VOLUME ONE HUNDRED AND EIGHTY NINE

PROGRESS IN MOLECULAR BIOLOGY AND TRANSLATIONAL SCIENCE

Large Animals as Models for Human Diseases

VOLUME ONE HUNDRED AND EIGHTY NINE

PROGRESS IN MOLECULAR BIOLOGY AND TRANSLATIONAL SCIENCE

Large Animals as Models for Human Diseases

Edited by

YA-XIONG TAO
*Department of Anatomy,
Physiology and Pharmacology,
College of Veterinary Medicine,
Auburn University, Auburn, AL, United States*

ELSEVIER

ACADEMIC PRESS
An imprint of Elsevier

Academic Press is an imprint of Elsevier
50 Hampshire Street, 5th Floor, Cambridge, MA 02139, United States
525 B Street, Suite 1650, San Diego, CA 92101, United States
The Boulevard, Langford Lane, Kidlington, Oxford OX5 1GB, United Kingdom
125 London Wall, London EC2Y 5AS, United Kingdom

First edition 2022

ISBN: 978-0-323-99441-5
ISSN: 1877-1173

For information on all Academic Press publications
visit our website at https://www.elsevier.com/books-and-journals

Publisher: Zoe Kruze
Acquisitions Editor: Ashlie M. Jackman
Developmental Editor: Jhon Michael Peñano
Production Project Manager: James Selvam
Cover Designer: Matthew Limbert

Typeset by STRAIVE, India

Contents

Contributors

Yongsong Chen
The First Affiliated Hospital of Shantou University, Shantou University Medical College, Shantou, China

Xiaoying Feng
The First Affiliated Hospital of Shantou University, Shantou University Medical College, Shantou, China

Chad D. Foradori
Department of Anatomy, Physiology and Pharmacology, College of Veterinary Medicine, Auburn University, Auburn, AL, United States

Weiren Huang
The First Affiliated Hospital of Shantou University, Shantou University Medical College, Shantou; Department of Urology, Shenzhen Institute of Translational Medicine, Shenzhen Second People's Hospital, The First Affiliated Hospital of Shenzhen University; Guangdong Key Laboratory of Systems Biology and Synthetic Biology for Urogenital Tumors; International Cancer Center of Shenzhen University; Shenzhen Institute of Synthetic Biology, Shenzhen Institutes of Advanced Technology, Chinese Academy of Sciences, Shenzhen, China

Chen-Che Jeff Huang
Department of Anatomy, Physiology and Pharmacology, College of Veterinary Medicine, Auburn University, Auburn, AL, United States

Ren-Lei Ji
Department of Anatomy, Physiology and Pharmacology, College of Veterinary Medicine, Auburn University, Auburn, AL, United States

Zhichao Li
Department of Urology, Shenzhen Institute of Translational Medicine, Shenzhen Second People's Hospital, The First Affiliated Hospital of Shenzhen University; Guangdong Key Laboratory of Systems Biology and Synthetic Biology for Urogenital Tumors; International Cancer Center of Shenzhen University, Shenzhen, China

Laci Mackay
Department of Anatomy, Physiology and Pharmacology, College of Veterinary Medicine, Auburn University, Auburn, AL, United States

Rebecca L. Nance
Scott-Ritchey Research Center; Department of Pathobiology, Auburn University College of Veterinary Medicine, Auburn, AL, United States

John F. Roberts
Department of Comparative, Diagnostic & Population Medicine, College of Veterinary Medicine, University of Florida, Gainesville, FL, United States

Hannah Ruetten
Department of Comparative Biosciences, School of Veterinary Medicine, University of Wisconsin-Madison, Madison, WI, United States

Abdul Mohin Sajib
Division of Cellular and Gene Therapies, Center for Biologics Evaluation and Research, US Food and Drug Administration, Silver Spring, MD, United States

Bruce F. Smith
Scott-Ritchey Research Center; Department of Pathobiology, Auburn University College of Veterinary Medicine, Auburn, AL, United States

Ya-Xiong Tao
Department of Anatomy, Physiology and Pharmacology, College of Veterinary Medicine, Auburn University, Auburn, AL, United States

Chad M. Vezina
Department of Comparative Biosciences, School of Veterinary Medicine, University of Wisconsin-Madison, Madison, WI, United States

Jia Wang
The First Affiliated Hospital of Shantou University, Shantou University Medical College, Shantou, China

Preface

Animal models have been used in biomedical research for more than a century. These studies contributed tremendously to advance our understanding of basic biology, including physiology, and pathogenesis of diseases. According to the Foundation for Biomedical Research (https://fbresearch.org/medical-advances/nobel-prizes/): "Lab animals have made important contributions to nearly every Nobel Prize in Medicine." Its tabulation shows that 188 of the 224 Nobel Prize recipients in Physiology or Medicine used animal models in their research (accessed on March 30, 2022).

Invertebrates such as fruit fly and nematode have been used as animal models. The studies on fruit fly led to six Nobel Prizes, including Thomas Hunt Morgan in 1933; Hermann Joseph Muller in 1946; Edward B. Lewis, Christiane Nüsslein-Volhard, and Eric F. Wieschaus in 1995; Richard Axel in 2004; Jules A. Hoffmann in 2011; and Jeffrey C. Hall, Michael Rosbash, and Michael W. Young in 2017. Work on the nematode led to three Nobel Prizes, including Robert Horvitz, John Sulston, and Sydney Brenner for programmed cell death in 2002; Andrew Fire and Craig Mello for RNA interference in 2006; and Martin Chalfie for green fluorescent protein in 2008.

Vertebrates have been more widely used as animal models for studying biology and human diseases. These vertebrates include fish, especially zebrafish, amphibians such as the African clawed frog, birds, especially chicken, and mammals. Of course, by far, the most commonly used lab animals are rodents, including genetically modified as well as naturally occurring disease models. For example, two strains of mice, *obese* (*ob/ob*) and *diabetes* (*db/db*), contributed tremendously to our current understanding of obesity and metabolic diseases.

However, rodents have their limitations. Large animals, including dog, cat, cow, sheep, pig, horse, and nonhuman primates, have also contributed dramatically to biomedical advance and led to 17 Nobel Prizes. Indeed, some Nobel laureates used multiple animal species in their Nobel Prize-winning work. For example, the 2003 Nobel Prize in Physiology or Medicine to Paul C. Lauterbur and Sir Peter Mansfield involved studies using clam, frog, mouse, rat, rabbit, pig, dog, and chimpanzee. The 2015 Nobel Prize in Physiology or Medicine to William C. Campbell, Satoshi

Omura, and Youyou Tu involved studies using chicken, mice, dogs, sheep, cattle, and monkeys.

Another concept that is gaining momentum during the past several decades is "One Medicine" or "One Health," bridging human and veterinary medicine, supported by both the American Medical Association and the American Veterinary Medical Association. Rudolf Virchow, the 19th century German physician and Father of Modern Pathology, first proposed that there should be no dividing line between human and veterinary medicine. Canadian physician, Sir William Osler, the Father of Modern Medicine, taught both medical and veterinary students, and first proposed the "One Medicine" concept. However, it was Calvin Schwabe, the Father of Veterinary Epidemiology, who revitalized this concept with his book entitled "Veterinary Medicine and Human Health," first published in 1969 and subsequently in 1984. He coined the term "One Medicine" and crystalized the concept (now called "One Health") in the 20th century (see https://onehealthinitiative.com/a-perspective-some-significant-historic-inspirations-for-advancing-the-one-medicine-one-health-concept-movement-19th-20th-and-21st-centuries/ for a more detailed history of this concept). Interestingly, he obtained his DVM at Auburn University in 1954. The One Health concept further incorporates the environmental aspect into human and animal health, and promotes health rather than treating diseases.

Another benefit of studying these large domestic animals is the dual benefit for both agriculture and human medicine, especially with production animals such as pig, sheep, cattle, goat, chicken, and fishes. For example, the activation of melanocortin-4 receptor, a critical regulator of energy homeostasis in vertebrates, is used to treat human obesity. But in production animals, our aim is to promote positive energy balance and higher feed efficiency so the farmers can make a bigger profit in a shorter period of time, with concurrent environmental benefit.

This volume provides a glimpse into using large animals as disease models for human medicine. Chad Foradori and Laci Mackay review the contribution of sheep to our understanding of neuroendocrinology, especially neuroendocrine regulation of reproduction. Hannah Ruetten and Chad Vezina summarize the relevance of dog for modeling urologic diseases, including urinary tract infection, benign prostatic hyperplasia, prostate cancer, erectile dysfunction, and incontinence. Rebecca Nance, Abdul Sajib, and Bruce Smith discuss how dogs can be used to study cancer and potential precision medicine for the treatment of both human and dog cancers. John Roberts and Jeff Huang discuss how cows can be used for studying human ovarian diseases. Ren-Lei Ji and I review naturally occurring mutations in several G protein-coupled receptor genes, including rhodopsin, type B endothelin receptor, CCR5 chemokine receptor, and melanocortin-1 receptor, in humans and animals, highlighting the similarities in the phenotypes between animals and humans. I also reviewed melanocortin-4 receptor mutations in fish, chicken, several species of large animals, and humans. Finally, although technically not an animal model, patient-derived organoids are widely used for modeling human diseases. Weiren Huang and his colleagues provide a comprehensive review on using patient-derived organoids for research on many different cancers. I thank all the authors for their excellent contributions to this volume.

Finally, I thank Dr. David Teplow, the Editor of the series, for his encouragement and support. I also thank Jhon Michael Peñano, Editor at Elsevier, for his guidance and patience during the prolonged period of germination of this volume, especially with the COVID-19 pandemic disrupting many scientists' schedules.

Ya-Xiong Tao
Auburn, AL, United States

CHAPTER ONE

Sheep as a model for neuroendocrinology research

Chad D. Foradori* and Laci Mackay
Department of Anatomy, Physiology and Pharmacology, College of Veterinary Medicine, Auburn University, Auburn, AL, United States
*Corresponding author: e-mail address: foradori@auburn.edu

Contents

Abstract

Animal models remain essential to understand the fundamental mechanisms of physiology and pathology. Particularly, the complex and dynamic nature of neuroendocrine cells of the hypothalamus make them difficult to study. The neuroendocrine systems of the hypothalamus are critical for survival and reproduction, and are highly conserved throughout vertebrate evolution. Their roles in controlling body metabolism, growth and body composition, stress, electrolyte balance, and reproduction, have been intensively studied, and have yielded groundbreaking discoveries. Many of these discoveries would not have been feasible without the use of the domestic sheep (*Ovis aries*). The sheep has been used for decades to study the neuroendocrine systems of the hypothalamus and has become a model for human neuroendocrinology. The aim of this chapter is to review some of the profound biomedical discoveries made possible by the use of sheep. The advantages and limitations of sheep as a neuroendocrine model will be discussed. While no animal model can perfectly recapitulate a human disease or condition, sheep are invaluable for enabling manipulations not possible in human subjects and isolating physiologic variables to garner insight into neuroendocrinology and associated pathologies.

Progress in Molecular Biology and Translational Science, Volume 189
ISSN 1877-1173
https://doi.org/10.1016/bs.pmbts.2022.01.001

1. Introduction

In all multicellular organisms, there is a requirement for coordination of bodily systems incorporating specialized cells distributed throughout the organism. Two of the systems performing this function are the nervous system and the endocrine system. The nervous system uses electrochemical signals, neurotransmitters, to directly transmit and receive information to and from other neurons and supporting cells. The endocrine system produces chemical agents, hormones, that are transported via the vascular system to receptive cells throughout the body. Portions of the nervous system produce not only neurotransmitters that act across synapses (electrochemical signals), but also release hormones that reach distant target cells via the bloodstream (endocrine signals). Because of their shared neural and endocrine characteristics, these cells are referred to as neuroendocrine cells. Neuroendocrine cells are found in multiple locations including the pancreas (islet cells), adrenal medulla (chromaffin cells), bifurcating airways (fetal pulmonary neuroendocrine cells), and gastrointestinal tract (enteroendocrine cells).[1] However, the most prominent and well-studied neuroendocrine cells are those located in the hypothalamus of the diencephalon.

The neuroendocrine cells of the hypothalamus are in a unique location. The hypothalamus resides at the intersection of several major neuronal pathways including the medial forebrain bundle, fornix, mammillothalamic tract, terminal stria, dorsal longitudinal fasciculus, and retinohypothalamic tract. These and other neural connections make the hypothalamus a major integration point for the central nervous system. The hypothalamus's privileged location is also buttressed by the presence of circumventricular organs (CVOs). The CVOs are characterized by the presence of fenestrated capillaries with higher permeability compared to common neural vasculature. The lack of the normal blood-brain-barrier in CVOs permits the surrounding cells to sample and interact with the contents of the systemic circulation and allows the secretions from those cells directly into the circulation. In mammals, CVOs include the median eminence, organum vasculosum lamina terminalis (OVLT), subfornical organ and the area postrema.[2] The rostral border of the hypothalamus is partially composed of the OVLT and a portion of hypothalamic ventral surface protrudes into the median eminence leading into the pituitary stalk. Hypothalamic neurons adjacent to or those which send projections to CVOs play a major role in many homeostatic functions. Some of these functions include regulation and control of blood pressure, temperature, respiration, appetite and energy balance, stress, reproduction, lactation, growth, and sleep.

Circuitry of the neuroendocrine systems of the hypothalamus generate complex patterns of hormone secretion that involve ultradian pulses of secretion superimposed upon daily (circadian) rhythms. Secretory patterns are generated by interactions between external stimuli, hypothalamic neural networks, pituitary cells, and the target organ(s), with positive and negative hormonal feedbacks at multiple levels with varying timescales (seconds, minutes, hours, days, seasons). While many neuroendocrine systems can act independently, the neuroendocrine systems of the hypothalamus commonly affect each other in a variety of ways. For example, the lack of adequate nutrients or energy stores activates satiety centers in the hypothalamus to increase feeding behavior but many of these same factors will inhibit neuroendocrine cells involved in reproduction and activate others involved in the stress response. The complex and dynamic nature of neuroendocrine circuits in the hypothalamus makes them difficult to study. A variety of animal models have been developed to better understand the interplay between brain, glandular, and various somatic tissue. One useful model is the domestic sheep (*Ovis aries*). The sheep has been used for decades to study the neuroendocrine systems of the hypothalamus and has become a model for human neuroendocrine functions and pathologies.

Sheep are docile, easy to handle, and relatively inexpensive with respect to housing and feeding. Their size (50 – 90 kg) is more like humans than laboratory rodents. Their larger size permits collection of larger volumes and more frequent blood samples without significant changes in blood chemistry or disturbance to the animal. This allows for the study of changes in hormones, metabolites, immune factors, or cellular components in the same animal over extended time periods. The sheep's size is also ideal for clinical imaging modalities designed for humans. For example, in the sheep, ovarian follicular development can be monitored, non-invasively, by ultrasound and samples from follicles can be taken after different treatments, without having to sacrifice the animal.

The use of sheep in biomedical research has allowed for some of the most tremendous scientific discoveries of the twentieth century. The first cloned mammal from a non-reproductive cell was a sheep.[3] Pig and sheep hypothalami were used to isolate and sequence thyrotropin releasing hormone and gonadotropin-releasing hormone (GnRH)—leading to the Nobel Prize in 1977.[4] Prostaglandins were first isolated from sheep which lead to the Nobel Prize in 1982.[5] Prions were first isolated from sheep and goats leading to the Nobel Prize in 1997.[6] Many of these studies revealed that the basic physiology of sheep better relates to humans than laboratory animals. Indeed, the human genome is more similar to the genome sequences of sheep when compared with rodents.[7] In addition to these similarities, sheep are much less inbred than laboratory rodents. Some would argue that it is advantageous to use animals that are highly inbred because this decreases the amount of variability in an experiment. However, each human has a unique genetic makeup and sometimes solutions for problems in inbred laboratory animals cannot be translated for use in humans. Therefore, in these instances, it is more beneficial to use a less inbred animal model, such as sheep. Similarly, naturally occurring, or spontaneous diseases in diverse animals remain a largely untapped resource and given the similarities in pathophysiology to humans, they not only allow for studying new treatment strategies but also disease etiology and prevention. Naturally occurring disease models, particularly for longer-lived, large animal species, allow for studying disorders at an age when the disease is most prevalent. In addition, improved awareness of the progress in animal models is mutually beneficial for producers of economically important animals, veterinarians, and human researchers. For these reasons and more, sheep have been instrumental in our understanding of physiology in general and neuroendocrine functions, specifically.

Considering the majority of work investigating neuroendocrine systems in the sheep have been focused on reproduction, that will be the primary focus of this chapter.

2. Reproduction

Sheep make an excellent model for the study of human reproduction. Although the ovine ovarian or estrous cycle can vary across breeds, most domesticated breeds used in neuroendocrine research present with 16–17-day estrous cycles.[8] Although shorter in length, sheep have a similar ovarian cycle to humans (approximately 28 days). The ovarian cycle can be divided into a follicular and luteal phase. A short follicular phase in the ewe accounts for the shorter estrous cycle while the luteal phase is similar in duration between humans and sheep. During the follicular phases, an ovarian follicle develops and produces increasing amounts of estrogen. The follicular phase ends with ovulation of the mature ovum from the follicle followed by a precipitous drop in estrogen. The remnants of the follicle in the ovary reorganize and differentiate into a new endocrine organ, the corpus luteum. The corpus luteum produces high levels of progesterone to prepare the uterus to establish and maintain pregnancy. If implantation does not occur, the corpus luteum regresses in the later part of the luteal phase resulting in a drop in progesterone and the cycle begins again with the start of the next follicular phase.

In all female mammals, the estrous cycle is maintained by hormonal signals conveyed between the brain, pituitary gland, and ovaries. These tissues function together as a hypothalamic-pituitary-gonadal (HPG) axis. Specialized hypothalamic neuroendocrine cells secrete GnRH into the hypophyseal portal vessel system to stimulate the release of gonadotropins (luteinizing hormone (LH) and follicle-stimulating hormone (FSH)) from the pituitary, which act on the ovary to regulate steroidogenesis (estradiol and progesterone), and follicle maturation. GnRH is the primary hypothalamic regulator of LH and FSH. FSH synthesis and release is additionally under the control of gonadal peptide hormones inhibin and activin. LH more directly recapitulates the release profile of GnRH, and its pulsatile release is pivotal in puberty onset, gonadal steroid synthesis, and ovulation. In turn, GnRH release triggers a regular or tonic secretion of LH and FSH which results in gonadal hormone synthesis and gamete maturation. The gonadal hormones feedback onto GnRH release from the hypothalamus and gonadotropin release from the pituitary gland. Ovulation is the result of a surge of high frequency, low amplitude release of GnRH and subsequent release of LH. The rise in GnRH/LH release is initiated by high estrogen concentrations at the end of the follicular phase. The tonic secretion of GnRH occurs in males and females and is primarily controlled by gonadal hormone negative feedback. The ability of estrogen to induce a GnRH (and subsequent LH) preovulatory surge is termed positive feedback and is only present in the female.

GnRH neuroendocrine cells are the final common mediators of the reproduction in all mammals. A critical feature of these neurons is their ability to secrete GnRH in discrete pulses; the frequency of these pulses is decoded by the pituitary to evoke preferential synthesis and release of LH or FSH.[9] In the early 1970s, Knobil and colleagues first demonstrated tonic (pulsatile) and cyclic (surge) gonadotropin release in female rhesus monkeys and predicted that the pulsatile nature of tonic LH release was due to GnRH stimulation to the pituitary gland.[10,11] This direct pulsatile GnRH-induced LH release was later described in sheep by measurements of GnRH in the hypophyseal portal system blood and LH in the systemic circulation.[12] These studies demonstrated for the first time that GnRH pulses measured in the pituitary portal circulation synchronized with LH pulses detected in the peripheral circulation.

The frequency and amplitude of GnRH release undergo cyclical changes across the ovarian cycle in response to gonadal steroid feedback.[13] In the follicular phase, estrogen feedback increases GnRH pulse frequency while reducing pulse amplitude.[14] The increased frequency drives steroidogenesis. The resulting elevation in estrogen concentration elicits a switch in steroid feedback from negative to positive. This culminates in a surge of GnRH and LH release that triggers ovulation. Following ovulation, increasing progesterone from the corpus luteum provides negative feedback to reduce GnRH pulse frequency. These dynamic changes in the pattern of GnRH release are critical to produce the proper levels of gonadotropins necessary for follicular development and ovulation.

The origin, location, and physiology of GnRH neurons make them some of the most unique cells in the brain. Born in the nasal placode, GnRH migrate into the telencephalon along the terminal nerve, and continue their migration to the diencephalon during mid- to late embryogenesis.[15] After migration, GnRH neurons are scattered throughout the preoptic area and hypothalamus. The majority of GnRH neurons send axonal projections to the median eminence at the base of the third ventricle and make specialized axon terminals termed dendrons which make contact or close associations with the vasculature of the primary portal plexus surrounding the median eminence.[16] When stimulated, action potentials propagate along the GnRH cell axon and results in an influx of calcium ions into the dendrons.[16] The increased calcium levels result in vesicle docking and release of the GnRH decapeptide into the median eminence extracellular space and into adjacent vasculature, which drains into the portal vein that empties into the secondary portal plexus surrounding the gonadotrophs of the anterior pituitary. The unique vasculature allows for extremely low concentrations of hypothalamic derived hormones to elicit large secondary hormone responses from pituitary cells. The ability to assess GnRH secretion in the hypophyseal portal blood of unanesthetized sheep has produced a wealth of information on the normal pattern of endogenous GnRH secretion.[12]

To generate synchronized pulses of secretion, GnRH neurons must be coordinated, presumably with synchronous bursts of electrical activity.[17] Some direct interaction seems likely to be involved, but GnRH neurons are not clustered together into one specific brain region or nuclei. They are scattered sparsely across the preoptic area and hypothalamus. It has long been postulated that there was a collection of neurons, upstream from GnRH that allowed for synchronization of GnRH release. The neuronal substance responsible for the coordinated pulsatile release of GnRH was termed the GnRH pulse generator.[18,19] The pulse generator hypothesis proposes that a network of cells in the mediobasal hypothalamus drives the coordinated release of GnRH. Because GnRH neurons do not express the classical nuclear gonadal steroid receptors required for the neuroendocrine control of reproduction, the GnRH pulse generator must be upstream of GnRH neurons and hormone sensitive. Therefore, these neurons should be able to convey the changing steroid environment onto GnRH cells.[20–22] Increasing evidence accumulated during the past two decades has demonstrated that a unique network of neurons in the mediobasal hypothalamus coexpressing three peptides—kisspeptin, neurokinin B, and dynorphin—serve as the GnRH pulse generator. Much of the work identifying and characterizing this unique network has been performed in the sheep.

3. Kisspeptin and KNDy neurons

In 2003, kisspeptin was identified as an unprecedented potent secretagogue of GnRH.[23–26] Mutation of the kisspeptin receptor (Kiss1R, also named GPR54) or of the kisspeptin gene (Kiss1) results in hypogonadotropic hypogonadism and infertility.[27–30] Kisspeptins are a group of peptides ranging from 54 to 10 amino acids (longest forms: Kisspeptin-54 in humans, Kisspeptin-53 in sheep, or Kisspeptin-52 in rodents). All kisspeptins share a C-terminal of 10 amino acids which is responsible for activation of Kiss1R.[31] The Kisppeptin-10 sequence is similar among species, suggesting a conserved physiological function.[32] While both central and peripheral administration of kisspeptins can rapidly and robustly induce gonadotropin release, kisspeptin directly stimulates GnRH neurons at the level of the hypothalamus.[25,33] The central action of kisspeptin was first demonstrated in the sheep demonstrating GnRH concentrations in the cerebrospinal fluid would quickly increase after intracerebroventricular delivery of kisspeptin.[25] In addition, kisspeptin-induced gonadotropin release is blocked by GnRH antagonist, indicating that kisspeptin stimulation of gonadotropin release is mediated by GnRH.[23,24,34] Kiss1R is expressed in a majority of GnRH neurons and electrophysiological studies demonstrate that kisspeptin exerts a potent direct depolarizing effect on GnRH neurons.[23,25,35–37] GnRH neuron-specific Kiss1R knockout mice are infertile, but expression of Kiss1R in GnRH neurons of global Kiss1R knockout mice restores fertility.[38] Taken together, these findings suggest that kisspeptin directly stimulates GnRH release via Kiss1R expressed in GnRH neurons leading to subsequent gonadotropin secretion.

In most mammals, there are two distinct populations of kisspeptin neurons. The largest number of kisspeptin neurons are located in the arcuate nucleus (human infundibular nucleus) with a smaller population in the preoptic area (POA).[39,40] In rodents, POA kisspeptin neurons consist of cells clustered in an area known as the anteroventricular nucleus (AVPV) and extend caudally into the periventricular preoptic nucleus (PeN).[24] In sheep and primates, kisspeptin cells are located in the POA but are more dispersed, not well-defined to specific nuclei, and are less numerous than in the rodent. Of note, the rostral population of kisspeptin cells are present in the female but little to no kisspeptin cells are found in the POA of the male.[39,40] The larger arcuate kisspeptin population is present in males and females.[39,41,42]

The major difference between the arcuate nucleus and POA populations of kisspeptin cells is the coexpression of additional neuropeptides in the arcuate population[43,44] A large majority of arcuate kisspeptin cells in the sheep colocalize with the neuropeptides, neurokinin B, and dynorphin.[44,45] The colocalization of kisspeptin, neurokinin B, and dynorphin peptides in a single group of cells has led to this cell population being termed "KNDy" neurons.[46] The KNDy neurons form an internal network of connections within the arcuate nucleus, as well as rostral projections to GnRH cell bodies and ventral/caudal projections to GnRH fibers and dendrons.[40,46–50] Reciprocal KNDy-KNDy network within this population is thought to play a key role in the synchronization of GnRH/LH pulses.[18]

In sheep, neurokinin B receptor (NK3R) activation stimulates LH release whereas antagonism of the receptor suppresses GnRH/LH pulses.[51,52] Kisspeptin infusion in animals pretreated with a neurokinin B antagonist restores GnRH/LH pulses, suggesting that kisspeptin action is downstream of neurokinin signaling.[52] GnRH neurons lack NK3R; however, NK3R is expressed in KNDy neurons.[53,54] Dynorphin is an endogenous opioid peptide that selectively binds the κ-opioid receptor (KOR) and inhibits GnRH/LH release.[45,55,56] KOR is expressed in GnRH and KNDy neurons in ewes.[57] Dynorphin terminates each GnRH pulse and limits the amount of GnRH released during each pulse.[58,59] These findings support the hypothesis that kisspeptin is the primary stimulatory output of KNDy neurons that drives GnRH secretion. Neurokinin B acts as an internal stimulator within the network and subsequent dynorphin release acts as a stop signal to the internal network and perhaps GnRH neurons directly.[45,46,60] Together, the expression and release of the three neuropeptides of the KNDy neurons act as a GnRH pulse generator.

A requirement for any proposed GnRH pulse generator is the ability to detect and respond to changing gonadal hormone concentrations and to convey those changes onto GnRH neurons since GnRH neurons do not express gonadal hormone receptors required for reproduction, specifically estrogen receptor-α (ERα).[61] KNDy neurons represent the link underlying feedback effects of steroids on GnRH secretion. The majority of KNDy neurons express not only ERα but also androgen receptor and progesterone receptor.[62,63] In sheep, dynorphin neurons in the KNDy neurons mediate the negative feedback effect of progesterone on GnRH/LH pulse secretion.[55,64] Arcuate dynorphin expression and release increases when circulating progesterone concentrations are high.[65] This is consistent with results in human studies, but not with data on manipulation by sex steroids in nonhuman primates.[66,67] Estrogen stimulates kisspeptin expression in the POA and decreases expression in the arcuate nucleus across multiple species.[32,39,68,69] Moreover, the immediate early gene, c-Fos, a measure of cellular activation, is induced in POA kisspeptin cells during the GnRH/LH preovulatory surge.[69,70] However, the role of KNDy neurons in integrating estrogen feedback is more complex, at least in ewes and possibly humans. Kisspeptin expression in the arcuate nucleus is elevated during the late follicular phase compared to the luteal phase, and there is c-Fos induction in arcuate kisspeptin neurons during GnRH/LH surge.[71–73] This has led to the theory that in the ewe and possibly in the human, the KNDy neurons mediate both the negative and positive feedback effects of estrogen.[69,73] As mentioned previously, towards the end of the follicular phase of the ovarian cycle, there is a transient switch from negative to positive estrogen feedback leading to the preovulatory GnRH/LH surge. The surge release of GnRH and LH is preceded by the activation of kisspeptin cells and an increase in Kiss1 expression in the arcuate nucleus in the ewe.[69–71] These findings are supported by previous work demonstrating that estrogen implants in the mediobasal hypothalamus could induce a surge of LH release.[74,75] Similarly, in the monkey, knife cuts isolating the mediobasal hypothalamus do not block estrogen positive feedback on LH release, suggesting that the KNDy neurons in the primate arcuate nucleus could also mediate estrogen positive and negative feedback onto GnRH release as postulated to be possible in the ewe.[76] The discovery and characterization of kisspeptin and the KNDy neurons has impacted virtually all aspects of neuroendocrinology.

4. Puberty

Timing of puberty has a great influence on animal and human health. Precocious or delayed puberty onset has been linked to multiple ailments in humans, including increased risk of obesity, cardiovascular disease, breast cancer, and depression.[77–80] Prior to puberty, LH secretion is pulsatile, as in the adult animal, but the frequency of LH pulses is reduced.[81] The low frequency LH pulses are unable to provide enough stimulus for follicular development or produce a sustained rise in estrogen concentration to trigger the preovulatory surge mechanism. In the pre-pubertal sheep, this is due to enhanced sensitivity to estrogen negative feedback, which declines as puberty begins.[82]

During this period, the hypothalamus becomes less sensitive to the negative feedback of estrogen, stimulating increased pulse frequency for both GnRH and LH.[82] The increase in GnRH/LH pulse frequency induces estrogen production by the growing ovarian follicles, eventually inducing an LH surge and ovulation.[82] The cause of the change in estrogen sensitively remains unclear. However, not surprisingly, kisspeptin and KNDy neurons have been implicated in puberty onset. Puberty fails to occur in animals lacking the kisspeptin receptor, Kiss1R. Conversely, gain-of-function mutations of Kiss1R results in precocious puberty in girls.[83] Kisspeptin treatment has been shown to increase LH secretion before puberty in all species examined.[84–88] Kisspeptin release increases during puberty in the mediobasal hypothalamus of female monkeys.[89] Kisspeptin gene expression and/or protein increases during pubertal onset in some but not all species.[51,85,86,90–93] Because the ovary is the main source of estrogen, this implies a negative effect of estrogen on this kisspeptin population. These data are consistent with the hypothesis that puberty is the result of a reduction in steroid negative feedback, leading to an increase in kisspeptin secretion.[82]

However, there are conflicting findings regarding estrogen sensitivity and KNDy neurons in puberty transition in the ewe. ERα mRNA expression has been shown to increase, not decrease, in arcuate kisspeptin neurons of ewes approaching puberty while kisspeptin expression does not change.[91,94,95] Similarly, the expression of the stimulatory neuropeptide in these same KNDy cells, NKB, which would be predicted to increase as puberty commences, does not change during puberty onset.[51,93] In addition, expression of the inhibitory dynorphin peptide in the arcuate has been shown to be very low before puberty and increase as puberty approaches in the ewe.[93,95,96] Thus, at least in sheep, neither KNDy protein nor mRNA expression changes in a predictable manner during pubertal development. The pubertal increase in LH secretion appears to occur absent of large changes in KNDy protein or mRNA expression.

In primates, hypothalamic kisspeptin and Kiss1R expression increases across pubertal onset and as GnRH pulses increased during puberty onset so did kisspeptin release at the level of the median eminence.[92,97] However, this does not elucidate the initial cause in the changes in the kisspeptin network. Garcia *et al.* (2019) postulated that the increase in kisspeptin release associated with puberty was dependent upon an escape from a not-as-yet completely characterized inhibitory input.[98] In higher primates, there is a "brake" on the secretion of GnRH and gonadotropins prior to puberty which is independent of negative feedback effects of gonadal steroids.[99] In primates, plasma gonadotropin profiles in gonadectomized animals were comparable to gonad-intact controls during juvenile and childhood. Gonadotropin secretion was suppressed until just before puberty in both gonad-intact and gonadectomized animals.[100] Interestingly, estrogen-dependent suppression of gonadotropin secretion is manifested just before puberty onset in monkeys.[101] Thus, the central mechanism inhibiting GnRH/LH secretion in primates appears to differ from that in rodents and sheep.[102] KNDy neurons are believed to represent a critical component of pulsatile GnRH release, i.e., the GnRH pulse generator, in puberty onset in the primate.[103] However, they may not be, in and of themselves, direct targets for signals that regulate puberty onset. Some support has been found for the tonic inhibition by γ-aminobutyric acid (GABA) neurotransmission in the suppression of GnRH in pre-pubertal monkeys and that KNDy neurons may mediate the inhibitory role of GABA in pre-pubertal restraint of GnRH neurons in primates.[102] Further studies are needed to clarify how these inhibitory actions disappear at puberty in primates. While sheep may not share similar neuroendocrine mechanisms in puberty onset to humans, it is interesting that GABA has been shown to inhibit GnRH and LH pulses in sheep and, of note, the transition from seasonal reproductive inactivity (anestrus) to breeding season involves changes to a hypothalamic GABAergic network input to KNDy neurons.[104,105]

5. Pregnancy

For obvious ethical reasons, many of the questions regarding the progression of human pregnancy and specifically the development of neuroendocrine systems cannot be answered directly, necessitating the use of appropriate animal models. A variety of animal models exist for the study of both normal and compromised pregnancies, including laboratory rodents, non-human primates, and domestic ruminants.[106] No animal model truly recapitulates human pregnancy, yet sheep have been used to study the normal and abnormal development of the fetus for almost a century. This is due to having a similar long gestation period, non-litter bearing, some comparable placental vascular structures, the ability to surgically place/maintain catheters in both the maternal and fetal vasculatures and overlapping *in utero* gestational brain development.

Anatomically, the sheep and human placenta can be considered quite divergent. However, there are important similarities in functional structures between human and sheep placentae. In humans, the conceptus deeply invades into the uterine lining, whereas in sheep, there are specialized regions of well-vascularized but non-glandular endometrium, known as caruncles, that protrude into the lumen of the uterus.[107] The sheep placenta has numerous discrete attachment sites comprised of the fetal cotyledon and maternal caruncle, rather than the large discoid placenta that occurs in humans. The villous tree of the sheep and human cotyledons are structurally similar, as both can be divided into stem, intermediate, and terminal villi, and the fetal vessels within the villi are comprised of stem arteries and veins, intermediate arterioles and venules, and terminal capillaries.[108] The relatively large size, docile nature, and the well-understood placental vasculature of sheep allows researchers the ability to surgically place and maintain catheters in both the maternal and fetal side of the placenta. This has led to studies using repeated sampling from non-anesthetized pregnant ewes. Repeated blood samples can be obtained throughout gestation under different experimental conditions.[107] While this research has been diverse, the bulk of the work has focused on regulation of placental oxygen levels, nutrient transfer, gestational growth, and endocrine function.[107,109]

Landmark work has been performed on the development of the cardiovascular system,[110] hypothalamic-pituitary-adrenal axis,[111] brain development,[112] effects of hypoxia/asphyxia,[113] therapeutic hypothermia,[114] inflammation,[115] anti-depressants,[116] broncho-pulmonary dysplasia (BPD),[117] and metabolism.[118] The life-saving clinical practices of glucocorticoid treatment in premature fetuses to promote lung development and the therapeutic hypothermia were developed in sheep and revolutionized perinatal medicine.[119] The chronically instrumented pregnant sheep have been used as a model of human fetal development and responses to pathophysiologic stimuli such as lipopolysaccharide.[120,121] The lesions following lipopolysaccharide exposure mimic what is seen in preterm infants with periventricular leukomalacia, which is due to a similar maturational profile of both species.[122,123]

Intrauterine growth restriction (IUGR) is defined as growth of fetus below its in-utero growth potential based on gestational age. It is one of the most common problems in obstetrics. Approximately 100,000 babies per year in the United States have IUGR, and an additional 250,000 are considered "small for gestational age".[124] IUGR is a silent cause of various morbidities and mortalities in many infants. Although modern perinatal care allows the majority of preterm and IUGR infants to survive the newborn period, it is now clear that there is a latent health "cost" for their survival.[125] Along with increased mortality rates early in life, IUGR causes various metabolic disorders that may lead to reduced quality and duration of life in affected individuals.[126] Sheep models of IUGR are similar to those in human IUGR fetuses.[127] The sheep IUGR model is performed through carunclectomy (removing plastination sites before the animal is pregnant) or hyperthermia which creates placental insufficiency. Continued use of the pregnant sheep model may lead to prenatal interventions or postnatal treatments to improve outcomes in IUGR-born offspring.[128]

In prenatal and perinatal brain developers, such as humans and sheep, the brain mass increases dramatically before birth, and the majority of central nervous system and autonomic development occurs *in utero*.[129] In contrast, rats, mice, dogs, and cats, are postnatal brain developers, with brain mass increasing and sleep states developing after birth.[123] The prenatal timing of brain development makes the sheep a valuable species in studies of brain development. Particularly, sheep have been used as an animal model for understanding how sex differences in the hypothalamic development occur and understanding the common human centers involving sex differences in behavior, metabolism, and hormone function.

Sexual dimorphism is associated with the activation of the fetal HPG axis. In sheep, fetal HPG activation occurs around gestational day (GD) 70 or about 40% of the way through gestation.[130] In humans, the HPG axis turns on at the beginning of the second trimester, which, similar to sheep, is about 40% of gestation.[131,132] In rats and mice, hypothalamic control of gonadotropic function occurs late in gestation around GD 18–19, approximately 90% through gestation.[133,134] Notably, in all species, the time when the HPG axis begins to function corresponds to the period of gestation when brain masculinization occurs under the influence of testosterone. In sheep and humans, brain masculinization is largely completed before birth, while in mice and rats the process extends into the first week of postnatal life.[135] The critical period for sexual differentiation in sheep occurs between GD30 and GD90 when testosterone is higher in males than in females. Masculinization of the genitalia by testosterone happens during the first half of this time interval, while appearance of sexually dimorphic brain structures and nuclei occur during the latter half.[136–138] Multiple studies have demonstrated similar sexually dimorphic brain structures and foci in humans and sheep.[139] Although consistent differences have been identified, the male and female brain are very similar. In general, using the term "sexual dimorphic" is correct only regarding a few brain characteristics. The specific sexually dimorphic differences identified in the KNDy neurons will be discussed below. Two sexually dimorphic traits that have garnered increased interest are sexual orientation and core gender identity. While out of the purview of direct neuroendocrine function, the sheep has also been used to study naturally occurring sexual behaviors and variations in the rams' sexual partner preferences.[140]

Sheep have significantly longer generation intervals and are seasonal breeders which can be problematic for some projects but advantageous to others. The advantages are evidenced by the recent development of a "biobag" and pumpless oxygenator to allow growth and development of the preterm lamb outside the womb for a month[141] and the possibilities for advances in clinical care.[142] Similarly, while sheep have several advantages when studying pregnancy in a non-litter bearing animal, the higher rate of multiple pregnancies compared to human twinning could be problematic for some projects.[143] This, however, is also a virtue of the model, as studying the effects of twinning on fetal development is an important biomedical task.[143–145] For example, genetic mutations in the sheep genes growth differentiation factor 9 (GDF9) and bone morphogenetic protein 15 (BMP15) were identified and found to impact twinning rates and fertility.[146,147] Soon after these discoveries, mutations in the orthologous human genes were shown to be associated with premature ovarian failure and spontaneous dizygotic twinning.[148–150]

All pregnancies, including ones presenting with no outward presenting pathologies may be exposed to a range of maternal health issues (e.g., asthma, mental illness, maternal stress, and drug use) and complications of pregnancy (e.g., gestational diabetes mellitus) that may require medications that impact the developing fetus. Therefore, understanding immediate and long-term effects of fetal exposure to potential interventions and environmental exposures is critical to establish the safety and efficacy of medications and agents before widespread use. More recently, sheep have been used to investigate the most commonly used analgesics to determine the potential impacts of each drug on maternal pain and fetal well-being.[151]

6. Hypothalamic and neuroendocrine sexual dimorphism

As mentioned previously, there is a sexual dimorphic difference in the expression pattern of the neuropeptide kisspeptin. Across species, males have little to no kisspeptin expression in the preoptic area whereas females have a high number of detectable kisspeptin cells. This sexual dimorphic expression is independent of gonadal steroid-induced changes to kisspeptin expression in adult animals. In addition, although there are conflicting findings, the number of kisspeptin expressing cells of the arcuate have also been shown to be sexually dimorphic in some species.[40,51,90,152–155] Results from neonatal gonadectomy and/or steroid treatments in the rat on the hormone control of this sexual dimorphism has implicated the well-established organizational effects of androgen exposure during hypothalamic organization.[156,157] This hypothesis states that testicular secretion of testosterone during a receptive developmental period is converted to estrogen, which then produces a permanent decrease in kisspeptin-expressing cells in the preoptic area. This conclusion is also supported by the effects of genetic manipulations in mice that do not express the aromatase enzyme or alpha-fetoprotein.[158,159]

There is no information on factors responsible for kisspeptin sexual dimorphism in humans, but in sheep, organizational effects of steroids likely play a role because the sex difference is seen in gonadectomized animals.[40,51,153] However, gestational testosterone treatment of females did not affect the number of kisspeptin cells, but it did decrease the number of NKB and dynorphin immunoreactive cells to levels seen in males.[153] Interestingly, the sexual differences in arcuate kisspeptin expression develops between 6 and 12 months of age as cell numbers in male sheep declines which is independent of gonadal steroids (because it is evident in gonadectomized animals).[51] In contrast to sheep, the sexual dimorphism in arcuate kisspeptin neurons in the mouse is lost during development.[160] Kisspeptin can elicit a robust release of LH that is accompanied by a delayed rise in serum testosterone secretion in GD85 lamb fetuses, an age that falls within the timeframe of the critical period of sexual differentiation.[161] In humans, like the sheep, it is thought that the conversion to estrogen is much less important than the permanent masculinizing effects of testosterone, both on the genitals and the brain.[162]

7. Polycystic ovarian syndrome

Polycystic ovary syndrome (PCOS) is the most common cause of female infertility. The estimated proportion of women affected varies with the criteria used for diagnosis. According to the NIH criteria, approximately 10% of women present with the clinical signs of oligo/anovulation and hyperandrogenism, with the exclusion of other causes of androgen excess.[163] PCOS encompasses a range of other common accompanying maladies, in particular, metabolic disorders of obesity, insulin resistance, abdominal adiposity, and in some cases glucose intolerance, which often progresses to type 2 diabetes mellitus.[164] Despite intense research efforts, the causes of PCOS remain unclear, due in part to diagnostic inconsistencies, the diversity of clinical phenotypes, as well as the complex endocrine profile in the pathophysiology. Existing evidence suggests that PCOS can be caused by both genetic and environmental factors. PCOS clusters in families, implying at least a partial genetic basis.[165] Several genomic variants have been identified as correlating with increased susceptibility to PCOS; these include mutations in genes related to obesity, insulin resistance, gonadotropin receptors, and steroidogenesis.[165] Additional genes and functional variants of interest are likely to be identified associated with PCOS, with the heritability currently accounting for less than 10% of cases. The limited linkage to PCOS suggests that other factors such as epigenetics and environmental interactions may play a role. The most widely implicated environmental factor associated with PCOS is the exposure to high levels of androgens during gestation.[166] In the human, exposure of female fetuses to excess androgens can occur during diabetic pregnancies and clinical conditions such as congenital adrenal hyperplasia and adrenal tumors.[167,168]

Prenatally androgenized sheep have been shown to be a clinically realistic model of PCOS.[169] Similar to monkeys, sheep exposed to androgens during early- to mid- gestation (GD30-90 out of approximately 150 days of pregnancy) develop oligo-ovulation, PCOS ovarian morphology, elevated LH levels and pulse frequency, and hyperandrogenism.[170–174] In addition, prenatally androgen treated sheep display multiple metabolic characteristics of PCOS, including excess adult body weight and insulin resistance.[169,175–179]

Prenatal androgen treatment results in the progressive deterioration of ovarian cyclicity, and reduced fertility, with most females becoming anovulatory by early adulthood.[180] The ability for repeated and sustained blood sampling in sheep has allowed detailed hormonal profiling studies showing that the progressive reproductive failure in prenatally androgenized sheep is due to hyperactivity of the reproductive neuroendocrine axis. The HPG hyperactivity appears to stem from defects in the steroid feedback mechanisms controlling GnRH and LH secretion, namely estradiol negative, estradiol positive, and progesterone negative feedback.[181–183] Further work in the sheep subsequently compared the effects of prenatal testosterone versus dihydrotestosterone (a non-aromatizable androgen). These findings suggests that prenatal androgen effects on negative estradiol feedback is mediated by androgen receptor activation and that alterations of positive estrogen feedback are dependent on the conversion of testosterone to estrogen.[174,184] Moreover, prenatal androgen treatment decreased the percentage of KNDy and Agouti-related protein (AgRP) expressing neurons that colocalized with insulin receptor B, consistent with reduced insulin sensitivity.[185]

Disrupted progesterone negative feedback has been demonstrated in prenatally androgen treated monkeys, sheep, and PCOS women.[186,187] KNDy neurons were previously shown to mediate progesterone negative feedback on GnRH release.[55] Prenatal androgen treatment does not alter the number of kisspeptin-positive cells in ewes, whereas the number of cells expressing either dynorphin, neurokinin B, or the progesterone receptor is reduced.[153] The reduction in progesterone receptor and dynorphin might explain the increased GnRH/LH pulse frequency/amplitude and diminished progesterone inhibition in the PCOS phenotype.

Other key neuroendocrine alterations seen in prenatal testosterone sheep include increased number of GABAergic inputs to KNDy and GnRH neurons.[188] Additional work is needed to determine if these neuroanatomical changes in prenatally androgenized females might explain the changes to estrogen positive and negative feedback identified in these animals. While not directly addressing the causality of human PCOS, the sheep model provides a unique resource for addressing underlying mechanisms and developing prevention and treatment options that are likely to be of translational value.

8. Growth hormone regulation

The neuroendocrine cells of the hypothalamus are integral to the regulation of energy homeostasis. Consequently, hypothalamic neural circuits have multi-purpose functions in regulating reproduction and growth/metabolism. One way this is accomplished is through the control of growth hormone (GH) release from somatotrophs in the pituitary. The name "growth hormone" is misleading, for GH has important actions throughout life, not only in the developmental increase in overall body size. GH and the subsequent induced release of insulin-like growth factor-1 in adult animals is responsible for changes in adiposity, bone strength, and glucose homeostasis. Pulsatile GH secretion from the pituitary is orchestrated by alternating stimulation by growth hormone releasing hormone (GHRH) neurons located in the arcuate nucleus and inhibition by somatostatin (SS) neurons in the periventricular nucleus.[189,190] The GHRH and SS neuroendocrine cells project to the median eminence and secrete their neuropeptides into the hypophyseal portal circulation to act upon the pituitary somatotrophs.[191] Neuropeptide Y (NPY) neurons located in the arcuate nucleus are upstream from GHRH and SS neurons. NPY neurons stimulate GHRH neurons and inhibit SS neurons.[190,192,193] NPY fibers are in close apposition or synapse with SS and GHRH neurons.[193,194] GHRH neurons express NPY receptors and central infusion of NPY results in increased GHRH in the third ventricle cerebrospinal fluid.[195,196] The dichotomy of NPY effects on these distinct cell groups (stimulation of GHRH/inhibition of SS) is thought to be due to NPY neurons' complex signaling through the release of NPY, AgRP, norepinephrine, and GABA.[197–199]

The pulse frequency and amplitude of GH is set by a multitude of factors. Work in the sheep has characterized some of these factors in conditions which, unlike in the rodent, better model effects identified in the human. Short-term fasting decreases the secretion of GH in rodents but increases secretion in other mammals such as sheep and humans.[200–205] The rise in GH levels results in increased gluconeogenesis to maintain blood glucose levels during short-term fasting.

In the fasted sheep, kisspeptin activates NPY neurons and induces a robust release of GH.[206,207] NPY cells of the arcuate nucleus express the kisspeptin receptor, and central infusion of kisspeptin increases NPY expression.[208,209] Both kisspeptin and NPY treatment induces GH release in the sheep, but kisspeptin-induced GH can be blocked by pretreatment with a NPY receptor antagonist.[192,207,210] Kisspeptin stimulates NPY neurons in the arcuate nucleus resulting in stimulation of GHRH cells and the inhibition of SS cells which leads to increased GH plasma levels in short-term fasted sheep.[207] This diversity between species is a result of differences in the hypothalamic response to altered nutritional status. In rodents, SS release is increased by undernutrition, whereas levels are lower in the hypophyseal portal circulation of food restricted sheep.[201,211] Alterations in the expression of NPY can also be linked to GH release in a species dependent manner. In both rodents and sheep, NPY expression is robustly increased by under-nutrition, but the effect on GH secretion is inhibitory in rodents and stimulatory in ruminants and humans.[212,213] These findings support the existence of a neural bridge between a pivotal reproductive factor, kisspeptin, and mediators of GH release. Accordingly, kisspeptin cells may have a dual function to regulate both reproduction and metabolic homeostasis.

9. Caveats and conclusions

This chapter has made the argument for the use of sheep to better understand the neuroendocrine functions involved in human physiology and diseases. Their nature, size, and physiology allow sheep a degree of precedency above most of the more commonly used research animals. However, there are drawbacks to the use of sheep in biomedical research. In general, there is a lack of antibodies and other reagents available for sheep tissue. More are becoming available, particularly for sheep, since their genome has been sequenced. As noted, most sheep breeds are not inbred to a point of complete phenotypic homology, so there is more genetic variation and "noise" in some experiments. Similarly, the ability to manipulate gene function in rodent models is far easier than in sheep. However, comparing conserved sequences and identifying conserved gene function across a variety of mammalian species may be beneficial in extrapolation to function in humans.

Housing facilities for large animals may also be an issue at some institutions, costs per animal are usually significantly greater than for rodents. However, in some cases multiple samples can be taken from the same animal without sacrifice, and a greater amount of sample material may be obtained per animal, which could allow research to be conducted with fewer numbers of animals than using rodents.

Sheep and humans share similar physiology which make them an excellent model for many human conditions. However, one of the important differences between mature humans and sheep is their source of energy. In humans, metabolic energy is directly derived from glucose, primarily supplied by ingested starch or from the intracellular breakdown of stored fatty acids and glycogen.[214,215] Ruminants, such as sheep, obtain energy primarily from short-chain volatile fatty acids, such as acetate, propionate, and butyrate, absorbed from the gut. These fatty acids are produced in the rumen by the action of rumen microorganisms metabolizing the complex polysaccharides of consumed fibrous vegetation.[216,217] The precursors for fatty acid synthesis in humans and ruminants are glucose and acetate, respectively.[216] In humans, fatty acid synthesis primarily occurs in the liver; in ruminants, it occurs in adipose tissue.[216] Furthermore, before birth the main energy source for the sheep fetus is glucose supplied across the placenta,[218] as it is for the human fetus. Transition to free fatty acids as the main energy source in sheep occurs after birth with establishment of the rumen.

Caveats should certainly be considered before developing or adopting any animal model. Experimental studies that are relevant to human health rely on appropriate animal models. Consideration of the strengths and weaknesses of any animal model of human disease is fundamental to effective and meaningful clinical translation of discoveries. Researchers must be able to identify the most appropriate animal model for research that will inform and translate to fundamental advances. Even more importantly, from a translational standpoint, the model provides a means to develop prevention and treatment strategies. As one animal model may have advantages in one study, it may have disadvantages in others. The careful choice of model is therefore crucial.

References

1. DeLellis RA. The neuroendocrine system and its tumors: an overview. *Am J Clin Pathol.* 2001;115(suppl):S5–16. https://doi.org/10.1309/7gr5-l7yw-3g78-ldj6.
2. Duvernoy HM, Risold P-Y. The circumventricular organs: an atlas of comparative anatomy and vascularization. *Brain Res Rev.* 2007;56(1):119–147. https://doi.org/10.1016/j.brainresrev.2007.06.002.
3. Campbell KH, McWhir J, Ritchie WA, Wilmut I. Sheep cloned by nuclear transfer from a cultured cell line. *Nature.* 1996;380(6569):64–66. https://doi.org/10.1038/380064a0.
4. Schally AV. Aspects of hypothalamic regulation of the pituitary gland. *Science.* 1978;202(4363):18–28. https://doi.org/10.1126/science.99816.
5. Flower RJ. Prostaglandins, bioassay and inflammation. *Br J Pharmacol.* 2006;147(suppl 1):S182–S192. https://doi.org/10.1038/sj.bjp.0706506.
6. Prusiner SB. Novel proteinaceous infectious particles cause scrapie. *Science.* 1982;216(4542):136–144. https://doi.org/10.1126/science.6801762.
7. Li X, Yang J, Shen M, et al. Whole-genome resequencing of wild and domestic sheep identifies genes associated with morphological and agronomic traits. *Nat Commun.* 2020;11(1):2815. https://doi.org/10.1038/s41467-020-16485-1.
8. Goodman R, Inskeep K. Control of the ovarian cycle of the sheep. Knobil Neill's Physiol Reprod. 2015;2:1259-1305. doi:10.1016/B978-0-12-397175-3.00027-2.

9. Wildt L, Hausler A, Marshall G, et al. Frequency and amplitude of gonadotropin-releasing hormone stimulation and gonadotropin secretion in the rhesus monkey. *Endocrinology*. 1981;109(2):376–385. https://doi.org/10.1210/endo-109-2-376.
10. Dierschke DJ, Bhattacharya AN, Atkinson LE, Knobil E. Circhoral oscillations of plasma LH levels in the ovariectomized rhesus monkey. *Endocrinology*. 1970;87(5):850–853. https://doi.org/10.1210/endo-87-5-850.
11. Monroe SE, Atkinson LE, Knobil E. Patterns of circulating luteinizing hormone and their relation to plasma progesterone levels during the menstrual cycle of the Rhesus monkey. *Endocrinology*. 1970;87(3):453–455. https://doi.org/10.1210/endo-87-3-453.
12. Moenter SM, Evans NP. Gonadotropin-releasing hormone (GnRH) measurements in pituitary portal blood: a history. *J Neuroendocrinol*. 2021;e13065. https://doi.org/10.1111/jne.13065.
13. Reame N, Sauder SE, Kelch RP, Marshall JC. Pulsatile gonadotropin secretion during the human menstrual cycle: evidence for altered frequency of gonadotropin-releasing hormone secretion. *J Clin Endocrinol Metab*. 1984;59(2):328–337. https://doi.org/10.1210/jcem-59-2-328.
14. Moenter SM, Caraty A, Locatelli A, Karsch FJ. Pattern of gonadotropin-releasing hormone (GnRH) secretion leading up to ovulation in the ewe: existence of a preovulatory GnRH surge. *Endocrinology*. 1991;129(3):1175–1182. https://doi.org/10.1210/endo-129-3-1175.
15. Wray S. Development of luteinizing hormone releasing hormone neurones. *J Neuroendocrinol*. 2001;13(1):3–11. https://doi.org/10.1046/j.1365-2826.2001.00609.x.
16. Herde MK, Iremonger KJ, Constantin S, Herbison AE. GnRH neurons elaborate a long-range projection with shared axonal and dendritic functions. *J Neurosci*. 2013;33(31):12689–12697. https://doi.org/10.1523/JNEUROSCI.0579-13.2013.
17. Krupa M, Vidal A, Clément F. A network model of the periodic synchronization process in the dynamics of calcium concentration in GnRH neurons. *J Math Neurosci*. 2013;3(1):4. https://doi.org/10.1186/2190-8567-3-4.
18. Lehman MN, Coolen LM, Goodman RL. Importance of neuroanatomical data from domestic animals to the development and testing of the KNDy hypothesis for GnRH pulse generation. *Domest Anim Endocrinol*. 2020;106441. https://doi.org/10.1016/j.domaniend.2020.106441.
19. Herbison AE. The gonadotropin-releasing hormone pulse generator. *Endocrinology*. 2018;159(11):3723–3736. https://doi.org/10.1210/en.2018-00653.
20. Vidal A, Clément F. A dynamical model for the control of the gonadotrophin-releasing hormone neurosecretory system. *J Neuroendocrinol*. 2010;22(12):1251–1266. https://doi.org/10.1111/j.1365-2826.2010.02055.x.
21. Shivers BD, Harlan RE, Morrell JI, Pfaff DW. Absence of oestradiol concentration in cell nuclei of LHRH-immunoreactive neurones. *Nature*. 1983;304(5924):345–347. https://doi.org/10.1038/304345a0.
22. Lehman MN, Karsch FJ. Do gonadotropin-releasing hormone, tyrosine hydroxylase-, and beta-endorphin-immunoreactive neurons contain estrogen receptors? A double-label immunocytochemical study in the Suffolk ewe. *Endocrinology*. 1993;133(2):887–895. https://doi.org/10.1210/endo.133.2.8102098.
23. Irwig MS, Fraley GS, Smith JT, et al. Kisspeptin activation of gonadotropin releasing hormone neurons and regulation of KiSS-1 mRNA in the male rat. *Neuroendocrinology*. 2004;80(4):264–272. https://doi.org/10.1159/000083140.
24. Gottsch ML, Cunningham MJ, Smith JT, et al. A role for kisspeptins in the regulation of gonadotropin secretion in the mouse. *Endocrinology*. 2004;145(9):4073–4077. https://doi.org/10.1210/en.2004-0431.

25. Messager S, Chatzidaki EE, Ma D, et al. Kisspeptin directly stimulates gonadotropin-releasing hormone release via G protein-coupled receptor 54. *Proc Natl Acad Sci U S A*. 2005;102(5):1761–1766. https://doi.org/10.1073/pnas.0409330102.
26. Kinoshita M, Tsukamura H, Adachi S, et al. Involvement of central metastin in the regulation of preovulatory luteinizing hormone surge and estrous cyclicity in female rats. *Endocrinology*. 2005;146(10):4431–4436. https://doi.org/10.1210/en.2005-0195.
27. de Roux N, Genin E, Carel JC, Matsuda F, Chaussain JL, Milgrom E. Hypogonadotropic hypogonadism due to loss of function of the KiSS1-derived peptide receptor GPR54. *Proc Natl Acad Sci U S A*. 2003;100(19):10972–10976. https://doi.org/10.1073/pnas.1834399100.
28. Seminara SB, Messager S, Chatzidaki EE, et al. The GPR54 gene as a regulator of puberty. *N Engl J Med*. 2003;349(17):1614–1627. https://doi.org/10.1056/NEJMoa035322.
29. d'Anglemont de Tassigny X, Fagg LA, Dixon JP, et al. Hypogonadotropic hypogonadism in mice lacking a functional Kiss1 gene. *Proc Natl Acad Sci U S A*. 2007;104(25):10714–10719. https://doi.org/10.1073/pnas.0704114104.
30. Dungan HM, Gottsch ML, Zeng H, et al. The role of kisspeptin-GPR54 signaling in the tonic regulation and surge release of gonadotropin-releasing hormone/luteinizing hormone. *J Neurosci*. 2007;27(44):12088–12095. https://doi.org/10.1523/JNEUROSCI.2748-07.2007.
31. Ohtaki T, Shintani Y, Honda S, et al. Metastasis suppressor gene KiSS-1 encodes peptide ligand of a G-protein-coupled receptor. *Nature*. 2001;411(6837):613–617. https://doi.org/10.1038/35079135.
32. Oakley AE, Clifton DK, Steiner RA. Kisspeptin signaling in the brain. *Endocr Rev*. 2009;30(6):713–743. https://doi.org/10.1210/er.2009-0005.
33. Caraty A, Lomet D, Sebert ME, Guillaume D, Beltramo M, Evans NP. Gonadotrophin-releasing hormone release into the hypophyseal portal blood of the ewe mirrors both pulsatile and continuous intravenous infusion of kisspeptin: an insight into kisspeptin's mechanism of action. *J Neuroendocrinol*. 2013;25(6):537–546. https://doi.org/10.1111/jne.12030.
34. Matsui H, Takatsu Y, Kumano S, Matsumoto H, Ohtaki T. Peripheral administration of metastin induces marked gonadotropin release and ovulation in the rat. *Biochem Biophys Res Commun*. 2004;320(2):383–388. https://doi.org/10.1016/j.bbrc.2004.05.185.
35. Herbison AE, de Tassigny X, Doran J, Colledge WH. Distribution and postnatal development of Gpr54 gene expression in mouse brain and gonadotropin-releasing hormone neurons. *Endocrinology*. 2010;151(1):312–321. https://doi.org/10.1210/en.2009-0552.
36. Pielecka-Fortuna J, Chu Z, Moenter SM. Kisspeptin acts directly and indirectly to increase gonadotropin-releasing hormone neuron activity and its effects are modulated by estradiol. *Endocrinology*. 2008;149(4):1979–1986. https://doi.org/10.1210/en.2007-1365.
37. Liu X, Lee K, Herbison AE. Kisspeptin excites gonadotropin-releasing hormone neurons through a phospholipase C/calcium-dependent pathway regulating multiple ion channels. *Endocrinology*. 2008;149(9):4605–4614. https://doi.org/10.1210/en.2008-0321.
38. Kirilov M, Clarkson J, Liu X, et al. Dependence of fertility on kisspeptin-Gpr54 signaling at the GnRH neuron. *Nat Commun*. 2013;4:2492. https://doi.org/10.1038/ncomms3492.
39. Lehman MN, Merkley CM, Coolen LM, Goodman RL. Anatomy of the kisspeptin neural network in mammals. *Brain Res*. 2010;1364:90–102. https://doi.org/10.1016/j.brainres.2010.09.020.

40. Hrabovszky E, Ciofi P, Vida B, et al. The kisspeptin system of the human hypothalamus: sexual dimorphism and relationship with gonadotropin-releasing hormone and neurokinin B neurons. *Eur J Neurosci*. 2010;31(11):1984–1998. https://doi.org/10.1111/j.1460-9568.2010.07239.x.
41. Smith JT, Cunningham MJ, Rissman EF, Clifton DK, Steiner RA. Regulation of Kiss1 gene expression in the brain of the female mouse. *Endocrinology*. 2005;146 (9):3686–3692. https://doi.org/10.1210/en.2005-0488.
42. Ramaswamy S, Guerriero KA, Gibbs RB, Plant TM. Structural interactions between kisspeptin and GnRH neurons in the mediobasal hypothalamus of the male rhesus monkey (Macaca mulatta) as revealed by double immunofluorescence and confocal microscopy. *Endocrinology*. 2008;149(9):4387–4395. https://doi.org/10.1210/en.2008-0438.
43. Foradori CD, Amstalden M, Goodman RL, Lehman MN. Colocalisation of dynorphin a and neurokinin B immunoreactivity in the arcuate nucleus and median eminence of the sheep. *J Neuroendocrinol*. 2006;18(7):534–541. https://doi.org/10.1111/j.1365-2826.2006.01445.x.
44. Goodman RL, Lehman MN, Smith JT, et al. Kisspeptin neurons in the arcuate nucleus of the ewe express both dynorphin A and neurokinin B. *Endocrinology*. 2007;148 (12):5752–5760. https://doi.org/10.1210/en.2007-0961.
45. Navarro VM, Gottsch ML, Chavkin C, Okamura H, Clifton DK, Steiner RA. Regulation of gonadotropin-releasing hormone secretion by kisspeptin/dynorphin/neurokinin B neurons in the arcuate nucleus of the mouse. *J Neurosci*. 2009;29 (38):11859–11866. https://doi.org/10.1523/JNEUROSCI.1569-09.2009.
46. Lehman MN, Coolen LM, Goodman RL. Minireview: kisspeptin/neurokinin B/dynorphin (KNDy) cells of the arcuate nucleus: a central node in the control of gonadotropin-releasing hormone secretion. *Endocrinology*. 2010;151(8):3479–3489. https://doi.org/10.1210/en.2010-0022.
47. Kallo I, Vida B, Deli L, et al. Co-localisation of kisspeptin with galanin or neurokinin B in afferents to mouse GnRH neurones. *J Neuroendocrinol*. 2012;24(3):464–476. https://doi.org/10.1111/j.1365-2826.2011.02262.x.
48. Herde MK, Geist K, Campbell RE, Herbison AE. Gonadotropin-releasing hormone neurons extend complex highly branched dendritic trees outside the blood-brain barrier. *Endocrinology*. 2011;152(10):3832–3841. https://doi.org/10.1210/en.2011-1228.
49. Clarkson J, Herbison AE. Postnatal development of kisspeptin neurons in mouse hypothalamus: sexual dimorphism and projections to gonadotropin-releasing hormone neurons. *Endocrinology*. 2006;147(12):5817–5825. https://doi.org/10.1210/en.2006-0787.
50. Uenoyama Y, Inoue N, Pheng V, et al. Ultrastructural evidence of kisspeptin-gonadotrophin-releasing hormone (GnRH) interaction in the median eminence of female rats: implication of axo-axonal regulation of GnRH release. *J Neuroendocrinol*. 2011;23(10):863–870. https://doi.org/10.1111/j.1365-2826.2011.02199.x.
51. Nestor CC, Briscoe AM, Davis SM, Valent M, Goodman RL, Hileman SM. Evidence of a role for kisspeptin and neurokinin B in puberty of female sheep. *Endocrinology*. 2012;153(6):2756–2765. https://doi.org/10.1210/en.2011-2009.
52. Clarke IJ, Li Q, Henry BA, Millar RP. Continuous kisspeptin restores luteinizing hormone pulsatility following cessation by a neurokinin B antagonist in female sheep. *Endocrinology*. 2018;159(2):639–646. https://doi.org/10.1210/en.2017-00737.
53. Billings HJ, Connors JM, Altman SN, et al. Neurokinin B acts via the neurokinin-3 receptor in the retrochiasmatic area to stimulate luteinizing hormone secretion in sheep. *Endocrinology*. 2010;151(8):3836–3846. https://doi.org/10.1210/en.2010-0174.

54. Amstalden M, Coolen LM, Hemmerle AM, et al. Neurokinin 3 receptor immunoreactivity in the septal region, preoptic area and hypothalamus of the female sheep: colocalisation in neurokinin B cells of the arcuate nucleus but not in gonadotrophin-releasing hormone neurones. *J Neuroendocrinol.* 2010;22(1):1–12. https://doi.org/10.1111/j.1365-2826.2009.01930.x.
55. Goodman RL, Coolen LM, Anderson GM, et al. Evidence that dynorphin plays a major role in mediating progesterone negative feedback on gonadotropin-releasing hormone neurons in sheep. *Endocrinology.* 2004;145(6):2959–2967.
56. Yen S, Quigley M, Reid R, Ropert J, Cetel N. Neuroendocrinology of opioid peptides and their role in the control of gonadotropin and prolactin secretion. *Am J Obstet Gynecol.* 1985;152(4):485–493.
57. Weems PW, Witty CF, Amstalden M, Coolen LM, Goodman RL, Lehman MN. kappa-Opioid receptor is colocalized in GnRH and KNDy cells in the female ovine and rat brain. *Endocrinology.* 2016;157(6):2367–2379. https://doi.org/10.1210/en.2015-1763.
58. Goodman RL, Parfitt DB, Evans NP, Dahl GE, Karsch FJ. Endogenous opioid peptides control the amplitude and shape of gonadotropin-releasing hormone pulses in the ewe. *Endocrinology.* 1995;136(6):2412–2420. https://doi.org/10.1210/endo.136.6.7750462.
59. Weems PW, Coolen LM, Hileman SM, et al. Evidence that dynorphin acts upon KNDy and GnRH neurons during GnRH pulse termination in the ewe. *Endocrinology.* 2018;159(9):3187–3199. https://doi.org/10.1210/en.2018-00435.
60. Rance NE, Krajewski SJ, Smith MA, Cholanian M, Dacks PA. Neurokinin B and the hypothalamic regulation of reproduction. *Brain Res.* 2010;1364:116–128. https://doi.org/10.1016/j.brainres.2010.08.059.
61. Herbison AE, Theodosis DT. Localization of oestrogen receptors in preoptic neurons containing neurotensin but not tyrosine hydroxylase, cholecystokinin or luteinizing hormone-releasing hormone in the male and female rat. *Neuroscience.* 1992;50(2):283–298. https://doi.org/10.1016/0306-4522(92)90423-y.
62. Smith JT, Dungan HM, Stoll EA, et al. Differential regulation of KiSS-1 mRNA expression by sex steroids in the brain of the male mouse. *Endocrinology.* 2005;146(7):2976–2984. https://doi.org/10.1210/en.2005-0323.
63. Smith JT, Clay CM, Caraty A, Clarke IJ. KiSS-1 messenger ribonucleic acid expression in the hypothalamus of the ewe is regulated by sex steroids and season. *Endocrinology.* 2007;148(3):1150–1157. https://doi.org/10.1210/en.2006-1435.
64. Goodman RL, Holaskova I, Nestor CC, et al. Evidence that the arcuate nucleus is an important site of progesterone negative feedback in the ewe. *Endocrinology.* 2011;152(9):3451–3460.
65. Foradori CD, Goodman RL, Adams VL, Valent M, Lehman MN. Progesterone increases dynorphin a concentrations in cerebrospinal fluid and preprodynorphin messenger ribonucleic acid levels in a subset of dynorphin neurons in the sheep. *Endocrinology.* 2005;146(4):1835–1842. https://doi.org/10.1210/en.2004-1326.
66. Rometo A, Rance NE. Changes in prodynorphin gene expression and neuronal morphology in the hypothalamus of postmenopausal women. *J Neuroendocrinol.* 2008;20(12):1376–1381.
67. Eghlidi DH, Haley GE, Noriega NC, Kohama SG, Urbanski HF. Influence of age and 17beta-estradiol on kisspeptin, neurokinin B, and prodynorphin gene expression in the arcuate-median eminence of female rhesus macaques. *Endocrinology.* 2010;151(8):3783–3794. https://doi.org/10.1210/en.2010-0198.
68. Garcia-Galiano D, Pinilla L, Tena-Sempere M. Sex steroids and the control of the Kiss1 system: developmental roles and major regulatory actions. *J Neuroendocrinol.* 2012;24(1):22–33. https://doi.org/10.1111/j.1365-2826.2011.02230.x.

69. Smith JT, Li Q, Pereira A, Clarke IJ. Kisspeptin neurons in the ovine arcuate nucleus and preoptic area are involved in the preovulatory luteinizing hormone surge. *Endocrinology*. 2009;150(12):5530–5538. https://doi.org/10.1210/en.2009-0712.
70. Hoffman GE, Le WW, Franceschini I, Caraty A, Advis JP. Expression of fos and in vivo median eminence release of LHRH identifies an active role for preoptic area kisspeptin neurons in synchronized surges of LH and LHRH in the ewe. *Endocrinology*. 2011;152 (1):214–222. https://doi.org/10.1210/en.2010-0066.
71. Estrada KM, Clay CM, Pompolo S, Smith JT, Clarke IJ. Elevated KiSS-1 expression in the arcuate nucleus prior to the cyclic preovulatory gonadotrophin-releasing hormone/lutenising hormone surge in the ewe suggests a stimulatory role for kisspeptin in oestrogen-positive feedback. *J Neuroendocrinol*. 2006;18(10):806–809. https://doi.org/10.1111/j.1365-2826.2006.01485.x.
72. Smith JT, Saleh SN, Clarke IJ. Seasonal and cyclical change in the luteinizing hormone response to kisspeptin in the ewe. *Neuroendocrinology*. 2009;90(3):283–291. https://doi.org/10.1159/000227806.
73. Merkley CM, Porter KL, Coolen LM, et al. KNDy (kisspeptin/neurokinin B/dynorphin) neurons are activated during both pulsatile and surge secretion of LH in the ewe. *Endocrinology*. 2012;153(11):5406–5414. https://doi.org/10.1210/en.2012-1357.
74. Blache D, Fabre-Nys CJ, Venier G. Ventromedial hypothalamus as a target for oestradiol action on proceptivity, receptivity and luteinizing hormone surge of the ewe. *Brain Res*. 1991;546(2):241–249. https://doi.org/10.1016/0006-8993(91)91488-m.
75. Caraty A, Fabre-Nys C, Delaleu B, et al. Evidence that the mediobasal hypothalamus is the primary site of action of estradiol in inducing the preovulatory gonadotropin releasing hormone surge in the ewe. *Endocrinology*. 1998;139(4):1752–1760. https://doi.org/10.1210/endo.139.4.5904.
76. Krey LC, Butler WR, Knobil E. Surgical disconnection of the medial basal hypothalamus and pituitary function in the rhesus monkey. I Gonadotropin secretion. *Endocrinology*. 1975;96(5):1073–1087. https://doi.org/10.1210/endo-96-5-1073.
77. Reinehr T, Roth CL. Is there a causal relationship between obesity and puberty? *Lancet Child Adolesc Health*. 2019;3(1):44–54. https://doi.org/10.1016/s2352-4642(18)30306-7.
78. Patton GC, Hibbert ME, Carlin J, et al. Menarche and the onset of depression and anxiety in Victoria. *Australia J Epidemiol Community Health*. 1996;50(6):661–666. https://doi.org/10.1136/jech.50.6.661.
79. Lakshman R, Forouhi NG, Sharp SJ, et al. Early age at menarche associated with cardiovascular disease and mortality. *J Clin Endocrinol Metab*. 2009;94(12):4953–4960. https://doi.org/10.1210/jc.2009-1789.
80. Britt K. Menarche, menopause, and breast cancer risk. *Lancet Oncol*. 2012;13 (11):1071–1072. https://doi.org/10.1016/s1470-2045(12)70456-4.
81. Foster DL, Lemons JA, Jaffe RB, Niswender GD. Sequential patterns of circulating luteinizing hormone and follicle-stimulating hormone in female sheep from early postnatal life through the first estrous cycles. *Endocrinology*. 1975;97(4):985–994. https://doi.org/10.1210/endo-97-4-985.
82. Foster DL, Ryan KD. Endocrine mechanisms governing transition into adulthood: a marked decrease in inhibitory feedback action of estradiol on tonic secretion of luteinizing hormone in the lamb during puberty. *Endocrinology*. 1979;105 (4):896–904. https://doi.org/10.1210/endo-105-4-896.
83. Teles MG, Bianco SD, Brito VN, et al. A GPR54-activating mutation in a patient with central precocious puberty. *N Engl J Med*. 2008;358(7):709–715. https://doi.org/10.1056/NEJMoa073443.

84. Plant TM, Ramaswamy S, Dipietro MJ. Repetitive activation of hypothalamic G protein-coupled receptor 54 with intravenous pulses of kisspeptin in the juvenile monkey (Macaca mulatta) elicits a sustained train of gonadotropin-releasing hormone discharges. *Endocrinology*. 2006;147(2):1007–1013. https://doi.org/10.1210/en.2005-1261.
85. Han SK, Gottsch ML, Lee KJ, et al. Activation of gonadotropin-releasing hormone neurons by kisspeptin as a neuroendocrine switch for the onset of puberty. *J Neurosci*. 2005;25(49):11349–11356. https://doi.org/10.1523/JNEUROSCI.3328-05.2005.
86. Navarro VM, Castellano JM, Fernandez-Fernandez R, et al. Developmental and hormonally regulated messenger ribonucleic acid expression of KiSS-1 and its putative receptor, GPR54, in rat hypothalamus and potent luteinizing hormone-releasing activity of KiSS-1 peptide. *Endocrinology*. 2004;145(10):4565–4574. https://doi.org/10.1210/en.2004-0413.
87. Castellano JM, Navarro VM, Fernandez-Fernandez R, et al. Ontogeny and mechanisms of action for the stimulatory effect of kisspeptin on gonadotropin-releasing hormone system of the rat. *Mol Cell Endocrinol*. 2006;257-258:75–83. https://doi.org/10.1016/j.mce.2006.07.002.
88. Redmond JS, Macedo GG, Velez IC, Caraty A, Williams GL, Amstalden M. Kisspeptin activates the hypothalamic-adenohypophyseal-gonadal axis in prepubertal ewe lambs. *Reproduction*. 2011;141(4):541–548. https://doi.org/10.1530/REP-10-0467.
89. Guerriero KA, Keen KL, Millar RP, Terasawa E. Developmental changes in GnRH release in response to kisspeptin agonist and antagonist in female rhesus monkeys (Macaca mulatta): implication for the mechanism of puberty. *Endocrinology*. 2012;153(2):825–836. https://doi.org/10.1210/en.2011-1565.
90. Takumi K, Iijima N, Ozawa H. Developmental changes in the expression of kisspeptin mRNA in rat hypothalamus. *J Mol Neurosci*. 2011;43(2):138–145. https://doi.org/10.1007/s12031-010-9430-1.
91. Redmond JS, Baez-Sandoval GM, Spell KM, et al. Developmental changes in hypothalamic Kiss1 expression during activation of the pulsatile release of luteinising hormone in maturing ewe lambs. *J Neuroendocrinol*. 2011;23(9):815–822. https://doi.org/10.1111/j.1365-2826.2011.02177.x.
92. Shahab M, Mastronardi C, Seminara SB, Crowley WF, Ojeda SR, Plant TM. Increased hypothalamic GPR54 signaling: a potential mechanism for initiation of puberty in primates. *Proc Natl Acad Sci U S A*. 2005;102(6):2129–2134. https://doi.org/10.1073/pnas.0409822102.
93. Li Q, Smith JT, Henry B, Rao A, Pereira A, Clarke IJ. Expression of genes for Kisspeptin (KISS1), Neurokinin B (TAC3), Prodynorphin (PDYN), and gonadotropin inhibitory hormone (RFRP) across natural puberty in ewes. *Phys Rep*. 2020;8(5):e14399. https://doi.org/10.14814/phy2.14399.
94. Bedenbaugh MN, D'Oliveira M, Cardoso RC, Hileman SM, Williams GL, Amstalden M. Pubertal escape from estradiol negative feedback in ewe lambs is not accounted for by decreased ESR1 mRNA or protein in kisspeptin neurons. *Endocrinology*. 2018;159(1):426–438. https://doi.org/10.1210/en.2017-00593.
95. Aerts EG, Harlow K, Griesgraber MJ, et al. Kisspeptin, neurokinin B, and dynorphin expression during pubertal development in female sheep. *Biology (Basel)*. 2021;10(10). https://doi.org/10.3390/biology10100988.
96. Lopez JA, Bedenbaugh MN, McCosh RB, et al. Does dynorphin play a role in the onset of puberty in female sheep? *J Neuroendocrinol*. 2016;28(12). https://doi.org/10.1111/jne.12445.

97. Keen KL, Wegner FH, Bloom SR, Ghatei MA, Terasawa E. An increase in kisspeptin-54 release occurs with the pubertal increase in luteinizing hormone-releasing hormone-1 release in the stalk-median eminence of female rhesus monkeys in vivo. *Endocrinology*. 2008;149(8):4151–4157. https://doi.org/10.1210/en.2008-0231.
98. Garcia JP, Keen KL, Seminara SB, Terasawa E. Role of kisspeptin and NKB in puberty in nonhuman primates: sex differences. *Semin Reprod Med*. 2019;37(2):47–55. https://doi.org/10.1055/s-0039-3400253.
99. Plant TM. Neuroendocrine control of the onset of puberty. *Front Neuroendocrinol*. 2015;38:73–88.
100. Plant TM, Barker-Gibb ML. Neurobiological mechanisms of puberty in higher primates. *Hum Reprod Update*. 2004;10(1):67–77.
101. Chongthammakun S, Terasawa E. Negative feedback effects of estrogen on luteinizing hormone-releasing hormone release occur in pubertal, but not prepubertal, ovariectomized female rhesus monkeys. *Endocrinology*. 1993;132(2):735–743.
102. Terasawa E, Fernandez DL. Neurobiological mechanisms of the onset of puberty in primates. *Endocr Rev*. 2001;22(1):111–151.
103. Nestor CC, Bedenbaugh MN, Hileman SM, Coolen LM, Lehman MN, Goodman RL. Regulation of GnRH pulsatility in ewes. *Reproduction*. 2018;156(3):R83–r99. https://doi.org/10.1530/rep-18-0127.
104. Jackson GL, Kuehl D. Gamma-aminobutyric acid (GABA) regulation of GnRH secretion in sheep. *Reprod Suppl*. 2002;59:15–24.
105. Goodman RL, Jansen HT, Billings HJ, Coolen LM, Lehman MN. Neural systems mediating seasonal breeding in the ewe. *J Neuroendocrinol*. 2010;22(7):674–681. https://doi.org/10.1111/j.1365-2826.2010.02014.x.
106. Carter AM. Animal models of human pregnancy and placentation: alternatives to the mouse. *Reproduction*. 2020;160(6):R129–R143. https://doi.org/10.1530/rep-20-0354.
107. Barry JS, Anthony RV. The pregnant sheep as a model for human pregnancy. *Theriogenology*. 2008;69(1):55–67. https://doi.org/10.1016/j.theriogenology.2007.09.021.
108. Leiser R, Krebs C, Ebert B, Dantzer V. Placental vascular corrosion cast studies: a comparison between ruminants and humans. *Microsc Res Tech*. 1997;38(1-2):76–87. https://doi.org/10.1002/(SICI)1097-0029(19970701/15)38:1/2<76::AID-JEMT9>3.0.CO;2-S.
109. Makowski EL, Battaglia FC, Meschia G, et al. Effect of maternal exposure to high altitude upon fetal oxygenation. *Am J Obstet Gynecol*. 1968;100(6):852–861. https://doi.org/10.1016/s0002-9378(15)33590-0.
110. Thakor AS, Giussani DA. Effects of acute acidemia on the fetal cardiovascular defense to acute hypoxemia. *Am J Phys Regul Integr Comp Phys*. 2009;296(1):R90–R99. https://doi.org/10.1152/ajpregu.90689.2008.
111. Unno N, Giussani DA, Hing WK, Ding XY, Collins JH, Nathanielsz PW. Changes in adrenocorticotropin and cortisol responsiveness after repeated partial umbilical cord occlusions in the late gestation ovine fetus. *Endocrinology*. 1997;138(1):259–263. https://doi.org/10.1210/endo.138.1.4880.
112. Müller T, Löhle M, Schubert H, et al. Developmental changes in cerebral autoregulatory capacity in the fetal sheep parietal cortex. *J Physiol*. 2002;539(pt 3):957–967. https://doi.org/10.1113/jphysiol.2001.012590.
113. Jacobs SE, Berg M, Hunt R, Tarnow-Mordi WO, Inder TE, Davis PG. Cooling for newborns with hypoxic ischaemic encephalopathy. *Cochrane Database Syst Rev*. 2013;2013(1). https://doi.org/10.1002/14651858.CD003311.pub3, Cd003311.

114. Gunn AJ, Gunn TR, de Haan HH, Williams CE, Gluckman PD. Dramatic neuronal rescue with prolonged selective head cooling after ischemia in fetal lambs. *J Clin Invest.* 1997;99(2):248–256. https://doi.org/10.1172/jci119153.
115. Gotsch F, Romero R, Kusanovic JP, et al. The fetal inflammatory response syndrome. *Clin Obstet Gynecol.* 2007;50(3):652–683. https://doi.org/10.1097/GRF.0b013e31811ebef6.
116. Morrison JL, Rurak DW, Chien C, et al. Maternal fluoxetine infusion does not alter fetal endocrine and biophysical circadian rhythms in pregnant sheep. *J Soc Gynecol Investig.* 2005;12(5):356–364. https://doi.org/10.1016/j.jsgi.2005.03.001.
117. Allison BJ, Crossley KJ, Flecknoe SJ, et al. Ventilation of the very immature lung in utero induces injury and BPD-like changes in lung structure in fetal sheep. *Pediatr Res.* 2008;64(4):387–392. https://doi.org/10.1203/PDR.0b013e318181e05e.
118. McMillen I, Robinson JS. Developmental origins of the metabolic syndrome: prediction, plasticity, and programming. *Physiol Rev.* 2005;85:571–633.
119. Liggins GC. Premature delivery of foetal lambs infused with glucocorticoids. *J Endocrinol.* 1969;45(4):515–523. https://doi.org/10.1677/joe.0.0450515.
120. Nitsos I, Rees SM, Duncan J, et al. Chronic exposure to intra-amniotic lipopolysaccharide affects the ovine fetal brain. *J Soc Gynecol Investig.* 2006;13(4):239–247. https://doi.org/10.1016/j.jsgi.2006.02.011.
121. Yan E, Castillo-Meléndez M, Nicholls T, Hirst J, Walker D. Cerebrovascular responses in the fetal sheep brain to low-dose endotoxin. *Pediatr Res.* 2004;55(5):855–863. https://doi.org/10.1203/01.Pdr.0000115681.95957.D4.
122. Dean JM, van de Looij Y, Sizonenko SV, et al. Delayed cortical impairment following lipopolysaccharide exposure in preterm fetal sheep. *Ann Neurol.* 2011;70(5):846–856. https://doi.org/10.1002/ana.22480.
123. Dobbing J, Sands J. Comparative aspects of the brain growth spurt. *Early Hum Dev.* 1979;3(1):79–83. https://doi.org/10.1016/0378-3782(79)90022-7.
124. Sankaran S. Creasy and Resnik's maternal–fetal medicine: principles and practice sixth edition. *Obstet Med.* 2012;5(2):88–89. https://doi.org/10.1258/om.2011.11E005.
125. Howson CP, Kinney MV, McDougall L, Lawn JE. Born too soon: preterm birth matters. *Reprod Health.* 2013;10(1):1–9.
126. Winterhager E, Bendix I, Miller SL. *Causes and Consequences of Intrauterine Growth Restriction.* vol 11. Front Endocrinol (Lausanne); 2020.
127. Morrison JL. Sheep models of intrauterine growth restriction: fetal adaptations and consequences. *Clin Exp Pharmacol Physiol.* 2008;35(7):730–743. https://doi.org/10.1111/j.1440-1681.2008.04975.x.
128. Beede KA, Limesand SW, Petersen JL, Yates DT. Real supermodels wear wool: summarizing the impact of the pregnant sheep as an animal model for adaptive fetal programming. *Animal Frontiers.* 2019;9(3):34–43. https://doi.org/10.1093/af/vfz018.
129. Halliwell C. Chapter 4—The role of animal models in developmental brain research. In: Gibb R, Kolb B, eds. *The Neurobiology of Brain and Behavioral Development.* Academic Press; 2018:81–95.
130. Brooks AN, McNeilly AS, Thomas GB. Role of GnRH in the ontogeny and regulation of the fetal hypothalamo-pituitary-gonadal axis in sheep. *J Reprod Fertil Suppl.* 1995;49:163–175.
131. Scott HM, Mason JI, Sharpe RM. Steroidogenesis in the fetal testis and its susceptibility to disruption by exogenous compounds. *Endocr Rev.* 2009;30(7):883–925. https://doi.org/10.1210/er.2009-0016.
132. Padmanabhan V, Sarma HN, Savabieasfahani M, Steckler TL, Veiga-Lopez A. Developmental reprogramming of reproductive and metabolic dysfunction in sheep: native steroids vs. environmental steroid receptor modulators. *Int J Androl.* 2010;33(2):394–404. https://doi.org/10.1111/j.1365-2605.2009.01024.x.

133. Kreisman MJ, Song CI, Yip K, Natale BV, Natale DR, Breen KM. Androgens mediate sex-dependent gonadotropin expression during late prenatal development in the mouse. *Endocrinology*. 2017;158(9):2884–2894. https://doi.org/10.1210/en.2017-00285.
134. Migrenne S, Pairault C, Racine C, Livera G, Geloso A, Habert R. Luteinizing hormone-dependent activity and luteinizing hormone-independent differentiation of rat fetal Leydig cells. *Mol Cell Endocrinol*. 2001;172(1-2):193–202. https://doi.org/10.1016/s0303-7207(00)00339-7.
135. Bouvattier C, Maione L, Bouligand J, Dode C, Guiochon-Mantel A, Young J. Neonatal gonadotropin therapy in male congenital hypogonadotropic hypogonadism. *Nat Rev Endocrinol*. 2011;8(3):172–182. https://doi.org/10.1038/nrendo.2011.164.
136. Roselli CE, Estill CT, Stadelman HL, Meaker M, Stormshak F. Separate critical periods exist for testosterone-induced differentiation of the brain and genitals in sheep. *Endocrinology*. 2011;152(6):2409–2415. https://doi.org/10.1210/en.2010-1445.
137. Ford JJ, D'Occhio MJ. Differentiation of sexual behavior in cattle, sheep and swine. *J Anim Sci*. 1989;67(7):1816–1823. https://doi.org/10.2527/jas1989.6771816x.
138. Foster DL, Jackson LM, Padmanabhan V. Novel concepts about normal sexual differentiation of reproductive neuroendocrine function and the developmental origins of female reproductive dysfunction: the sheep model. *Soc Reprod Fertil Suppl*. 2007;64:83–107. https://doi.org/10.5661/rdr-vi-83.
139. Pallayova M, Brandeburova A, Tokarova D. Update on sexual dimorphism in brain structure-function interrelationships: a literature review. *Appl Psychophysiol Biofeedback*. 2019;44(4):271–284. https://doi.org/10.1007/s10484-019-09443-1.
140. Roselli CE. Programmed for preference: the biology of same-sex attraction in rams. *Neurosci Biobehav Rev*. 2020;114:12–15. https://doi.org/10.1016/j.neubiorev.2020.03.032.
141. Partridge EA, Davey MG, Hornick MA, et al. An extra-uterine system to physiologically support the extreme premature lamb. *Nat Commun*. 2017;8:15112. https://doi.org/10.1038/ncomms15112.
142. Berry MJ. EXTENDing fetal life: postnatal placentae for preterm babies. *J Physiol*. 2018;596(9):1547–1548. https://doi.org/10.1113/jp275922.
143. Re Frasch MG. The perinatal development of arterial pressure in sheep: effects of low birth weight due to twinning. *Reprod Sci*. 2008;15(9):863–864. author reply 865 https://doi.org/10.1177/1933719108324894.
144. Frasch MG, Müller T, Wicher C, et al. Fetal body weight and the development of the control of the cardiovascular system in fetal sheep. *J Physiol*. 2007;579 (pt 3):893–907. https://doi.org/10.1113/jphysiol.2006.124800.
145. Hancock SN, Oliver MH, McLean C, Jaquiery AL, Bloomfield FH. Size at birth and adult fat mass in twin sheep are determined in early gestation. *J Physiol*. 2012;590 (5):1273–1285. https://doi.org/10.1113/jphysiol.2011.220699.
146. Galloway SM, McNatty KP, Cambridge LM, et al. Mutations in an oocyte-derived growth factor gene (BMP15) cause increased ovulation rate and infertility in a dosage-sensitive manner. *Nat Genet*. 2000;25(3):279–283. https://doi.org/10.1038/77033.
147. Hanrahan JP, Gregan SM, Mulsant P, et al. Mutations in the genes for oocyte-derived growth factors GDF9 and BMP15 are associated with both increased ovulation rate and sterility in Cambridge and Belclare sheep (Ovis aries). *Biol Reprod*. 2004;70 (4):900–909. https://doi.org/10.1095/biolreprod.103.023093.
148. Di Pasquale E, Beck-Peccoz P, Persani L. Hypergonadotropic ovarian failure associated with an inherited mutation of human bone morphogenetic protein-15 (BMP15) gene. *Am J Hum Genet*. 2004;75(1):106–111. https://doi.org/10.1086/422103.

149. Dixit H, Rao LK, Padmalatha VV, et al. Missense mutations in the BMP15 gene are associated with ovarian failure. *Hum Genet.* 2006;119(4):408–415. https://doi.org/10.1007/s00439-006-0150-0.
150. Montgomery GW, Zhao ZZ, Marsh AJ, et al. A deletion mutation in GDF9 in sisters with spontaneous DZ twins. *Twin Res.* 2004;7(6):548–555. https://doi.org/10.1375/1369052042663823.
151. Varcoe TJ, Darby JRT, Gatford KL, et al. Considerations in selecting postoperative analgesia for pregnant sheep following fetal instrumentation surgery. *Animal Frontiers.* 2019;9(3):60–67. https://doi.org/10.1093/af/vfz019.
152. Adachi S, Yamada S, Takatsu Y, et al. Involvement of anteroventral periventricular metastin/kisspeptin neurons in estrogen positive feedback action on luteinizing hormone release in female rats. *J Reprod Dev.* 2007;53(2):367–378. https://doi.org/10.1262/jrd.18146.
153. Cheng G, Coolen LM, Padmanabhan V, Goodman RL, Lehman MN. The kisspeptin/neurokinin B/dynorphin (KNDy) cell population of the arcuate nucleus: sex differences and effects of prenatal testosterone in sheep. *Endocrinology.* 2010;151(1):301–311. https://doi.org/10.1210/en.2009-0541.
154. Desroziers E, Mikkelsen JD, Duittoz A, Franceschini I. Kisspeptin-immunoreactivity changes in a sex- and hypothalamic-region-specific manner across rat postnatal development. *J Neuroendocrinol.* 2012;24(8):1154–1165. https://doi.org/10.1111/j.1365-2826.2012.02317.x.
155. Xu Z, Kaga S, Mochiduki A, et al. Immunocytochemical localization of kisspeptin neurons in the rat forebrain with special reference to sexual dimorphism and interaction with GnRH neurons. *Endocr J.* 2012;59(2):161–171. https://doi.org/10.1507/endocrj.ej11-0193.
156. Kauffman AS, Gottsch ML, Roa J, et al. Sexual differentiation of Kiss1 gene expression in the brain of the rat. *Endocrinology.* 2007;148(4):1774–1783. https://doi.org/10.1210/en.2006-1540.
157. Homma T, Sakakibara M, Yamada S, et al. Significance of neonatal testicular sex steroids to defeminize anteroventral periventricular kisspeptin neurons and the GnRH/LH surge system in male rats. *Biol Reprod.* 2009;81(6):1216–1225. https://doi.org/10.1095/biolreprod.109.078311.
158. Bakker J, Pierman S, Gonzalez-Martinez D. Effects of aromatase mutation (ArKO) on the sexual differentiation of kisspeptin neuronal numbers and their activation by same versus opposite sex urinary pheromones. *Horm Behav.* 2010;57(4-5):390–395. https://doi.org/10.1016/j.yhbeh.2009.11.005.
159. Gonzalez-Martinez D, De Mees C, Douhard Q, Szpirer C, Bakker J. Absence of gonadotropin-releasing hormone 1 and Kiss1 activation in alpha-fetoprotein knockout mice: prenatal estrogens defeminize the potential to show preovulatory luteinizing hormone surges. *Endocrinology.* 2008;149(5):2333–2340. https://doi.org/10.1210/en.2007-1422.
160. Kauffman AS, Navarro VM, Kim J, Clifton DK, Steiner RA. Sex differences in the regulation of Kiss1/NKB neurons in juvenile mice: implications for the timing of puberty. *Am J Physiol Endocrinol Metab.* 2009;297(5):E1212–E1221. https://doi.org/10.1152/ajpendo.00461.2009.
161. Amodei R, Gribbin K, He W, et al. Role for kisspeptin and neurokinin B in regulation of luteinizing hormone and testosterone secretion in the fetal sheep. *Endocrinology.* 2020;161(4). https://doi.org/10.1210/endocr/bqaa013.
162. Arnold AP, McCarthy MM. Sexual differentiation of the brain and behavior: a primer. In: Pfaff DW, Volkow ND, eds. *Neuroscience in the 21st Century: From Basic to Clinical.* New York: Springer; 2016:2139–2168.

163. Fauser BC, Tarlatzis BC, Rebar RW, et al. Consensus on women's health aspects of polycystic ovary syndrome (PCOS): the Amsterdam ESHRE/ASRM-Sponsored 3rd PCOS Consensus Workshop Group. *Fertil Steril.* 2012;97(1):28–38. e25 https://doi.org/10.1016/j.fertnstert.2011.09.024.
164. Escobar-Morreale HF. Polycystic ovary syndrome: definition, aetiology, diagnosis and treatment. *Nat Rev Endocrinol.* 2018;14(5):270–284. https://doi.org/10.1038/nrendo.2018.24.
165. Bruni V, Capozzi A, Lello S. The role of genetics, epigenetics and lifestyle in polycystic ovary syndrome development: the state of the art. *Reprod Sci.* 2021. https://doi.org/10.1007/s43032-021-00515-4.
166. Risal S, Pei Y, Lu H, et al. Prenatal androgen exposure and transgenerational susceptibility to polycystic ovary syndrome. *Nat Med.* 2019;25(12):1894–1904.
167. Barnes RB, Rosenfield RL, Ehrmann DA, et al. Ovarian hyperandrogynism as a result of congenital adrenal virilizing disorders: evidence for perinatal masculinization of neuroendocrine function in women. *J Clin Endocrinol Metab.* 1994;79(5):1328–1333. https://doi.org/10.1210/jcem.79.5.7962325.
168. Yildiz BO, Azziz R. The adrenal and polycystic ovary syndrome. *Rev Endocr Metab Disord.* 2007;8(4):331–342. https://doi.org/10.1007/s11154-007-9054-0.
169. Cardoso RC, Padmanabhan V. Developmental Programming of PCOS Traits: Insights from the Sheep. *Med Sci (Basel).* 2019;7(7). https://doi.org/10.3390/medsci7070079.
170. Cardoso RC, Padmanabhan V. Prenatal steroids and metabolic dysfunction: lessons from sheep. *Annu Rev Anim Biosci.* 2019;7:337–360. https://doi.org/10.1146/annurev-animal-020518-115154.
171. Padmanabhan V, Veiga-Lopez A. Sheep models of polycystic ovary syndrome phenotype. *Mol Cell Endocrinol.* 2013;373(1-2):8–20. https://doi.org/10.1016/j.mce.2012.10.005.
172. Abbott DH, Dumesic DA, Eisner JR, Colman RJ, Kemnitz JW. Insights into the development of polycystic ovary syndrome (PCOS) from studies of prenatally androgenized female rhesus monkeys. *Trends Endocrinol Metab.* 1998;9(2):62–67. https://doi.org/10.1016/s1043-2760(98)00019-8.
173. Ortega HH, Salvetti NR, Padmanabhan V. Developmental programming: prenatal androgen excess disrupts ovarian steroid receptor balance. *Reproduction.* 2009;137(5):865–877. https://doi.org/10.1530/REP-08-0491.
174. Padmanabhan V, Veiga-Lopez A. Developmental origin of reproductive and metabolic dysfunctions: androgenic versus estrogenic reprogramming. *Semin Reprod Med.* 2011;29(3):173–186. https://doi.org/10.1055/s-0031-1275519.
175. Dou J, Puttabyatappa M, Padmanabhan V, Bakulski KM. Developmental programming: adipose depot-specific transcriptional regulation by prenatal testosterone excess in a sheep model of PCOS. *Mol Cell Endocrinol.* 2021;523:111137. https://doi.org/10.1016/j.mce.2020.111137.
176. Carrasco A, Recabarren MP, Rojas-García PP, et al. Insulin sensitivity in male sheep born to ewes treated with testosterone during pregnancy. *J Dev Orig Health Dis.* 2021;12(3):456–464. https://doi.org/10.1017/s2040174420000665.
177. Cardoso RC, Burns A, Moeller J, Skinner DC, Padmanabhan V. Developmental programming: insulin sensitizer prevents the GnRH-stimulated LH hypersecretion in a sheep model of PCOS. *Endocrinology.* 2016;157(12):4641–4653. https://doi.org/10.1210/en.2016-1613.
178. Ramaswamy S, Grace C, Mattei A, et al. Developmental programming of polycystic ovary syndrome (PCOS): prenatal androgens establish pancreatic islet α/β cell ratio and subsequent insulin secretion. *Sci Rep.* 2016;6(1):1–10.

179. Siemienowicz KJ, Coukan F, Franks S, Rae MT, Duncan WC. Aberrant subcutaneous adipogenesis precedes adult metabolic dysfunction in an ovine model of polycystic ovary syndrome (PCOS). *Mol Cell Endocrinol.* 2021;519: 111042.
180. Birch RA, Padmanabhan V, Foster DL, Unsworth WP, Robinson JE. Prenatal programming of reproductive neuroendocrine function: fetal androgen exposure produces progressive disruption of reproductive cycles in sheep. *Endocrinology.* 2003;144 (4):1426–1434. https://doi.org/10.1210/en.2002-220965.
181. Sarma HN, Manikkam M, Herkimer C, et al. Fetal programming: excess prenatal testosterone reduces postnatal luteinizing hormone, but not follicle-stimulating hormone responsiveness, to estradiol negative feedback in the female. *Endocrinology.* 2005;146(10):4281–4291. https://doi.org/10.1210/en.2005-0322.
182. Sharma TP, Herkimer C, West C, et al. Fetal programming: prenatal androgen disrupts positive feedback actions of estradiol but does not affect timing of puberty in female sheep. *Biol Reprod.* 2002;66(4):924–933. https://doi.org/10.1095/biolreprod66.4.924.
183. Steckler TL, Roberts EK, Doop DD, Lee TM, Padmanabhan V. Developmental programming in sheep: administration of testosterone during 60-90 days of pregnancy reduces breeding success and pregnancy outcome. *Theriogenology.* 2007;67(3):459–467. https://doi.org/10.1016/j.theriogenology.2006.08.010.
184. Veiga-Lopez A, Astapova OI, Aizenberg EF, Lee JS, Padmanabhan V. Developmental programming: contribution of prenatal androgen and estrogen to estradiol feedback systems and periovulatory hormonal dynamics in sheep. *Biol Reprod.* 2009;80 (4):718–725. https://doi.org/10.1095/biolreprod.108.074781.
185. Cernea M, Phillips R, Padmanabhan V, Coolen LM, Lehman MN. Prenatal testosterone exposure decreases colocalization of insulin receptors in kisspeptin/neurokinin B/dynorphin and agouti-related peptide neurons of the adult ewe. *Eur J Neurosci.* 2016;44(8):2557–2568. https://doi.org/10.1111/ejn.13373.
186. Robinson JE, Forsdike RA, Taylor JA. In utero exposure of female lambs to testosterone reduces the sensitivity of the gonadotropin-releasing hormone neuronal network to inhibition by progesterone. *Endocrinology.* 1999;140(12):5797–5805. https://doi.org/10.1210/endo.140.12.7205.
187. Pastor CL, Griffin-Korf ML, Aloi JA, Evans WS, Marshall JC. Polycystic ovary syndrome: evidence for reduced sensitivity of the gonadotropin-releasing hormone pulse generator to inhibition by estradiol and progesterone. *J Clin Endocrinol Metab.* 1998;83 (2):582–590. https://doi.org/10.1210/jcem.83.2.4604.
188. Porter DT, Moore AM, Cobern JA, et al. Prenatal testosterone exposure alters GABAergic synaptic inputs to GnRH and KNDy neurons in a sheep model of polycystic ovarian syndrome. *Endocrinology.* 2019;160(11):2529–2542. https://doi.org/10.1210/en.2019-00137.
189. Frohman LA, Downs TR, Chomczynski P. Regulation of growth hormone secretion. *Front Neuroendocrinol.* 1992;13(4):344–405.
190. Bluet-Pajot MT, Epelbaum J, Gourdji D, Hammond C, Kordon C. Hypothalamic and hypophyseal regulation of growth hormone secretion. *Cell Mol Neurobiol.* 1998;18 (1):101–123.
191. McMahon CD, Radcliff RP, Lookingland KJ, Tucker HA. Neuroregulation of growth hormone secretion in domestic animals. *Domest Anim Endocrinol.* 2001;20(2):65–87.
192. Gladysz A, Krejci P, Simunek J, Polkowska J. Effects of central infusions of neuropeptide Y on the somatotropic axis in sheep fed on two levels of protein. *Acta Neurobiol Exp (Wars).* 2001;61(4):255–266.
193. Tillet Y, Picard S, Bruneau G, et al. Hypothalamic arcuate neuropeptide Y-neurons decrease periventricular somatostatin-neuronal activity before puberty in the female lamb: morphological arguments. *J Chem Neuroanat.* 2010;40(4):265–271. https://doi.org/10.1016/j.jchemneu.2010.07.003.

194. Deltondo J, Por I, Hu W, et al. Associations between the human growth hormone-releasing hormone- and neuropeptide-Y-immunoreactive systems in the human diencephalon: a possible morphological substrate of the impact of stress on growth. *Neuroscience*. 2008;153(4):1146–1152. https://doi.org/10.1016/j.neuroscience.2008.02.072.
195. Thomas MG, Amstalden M, Hallford DM, et al. Dynamics of GHRH in third-ventricle cerebrospinal fluid of cattle: relationship with serum concentrations of GH and responses to appetite-regulating peptides. *Domest Anim Endocrinol*. 2009;37(4):196–205. https://doi.org/10.1016/j.domaniend.2009.05.006.
196. Lin S, Lin EJ, Boey D, et al. Fasting inhibits the growth and reproductive axes via distinct Y2 and Y4 receptor-mediated pathways. *Endocrinology*. 2007;148(5):2056–2065. https://doi.org/10.1210/en.2006-1408.
197. Horvath TL, Bechmann I, Naftolin F, Kalra SP, Leranth C. Heterogeneity in the neuropeptide Y-containing neurons of the rat arcuate nucleus: GABAergic and non-GABAergic subpopulations. *Brain Res*. 1997;756(1-2):283–286. https://doi.org/10.1016/s0006-8993(97)00184-4.
198. Allen LG, Crowley WR, Kalra SP. Interactions between neuropeptide Y and adrenergic systems in the stimulation of luteinizing hormone release in steroid-primed ovariectomized rats. *Endocrinology*. 1987;121(6):1953–1959. https://doi.org/10.1210/endo-121-6-1953.
199. Tong Q, Ye CP, Jones JE, Elmquist JK, Lowell BB. Synaptic release of GABA by AgRP neurons is required for normal regulation of energy balance. *Nat Neurosci*. 2008;11(9):998–1000. https://doi.org/10.1038/nn.2167.
200. Ho KY, Veldhuis JD, Johnson ML, et al. Fasting enhances growth hormone secretion and amplifies the complex rhythms of growth hormone secretion in man. *J Clin Invest*. 1988;81(4):968–975. https://doi.org/10.1172/JCI113450.
201. Thomas GB, Cummins JT, Francis H, Sudbury AW, McCloud PI, Clarke IJ. Effect of restricted feeding on the relationship between hypophysial portal concentrations of growth hormone (GH)-releasing factor and somatostatin, and jugular concentrations of GH in ovariectomized ewes. *Endocrinology*. 1991;128(2):1151–1158. https://doi.org/10.1210/endo-128-2-1151.
202. Glad CA, Kitchen EE, Russ GC, et al. Reverse feeding suppresses the activity of the GH axis in rats and induces a preobesogenic state. *Endocrinology*. 2011;152(3):869–882. https://doi.org/10.1210/en.2010-0713.
203. Steyn FJ, Leong JW, Huang L, et al. GH does not modulate the early fasting-induced release of free fatty acids in mice. *Endocrinology*. 2012;153(1):273–282. https://doi.org/10.1210/en.2011-1681.
204. Driver PM, Forbes JM. Episodic growth hormone secretion in sheep in relation to time of feeding, spontaneous meals and short term fasting. *J Physiol*. 1981;317:413–424. https://doi.org/10.1113/jphysiol.1981.sp013833.
205. Tannenbaum GS, Rorstad O, Brazeau P. Effects of prolonged food deprivation on the ultradian growth hormone rhythm and immunoreactive somatostatin tissue levels in the rat. *Endocrinology*. 1979;104(6):1733–1738. https://doi.org/10.1210/endo-104-6-1733.
206. Whitlock BK, Daniel JA, Wilborn RR, Maxwell HS, Steele BP, Sartin JL. Interaction of kisspeptin and the somatotropic axis. *Neuroendocrinology*. 2010;92(3):178–188. https://doi.org/10.1159/000318049.
207. Foradori CD, Whitlock BK, Daniel JA, et al. Kisspeptin stimulates growth hormone release by utilizing neuropeptide Y pathways and is dependent on the presence of ghrelin in the ewe. *Endocrinology*. 2017;158(10):3526–3539. https://doi.org/10.1210/en.2017-00303.
208. Backholer K, Smith JT, Rao A, et al. Kisspeptin cells in the ewe brain respond to leptin and communicate with neuropeptide Y and proopiomelanocortin cells. *Endocrinology*. 2010;151(5):2233–2243. https://doi.org/10.1210/en.2009-1190.

209. Kim GL, Dhillon SS, Belsham DD. Kisspeptin directly regulates neuropeptide Y synthesis and secretion via the ERK1/2 and p38 mitogen-activated protein kinase signaling pathways in NPY-secreting hypothalamic neurons. *Endocrinology*. 2010;151 (10):5038–5047. https://doi.org/10.1210/en.2010-0521.
210. Morrison CD, Daniel JA, Hampton JH, et al. Luteinizing hormone and growth hormone secretion in ewes infused intracerebroventricularly with neuropeptide Y. *Domest Anim Endocrinol*. 2003;24(1):69–80. https://doi.org/10.1016/s0739-7240(02)00206-0.
211. Watanobe H, Habu S. Leptin regulates growth hormone-releasing factor, somatostatin, and alpha-melanocyte-stimulating hormone but not neuropeptide Y release in rat hypothalamus in vivo: relation with growth hormone secretion. *J Neurosci*. 2002;22 (14):6265–6271. 20026615.
212. Marson L, Lauterio TJ, Della-Fera MA, Baile CA. Immunohistochemical distribution of cholecystokinin, dynorphin A and Met-enkephalin neurons in sheep hypothalamus. *Neurosci Lett*. 1987;81(1-2):35–40. https://doi.org/10.1016/0304-3940(87)90336-3.
213. McShane TM, May T, Miner JL, Keisler DH. Central actions of neuropeptide-Y may provide a neuromodulatory link between nutrition and reproduction. *Biol Reprod*. 1992;46(6):1151–1157. https://doi.org/10.1095/biolreprod46.6.1151.
214. Adeva-Andany MM, González-Lucán M, Donapetry-García C, Fernández-Fernández C, Ameneiros-Rodríguez E. Glycogen metabolism in humans. *BBA Clin*. 2016; 5:85–100. https://doi.org/10.1016/j.bbacli.2016.02.001.
215. Wong JM, Jenkins DJ. Carbohydrate digestibility and metabolic effects. *J Nutr*. 2007;137(suppl 11):2539s–2546s. https://doi.org/10.1093/jn/137.11.2539S.
216. Annison EF, Bryden WL. Perspectives on ruminant nutrition and metabolism. II. Metabolism in ruminant tissues. *Nutr Res Rev*. 1999;12(1):147–177. https://doi.org/10.1079/095442299108728866.
217. Nozière P, Ortigues-Marty I, Loncke C, Sauvant D. Carbohydrate quantitative digestion and absorption in ruminants: from feed starch and fibre to nutrients available for tissues. *Animal*. 2010;4(7):1057–1074. https://doi.org/10.1017/s1751731110000844.
218. Battaglia F, Meschia G. Principal substrates of fetal metabolism. *Physiol Rev*. 1978; 58(2):499–527.

CHAPTER TWO

Relevance of dog as an animal model for urologic diseases

Hannah Ruetten and Chad M. Vezina*
Department of Comparative Biosciences, School of Veterinary Medicine, University of Wisconsin-Madison, Madison, WI, United States
*Corresponding author: e-mail address: chad.vezina@wisc.edu

Contents

Abstract

We utilize animal models in urologic research to improve understanding of urinary physiology, determine the etiology of many urologic diseases, and discover and test novel therapeutic interventions. Dogs have a similar urinary tract anatomy and physiology to human and they develop many urologic diseases spontaneously. This chapter offers detailed comparisons of urinary tract anatomy, physiology, and the most common urologic diseases between humans and dogs. Dogs offer a unique opportunity for urologic research because they can be studied in research colonies and in client

Progress in Molecular Biology and Translational Science, Volume 189
ISSN 1877-1173
https://doi.org/10.1016/bs.pmbts.2021.12.002

owned cohorts. Dogs also are among a limited number of non-human species that require continence and socially appropriate urinary behaviors (ex. going to the bathroom outside, training to not have submissive urination, etc.). These features make dogs unique in the animal kingdom and make them an ideal animal model for urologic research.

1. Introduction

Urologic diseases can affect a range of organs including the bladder, urethra, kidney, and/or reproductive organs. The most common urologic diseases for which human patients seek treatment are urinary tract infections, prostate cancer, benign prostatic hyperplasia, erectile dysfunction, and urinary incontinence.[1] A non-comprehensive list of human urologic diseases can be found in Table 1, ordered in increasing prevalence based on number of physician office visits in 2007.[1] Also included in Table 1 are select references describing the spontaneous occurrence of each urologic disease in dogs or studies that use dogs as a large animal model.

Progress in urologic research has lagged that of other research fields. We are just beginning to unravel the complex etiologies of many urologic diseases. This is in part due to the stigma around urinary voiding- leading to underdiagnosis of urologic disorders, patient and community lack of awareness of urologic diseases and treatment options, and lower funding for urologic research than for other diseases.[2–4] Also, urologic diseases tend to be referred to as bothersome rather than life threatening, which further decreases their apparent significance. The truth is that urologic diseases are not just bothersome, but they can also progress to more severe disease, associated with depression, and have a considerable impact on "major life activities."[2,5–7]

We utilize animal models in urologic research to improve understanding of urinary physiology, determine the etiology of many urologic diseases (for which causes are currently unknown), and discover and test novel therapeutic interventions. Numerous invertebrates and mammalian species have been used in urologic research, including *Drosophila*, non-human primates, pigs, sheep, rabbits, guinea pigs, rats, mice, cats, and dogs. Each species has its pros and cons including cost of housing, ease of genetic manipulations, size, and resemblance to human—anatomy, physiology, pathologic features of disease, and response to treatment.

Dogs offer a unique opportunity for urologic research because they can be studied in research colonies and in client owned cohorts. Traditional canine studies utilize research beagles as study subjects. Research beagles

Table 1 Human urologic diseases and references for similar diseases in dogs.

Urologic diseases in humans	Selected reference(s) on urologic disease in dogs
Urinary tract infections (adult female)	Thompson MF, Litster AL, Platell JL, Trott DJ. Canine bacterial urinary tract infections: new developments in old pathogens. Vet J. 2011 190:22–27. doi: https://doi.org/10.1016/j.tvjl.2010.11.013. PMID: 21239193. Seguin MA, Vaden SL, Altier C, Stone E, Levine JF. Persistent urinary tract infections and reinfections in 100 dogs (1989–1999). J Vet Intern Med. 2003 17:622–631. doi: https://doi.org/10.1111/j.1939-1676.2003.tb02492.x. PMID: 14529127. Chew DJ. Diagnosing Initial and Recurrent Urinary Tract Infections in Dogs. Orlando, FL: North American Veterinary Conference; 2001.
Prostate cancer (adult male)	Leroy BE, Northrup N. Prostate cancer in dogs: comparative and clinical aspects. Vet J. 2009 180:149–162. doi: https://doi.org/10.1016/j.tvjl.2008.07.012. PMID: 18786842. Argyle DJ. Prostate cancer in dogs and men: a unique opportunity to study the disease. Vet J. 2009 180:137–138. doi: https://doi.org/10.1016/j.tvjl.2008.08.012. PMID: 18835203. Smith J. Canine prostatic disease: a review of anatomy, pathology, diagnosis, and treatment. Theriogenology. 2008 70:375–383. doi: https://doi.org/10.1016/j.theriogenology.2008.04.039. PMID: 18514299.
Benign prostatic hyperplasia (adult male)	Zambelli D, Cunto M, Gentilini F. Validation of a model to develop a symptom index for benign prostatic hyperplasia in dogs. Reprod Domest Anim. 2012 47 Suppl 6:229–231. doi: https://doi.org/10.1111/rda.12084. PMID: 23279506. Pinheiro D, Machado J, Viegas C, Baptista C, Bastos E, Magalhães J, Pires MA, Cardoso L, Martins-Bessa A. Evaluation of biomarker canine-prostate specific arginine esterase (CPSE) for the diagnosis of benign prostatic hyperplasia. BMC Vet Res. 2017 13:76. doi: https://doi.org/10.1186/s12917-017-0996-5. PMID: 28335775. Smith J. Canine prostatic disease: a review of anatomy, pathology, diagnosis, and treatment. Theriogenology. 2008 70:375–83. doi: https://doi.org/10.1016/j.theriogenology.2008.04.039. PMID: 18514299.

Continued

Table 1 Human urologic diseases and references for similar diseases in dogs.—cont'd

Urologic diseases in humans	Selected reference(s) on urologic disease in dogs
Erectile dysfunction (adult male)	(Spontaneous erectile dysfunction has not been reported in canine literature. However, dogs have still been utilized for erectile dysfunction research.) Hellstrom WJ. Functional measurements of penile erection in feline, canine and primate animal models. Int J Impot Res. 2001 13:149–150. doi: https://doi.org/10.1038/sj.ijir.3900682. PMID: 11525312. Takahashi Y, Ishii N, Lue TF, Tanagho EA. Pharmacological effects of adenosine on canine penile erection. Tohoku J Exp Med. 1991 165:49–58. doi: https://doi.org/10.1620/tjem.165.49. PMID: 1798976.
Urinary incontinence (adult female)	Acierno MJ, Labato MA. Canine incontinence. Vet Clin North Am Small Anim Pract. 2019 49:125–140. doi: https://doi.org/10.1016/j.cvsm.2018.11.003. PMID: 30736891.
Interstitial cystitis (non-specific)	(Not common in dogs however cats present with a similar disease to human—Feline Idiopathic Cystitis)
Urinary tract infections (children)	(See urinary tract infections—female adult citations above)
Urinary tract infections (adult male)	(See urinary tract infections—female adult citations above)
Lower tract transitional cell carcinoma (adult)	Fulkerson CM, Knapp DW. Management of transitional cell carcinoma of the urinary bladder in dogs: a review. Vet J. 2015 205:217–225. doi: https://doi.org/10.1016/j.tvjl.2015.01.017. PMID: 25747698. Knapp DW, Waters DJ. Naturally occurring cancer in pet dogs: important models for developing improved cancer therapy for humans. Mol Med Today. 1997 3:8–11. doi: https://doi.org/10.1016/s1357-4310(96)20031-0. PMID: 9021736. Luethcke KR, Ekena J, Chun R, Trepanier LA. Glutathione S-transferase theta genotypes and environmental exposures in the risk of canine transitional cell carcinoma. J Vet Intern Med. 2019 33:1414–1422. doi: https://doi.org/10.1111/jvim.15504. PMID: 31008543.
Urinary incontinence (adult male)	(see urinary incontinence- Female Adult citations above)

Prostatitis (adult male)	Barsanti JA, Finco DR. Canine bacterial prostatitis. Vet Clin North Am Small Anim Pract. 1980 9:679–700. doi: https://doi.org/10.1016/s0195-5616(79)50081-3. PMID: 6996296. Smith J. Canine prostatic disease: a review of anatomy, pathology, diagnosis, and treatment. Theriogenology. 2008 Aug;70(3):375–83. doi: https://doi.org/10.1016/j.theriogenology.2008.04.039. PMID: 18514299.
Upper urinary tract stones (non-specific)	Ross SJ, Osborne CA, Lulich JP, Polzin DJ, Ulrich LK, Koehler LA, Bird KA, Swanson LL. Canine and feline nephrolithiasis. Epidemiology, detection, and management. Vet Clin North Am Small Anim Pract. 1999 29:231–250, xiii-xiv. Doi: https://doi.org/10.1016/s0195-5616(99)50013-2. PMID: 10028160. Furrow E, Pfeifer RJ, Osborne CA, Lulich JP. An APRT mutation is strongly associated with and likely causative for 2,8-dihydroxyadenine urolithiasis in dogs. Mol Genet Metab. 2014 111:399–403. doi: https://doi.org/10.1016/j.ymgme.2013.12.002. PMID: 24359665.
Kidney cancer (non-specific)	Burgess KE, DeRegis CJ. Urologic oncology. Vet Clin North Am Small Anim Pract. 2019 49:311–323. doi: https://doi.org/10.1016/j.cvsm.2018.11.006. PMID: 30635132.
Urinary incontinence (children)	(See urinary incontinence—female adult citations above) Borns-Weil S. Inappropriate urination. Vet Clin North Am Small Anim Pract. 2019 49:141–155. doi: https://doi.org/10.1016/j.cvsm.2018.10.003. PMID: 30591188.
Interstitial cystitis (specific)	(Not common in dogs; however, cats can present with Feline Idiopathic Cystitis)
Urethral stricture (adult male)	Hill TL, Berent AC, Weisse CW. Evaluation of urethral stent placement for benign urethral obstructions in dogs. J Vet Intern Med. 2014 28:1384–1390. doi: https://doi.org/10.1111/jvim.12412. PMID: 25066355. Zhang K, Fang X, Zhu J, Yang R, Wang Y, Zhao W, Mo X, Fu Q. Effective reconstruction of functional urethra promoted with ICG-001 delivery using core-shell collagen/poly(Lactide-*co*-caprolactone) [P(LLA-CL)] nanoyarn-based scaffold: A study in dog model. Front Bioeng Biotechnol. 2020 8:774. doi: https://doi.org/10.3389/fbioe.2020.00774. PMID: 32754582.

Continued

Table 1 Human urologic diseases and references for similar diseases in dogs.—cont'd

Urologic diseases in humans	Selected reference(s) on urologic disease in dogs
Urethral stricture (adult female)	Hill TL, Berent AC, Weisse CW. Evaluation of urethral stent placement for benign urethral obstructions in dogs. J Vet Intern Med. 2014 28:1384–1390. doi: https://doi.org/10.1111/jvim.12412. PMID: 25066355. Wood MW, Vaden S, Cerda-Gonzalez S, Keene B. Cystoscopic-guided balloon dilation of a urethral stricture in a female dog. Can Vet J. 2007 48:731–733. PMID: 17824158.
Testicular cancer (adult male)	Grieco V, Riccardi E, Greppi GF, Teruzzi F, Iermanò V, Finazzi M. Canine testicular tumours: a study on 232 dogs. J Comp Pathol. 2008 138:86–89. doi: https://doi.org/10.1016/j.jcpa.2007.11.002. PMID: 18295787.
Undescended testis (male)	Khan FA, Gartley CJ, Khanam A. Canine cryptorchidism: an update. Reprod Domest Anim. 2018 Dec;53(6):1263–1270. doi: https://doi.org/10.1111/rda.13231. Epub 2018 Jun 28. PMID: 29956390.
Vesicoureteral reflux (non-specific)	B.E. Goulden B.V.Sc. M.R.C.V.S. Ph.D. Vesico-ureteral reflux in the dog. New Zealand Vet J. 1968; 16(12):167–175, DOI: https://doi.org/10.1080/00480169.1968.33769
Hypospadias (male)	Jurka P, Galanty M, Zielinska P, Max A, Sysa P. Hypospadias in six dogs. Vet Rec. 2009 Mar 14;164(11):331–3. doi: https://doi.org/10.1136/vr.164.11.331. PMID: 19287029.
Lower urinary tract/bladder stones (non-specific)	Syme HM. Stones in cats and dogs: what can be learnt from them? Arab J Urol. 2012 10:230–239. doi: https://doi.org/10.1016/j.aju.2012.06.006. PMID: 26558031. Bartges JW, Callens AJ. Urolithiasis. Vet Clin North Am Small Anim Pract. 2015 45:747–768. doi: https://doi.org/10.1016/j.cvsm.2015.03.001. PMID: 26002797.
Posterior urethral valves (juvenile male)	
Ureteropelvic junction obstruction (children)	Lam NK, Berent AC, Weisse CW, Bryan C, Mackin AJ, Bagley DH. Endoscopic placement of ureteral stents for treatment of congenital bilateral ureteral stenosis in a dog. J Am Vet Med Assoc. 2012 240:983–990. doi: https://doi.org/10.2460/javma.240.8.983. PMID: 22471828.

tend to have higher animal care costs than smaller animal models such as mice and rats. It is difficult to manipulate the dog genome to create de novo disease models; however, some spontaneous mutations in the canine genome are associated with an increased risk of urologic disease such as urolithiasis, and dog colonies have been expanded to examine disease mechanisms. The larger size of dogs compared to rodents make them ideal for refining surgical approaches for urologic diseases.

Veterinarians frequently reference human subjects research to determine probable etiologies and select treatments for their patients. This practice is possible because human anatomy, physiology, pathologic features of disease, and response to treatment is often conserved across mammalian species. It is less common for human physicians to reference veterinary literature for probable etiologies and treatment selection for human patients, although it is possible this practice will gain momentum with the recent emphasis on a "One Health" approach for research and medicine.

This "One Health" concept opens doors for client owned dogs as research models. Opportunities include the expansion of clinical studies and the development of biobanks and tissue banks for cohort and population-based screening. Client owned canines are genetically diverse and derive from diverse environmental backgrounds. They also have similar challenges for treatment including access to medical care and treatment protocol adherence. In this way, conducting clinical trials in client owned dogs better recapitulate variables present in human clinical medicine and successful trials in dogs are more likely to be successful in human patients compared to traditional animal studies. Dogs exist in the same environments as their human owners, providing further opportunities to study the influence of environmental toxicants and gene-environment interactions on urinary health.

Urinary diseases often have a complex multifactorial etiology with the possibility of both a physiologic and psychologic component. Dogs can spontaneously develop urinary dysfunction with both components. Further, dogs are among a limited number of non-human species that require continence and socially appropriate urinary behaviors (ex. going to the bathroom outside, training to not have submissive urination, etc.). This feature makes dogs unique in the animal kingdom and makes them an ideal animal model for urologic research. For a summary comparing features of mice/rats, research dogs, and client owned dogs, see Table 2.

Table 2 Strengths and limitations of using rodents, research colony dogs, and client owned dogs for urologic research studies.

	Mice/rats	Research colony dogs	Client owned dogs
Spontaneously develop common urologic diseases	Some diseases	Most diseases	Most diseases
Similarity of anatomy and physiology	Somewhat similar	Very similar	Very similar
Animal housing costs	Cheaper	Expensive	None
Genetic manipulation	Readily available	Some cohorts have been established with specific genetic mutations	No, but many have detailed pedigrees enabling tracing of heritability
Size	Small	Typically, 18–35 lb	Highly variable
Environment	Highly regulated	Highly regulated	Highly variable

2. Comparison of human and dog lower urinary tract anatomy and physiology

2.1 Kidneys

Two kidneys are located in the back part of the abdomen, behind the peritoneum, one on each side of the spine.[8,9] According to Henry Gray, kidneys in men and women span from the twelfth thoracic vertebra to the third lumbar vertebra.[9] The right kidney is positioned slightly lower than the left, and is about 10 cm long, 5 cm wide, and 2.5 cm thick.[9] The left kidney, therefore, is positioned slightly higher than the right and is longer and narrower.[9] The kidneys weigh 128–170 g each in men and 113–156 g in women.[9] According to Howard Evans, the kidney of an average-sized dog lies in a similar location, is 6–9 cm long, 4–5 cm wide, 3–4 cm thick, and weighs 25–35 g.[8] Evans states that in dog and Gray states that in humans each kidney is comprised of a cranial and caudal pole, a medial and lateral border, and dorsal and ventral surface.[8,9] The medial border contains an indented region, the hilus, where, in order of dorsal to ventral, the renal artery, the ureter and renal vein enter (Fig. 1A). The kidney is covered in a thin fibrous capsule and internally is comprised of the medulla and cortex.[8,9] Dogs have an unilobular unipapillary kidney structure (Fig. 1B)

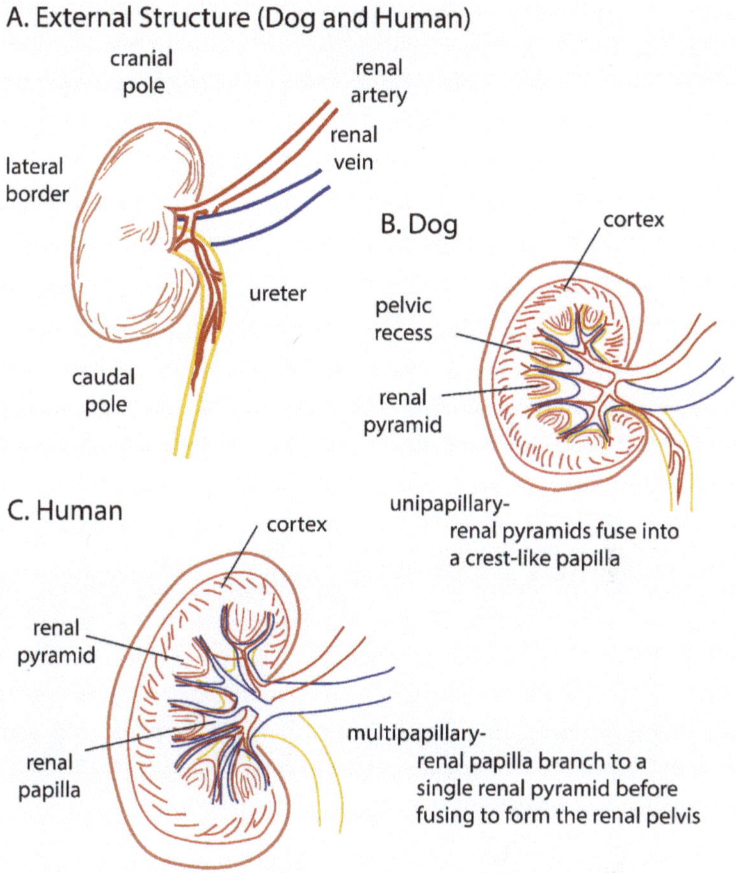

Fig. 1 Kidney structure in dog and human. (A) External structure of both dog and human. (B) Cross-section of unipapillary kidney structure of dog. (C) Cross-section of multipapillary kidney structure of human.

whereas humans have a multipapillary kidney (Fig. 1C).[10] The dog has significantly contributed to the development of dialysis and kidney transplants in humans,[10,11] and dogs are considered ideal models for studying chronic kidney disease and kidney stones.[12–15]

2.2 Ureters

The ureters are cylindrical tubes that carry urine from the kidney to the bladder. In humans, they are about 41 cm in length and the diameter of a "goose quill."[9] In dogs, they are 0.6–0.9 cm in diameter but their length varies greatly with the size of the dog.[8] In a 16 kg dog, the ureters are 12–16 cm long.[8] In both humans and dogs, according to Gray and Evans, respectively, the tube is comprised of 3 layers, an external fibrous layer, a muscular layer, and a mucus layer. The mucosal layer contains longitudinal folds and is comprised of transitional epithelium. The ureters drain into the bladder obliquely through slit like orifices just caudal to the bladder neck on the dorsolateral surface of the bladder.[8,9] Conditions such as ectopic ureters, ureterocele, and ureteral inflammation and scarring occur in both human and dog.[16]

2.3 Bladder

The bladder is a musculo-membranous organ that stores urine. According to Gray, the shape and position of the bladder vary greatly with age, sex, and quantity of urine it is holding.[9] In men, the bladder lies between the pubic symphysis and the rectum. In women, the bladder lies between the pubic symphysis and the vagina/cervix/uterus. Gray also states that in adult men and women, the bladder forms a cup-like shape when empty; the cranial surface contracts down to form the "V" shape of a "Y," with the stem of the "Y" being the urethra.[9] When slightly full the bladder has a rounded shape but is completely contained in the pelvic cavity. When very full, it is ovoid in shape and the cranial aspect extends beyond the pelvic cavity into the abdominal cavity and can reach as high as the umbilicus when distended. This change in bladder shape during urination was recently confirmed in men, using modern imaging techniques, with real-time MRI during patient urination.[17] In women, the bladder is larger in transverse diameter than in vertical diameter. At a typical capacity of 473 mL (US pint) of urine, the human bladder is 13 cm long and 8 cm wide.[9] According to Evans, the bladder of a 11 kg dog can hold 100–120 mL of urine without distension. At this capacity, the bladder measures 18 cm long and 17.5 cm wide. When contracted, it is 3.2 cm long and 2 cm wide.[8] In the intact male dog, the bladder lies between the pubis ventrally and the ductus deferens, small intestines, and rectum dorsally. In the intact female dog, the bladder lies between the pubis ventrally and the body of the uterus and cervix and small intestines dorsally.[8]

The musculature of the bladder, the detrusor, exists in three layers of smooth muscle in both dog and human: the internal longitudinal, the circular layer, and the external longitudinal.[18–23] The inner longitudinal layer is the thinnest and comprised of spindle shaped bundles that end as connective tissue at the bladder trigone.[19] The middle layer is the strongest muscle structure in the bladder.[19] The outer longitudinal layer is complex and includes a caudal collar-like structure that holds the muscles of the bladder together.[19] This collar, the *collare vesicae*, is fixed with two ventral extensions to the pubic symphysis called *musculi pubovesicales*.[19] The dorsal bundle of this outer layer forms the back of the bladder; projections of this bundle extend into the prostate in man and into the *septum vesicovaginale* in women.[19]

2.4 Internal urethral sphincter

Dorschner *et al.*, after evaluation of 3000 serial sections of urinary tract from 65 patients of various age and sex, found that the detrusor ends at the internal urethral orifice in humans and do not contribute to the internal urethral sphincter.[19] Others contest that the detrusor musculature contributes to the ventral aspect of the internal urethral sphincter in what is known as the base-plate model.[24–27] There is contention over whether an internal sphincter exists in dogs.[23] Miller *et al.* and Cullen *et al.* conclude that there is no pre-prostatic urethra in the dog, whereas Bruschini *et al.* conclude that the internal sphincter in male dogs resides within the prostate.[28,29] Stolzenburg *et al.* conclude that the detrusor extends into the urethra in female dogs and all three detrusor layers extend through the proximal third of the urethra, while individual bundles from the inner layer extend through the distal urethra.[30] In this way, no true internal urethral sphincter exists in female dogs.[30] Dorschner *et al.* demonstrate that the bladder trigone in men and women is comprised of a single layer of muscle with two parts: the *musculus interuretericus*, which is made of the musculature of the ureters and comes together to form the interureteric ridge, and the *musculus sphincter vesicae*, which elliptically wraps the internal urethral orifice and is the only muscle system capable of occluding the internal urethral orifice.[20] The interureteric ridge prevents reflux of urine into the ureters and also prevents draining of urine from the ureters directly into the urethra.[20] In men, the musculus sphincter vesicae is infiltrated with prostate glandular tissue and likely has a role in continence, preventing retrograde ejaculation, and placing pressure on the prostate to release prostatic fluid.[20]

2.5 Male urethra

The male urethra extends from the bladder neck to the tip of the penis. The human male urethra has a double curve when the penis is in the flaccid state but a single curve in the erect state. The urethra is 20–23 cm long.[9] There are three regions of urethra in men: prostatic, membranous, and spongy. In the resting state, the urethra is a slit that in the prostatic portion is horizontal and arched and at the tip of the penis the slit is vertical and flat.[9] The intact male dog urethra is divided into a pelvic urethra (preprostatic and prostatic portions) and a cavernous urethra. In a 11 kg dog, the urethra is 25 cm in length.[8]

2.5.1 Prostatic urethra (men vs. dogs)

In men, Gray states that the prostatic urethra is the widest and most dilate-able portion of the urethra extending from the apex to the base of the prostate.[9] It is about 3 cm long and spindle-like in shape with the middle being the widest and the beginning at the bladder neck and the exit at the membranous urethra being the narrowest.[9] The verumontanum is located on the floor of the prostatic urethra near the bladder neck. When ejaculation occurs, musculature surrounding the verumontanum contracts, causing it to block the passage of semen into the bladder and the passage of urine out of the bladder.[31] A prostatic sinus is located on each side of the verumontanum and is where prostatic fluid enters the urethra during ejaculation. In the middle line of the verumontanum is a depression, the *sinus pocularis* or *uterus masculinus*, through which the ejaculatory ducts drain.[9] There are five formations the verumontanum can take: "Volcano," "Lighthouse," "Whale Tail," "Castle Door," and "Hood." Of the five formations, "Volcano" is the most common. Formation does not change with benign prostatic hypertrophy or body mass index; however, the diameter of the verumontanum does decrease with benign prostatic hypertrophy and increase with body mass index.[32]

In dogs, Evans states that the prostatic urethra is also an arched slit in its resting state.[8] The urethral crest is located on the floor of the prostatic urethra. In the center of the urethral crest is an oval enlargement, the seminal hillock, which contains the *uterus masculinus*. The deferent ducts open onto either side of the seminal hillock and the prostatic ducts drain into the urethra on either side of the urethral crest.[8]

2.5.2 Membranous and spongy urethra (men) vs. cavernous urethra (dogs)

In men, Gray states that the membranous urethra is the shortest urethral segment, just 1–2 cm long, and is the narrowest portion of urethra. The spongy urethra is the longest, about 15 cm long, and lies within the corpus spongiosum. It is about 0.6 cm in diameter throughout its entire length. The lining contains many orifices of mucus glands and follicles. The portion contained within the bulb is often called the *bulbous portion* and contains the draining ducts of the Cowper's glands.[9]

In dogs, Evans states that the cavernous urethra begins at the ischial arch and extends to the tip of the penis.[8]

2.5.3 Longitudinal urethral muscles

Anatomical similarities between dorsal longitudinal urethral muscles suggests a conserved function in human and dog.[23] The muscles in both species course from the *colliculus seminalis* to the bulb of the penis and form the raised region in the urethra, the *crista urethralis*.[23] The ventral longitudinal urethral muscles differ between humans and dogs. In male dogs, the muscle is present only caudal to the prostate; however, in man, the muscle originates closer to the bladder neck and the muscle bundles extend laterally.[23]

2.5.4 External urethral sphincter

In both human and dog, it has been debated whether the external urethral sphincter or voluntary sphincter is a component of the pelvic floor. Dorschner *et al.* and Stolzenburg *et al.* found that the external urethral sphincter musculature in humans and dogs exists as an independent unit containing a smooth and striated muscle component.[18,23] In man, this sphincter is horse-shoe shaped but in dog, it is circular.[23] The striated musculature of the sphincter comprises 50% of the urethral wall in male dogs but only 35.5% of the urethral wall in men.[28,33] In dog, the smooth muscle component is underdeveloped and thin.[23] The human external sphincter has sympathetic, parasympathetic, and somatic innervation, leading researchers to believe that it has both involuntary and voluntary functions in continence.[34] Innervation of the dog sphincter is understudied.

2.6 Female urethra

The female human urethra is 2.5–4 cm long and 0.6 cm in diameter, is embedded in the anterior wall of the vagina, and extends from the bladder neck to the *meatus urinarius*.[9] The female human urethra contains a muscular, erectile, and mucus layer.[9] The female dog urethra is 7–10 cm long and 0.5 cm in diameter.[8] The dorsal wall of the urethra is immediately adjacent to the ventral wall of the vagina.[8]

2.6.1 Longitudinal urethral muscles and external urethral sphincter

The female human external urethral sphincter is comprised of circular smooth muscle bundles that wrap the proximal half of the urethra as well as a striated muscle component that contributes roughly half of urethral tone.[35–37] Stolzenburg *et al.* found that the female dog detrusor extends into the urethra and all three detrusor layers extend through the proximal third of the urethra, while individual bundles from the inner layer extend through the distal urethra.[30] In both female dogs and female humans there are smooth and striated muscle components to the external urethral sphincter.[30] The circular layer of female dog detrusor completely surrounds the urethra, forming the smooth muscle portion of the urethral sphincter.[30] The smooth muscle component begins to thin at the level where striated muscle begins to wrap the urethra.[30] Until at the caudal portion there is exclusively striated muscle wrapping the vagina and urethra as one continuous muscle.[30] The main difference between the human and dog female external urethral sphincter is that in humans the striated muscles take the form of a two horse-shoes stacked on top of each other, whereas the dog muscle is circular.[30] The individual striated muscle bundles are also smaller in female humans than in female dogs (Table 3).[30,38]

Table 3 A comparison of select anatomical features.

	Male human	**Female human**	**Male dog**	**Female dog**
Kidney	10 cm (L) × 5 cm (W) × 2.5 cm (H) 128–170 g (Wt)	10 cm (L) × 5 cm (W) × 2.5 cm (H) 113–156 g (Wt)	6-9c m (L) × 4–5 cm (W) × 3–4 cm (H) 25–35 g (Wt)	
Ureter	41 cm (L) × “goose quill” (dia.)		12–16 cm (L) × 0.6–0.9 cm (dia.)	
Bladder	13 cm (L) × 8 cm (dia.) @ 473 mL	shorter and wider	18 cm (L) × 17.5 cm (dia.) @100–120 mL	
Urethra	20–23 (L) variable dia.	2.5–4 (L) × 0.6 (dia.)	25 (L) variable dia.	7–10 (L) × 0.5 (dia.)
Internal urethral sphincter	Yes, but infiltrated with prostate glandular tissue	Yes	no true internal urethral sphincter	
External urethral sphincter	Yes Smooth muscle and striated muscle component Horseshoe shaped		Yes Smooth muscle and striated muscle component Greater ratio of striated muscle to smooth muscle than human Circular shaped	
Longitudinal muscle of urethra are an extension of detrusor	Conflicting reports		Yes	

3. Comparison of human and dog urine production, storage, and excretion

Urine production occurs in an involuntary manner and is well understood. However, the mechanisms of urine storage and emptying are complex and include involuntary and voluntary responses largely influenced by social factors in both humans and dogs. Because of the complex interplay of factors influencing urine storage and emptying, the process of urination remains incompletely understood to this day. Herein we will compare human to dog by summarizing existing evidence but will also reveal knowledge gaps and describe popular hypotheses.

Men urinate on average 1718 mL per day and women 1762 mL per day.[39] The dog urinates 20–100 mL of urine per kg body weight per day.[16] Human urine specific gravity is 1.001–1.035 and pH is 4.6–8.0[40] and dog urine specific gravity is 1.015–1.045 and pH is 6–7.5.[41]

3.1 Urine production

Production of urine begins in the kidney. The human kidney consists of 200,000 to 1,800,000 nephrons[42–45] and the dog has 341,000 to 655,000 nephrons,[46] each composed of a glomerulus and tubule. It is within the nephron that an ultrafiltrate of the blood plasma is formed in both species.[42,47] The first step of urine formation is glomerular filtration.[47] Blood enters the glomerulus through afferent arterioles.[47] Within the glomerulus water and many solutes diffuse across the glomerular capillary walls and across the glomerular capsule into the renal tubules (glomerular filtration).[47] As the filtrate moves through the tubules, the majority of water and solutes are resorbed (tubular reabsorption).[47] The tubule segments, named from glomerulus to collecting duct, are proximal convoluted tubule, descending limb of the loop of Henle, ascending limb of the loop of Henle, and distal convoluted tubule.[47]

The rate of urine production is influenced by the glomerular filtration rate, which is influenced by renal, neural, and hormonal factors.[47] Renal factors include tubuloglomerular feedback and the myogenic mechanism.[47] In the nephron, the juxtaglomerular apparatus is comprised of the macula densa, glomerular cells, and the extraglomerular mesangial cells. NaCl in the fluid reaching the macula densa decreases glomerular filtration rate (GFR) by a mechanism involving nitric oxide release and constriction of afferent arterioles (tubuloglomerular feedback).[47] Average GFR is 92.0 mL/min/1.73 m^2 in men and 88.1 mL/min/1.73 m^2 in women.[48]

GFR in dogs is reported as per kg body weight and ranges from 3.05 to 4.85 mL/min/kg depending on mode of analysis.[49] Decreased pressure of arteries feeding the afferent arterioles reduces the stretching of the arterial walls and leads to the release of renin from glomerular cells.[47] This release of renin feeds into the renin-angiotensin-aldosterone axis, which regulates systemic arteriole pressure and thereby controls the rate of plasma entering the kidney in both species.[47]

For humans and dogs, sympathetic innervation of the kidney also plays a role in the rate of filtration with the release of catecholamines causing vasoconstriction, increasing sodium reabsorption, and increasing renin secretion.[47] Hormonal factors influencing glomerular filtration include the aforementioned angiotensin, which acts through the renin-angiotensin--aldosterone system, as well as atrial natruiuretic peptide (ANP), which is released when the heart's atria are stretched.[47] Release of ANP relaxes cells in the glomerulus, increasing capillary surface area and rate of filtration.[47]

The volume and composition of urine is influenced by tubular reabsorption and tubular secretion.[47] In both human and dog, the volume and concentration of urine are controlled by several factors, but key players are the proximal convoluted tubule and the loop of Henle.[42] Filtrate enters the proximal convoluted tubule at about 80 mL/min and the majority of water and nutrients are quickly resorbed (65% water, 100% glucose and amino acids, 50% chloride ions, 80–90% bicarbonate ions, 50% urea).[47] Filtrate enters the loop of Henle at 40–45 mL/min. In the loop of Henle, 15% of water, 20–30% of sodium ions, 20–30% of potassium ions, 35% of chloride ions, and 10–20% of bicarbonate ions, are resorbed.[47] Dogs tend to excrete a smaller volume of more concentrated urine than humans.[16,39–41] The ratio of long to short loops of Henle is often referenced and compared between species where increased long loops correspond to a higher concentrating ability.[50] Humans have a short to long loop ratio of 85:15 and dogs only have long loops.[51] Filtrate enters the distal convoluted tubule at about 25 mL/min.[47] In the distal convoluted tubule, 10–15% of water, 5% of sodium ions, and 5% of chloride ions are resorbed.[47]

Five hormones determine the amount of water and solutes that are resorbed.[47] Angiotensin II and aldosterone increase water, sodium, and chloride ion resorption in the proximal convoluted tubule.[47] Antidiuretic hormone (or vasopressin) increases water resorption in the distal convoluted tubule.[47] ANP inhibits sodium ion resorption and suppresses angiotensin II and aldosterone.[47] Finally, parathyroid hormone increases calcium ion resorption and decreases bicarbonate ion resorption.[47]

3.2 Urine storage

Urine is moved through the ureters in peristaltic waves.[42] Calmodulin activates myosin light-chain kinase in ureteral smooth muscle, causing contraction and electrical pacemaker cells in the proximal renal pelvis, controlling the frequency of these waves (in humans 1–6 waves per minute).[42,47] The ureteral smooth muscle receives both sympathetic and parasympathetic innervation and this innervation can modulate the contractility of the ureters; however, neither sympathetic nor parasympathetic innervation is required for ureter peristalsis.[42,52] Similarly, the volume of urine production can modulate the amplitude and frequency of peristaltic contractions but when decoupled, spontaneous peristalsis occurs independent of urine flow.[52] Urine exists in the ureters and enters the bladder through ureteral orifices in the bladder trigone.[42] A flap of mucosa covers each ureteral orifice that opens to allow filling but prevents backflow of urine from the bladder during the urine storage phase.[42]

The bladder and urinary sphincter receive sympathetic innervation (from intermediolateral cell column of T10 to L2), parasympathetic innervation (from the intermediolateral cell column of S2 to S4), and somatic innervation (from motor neurons of S2 to S4).[42] During the storage phase, the cortical and suprapontine centers in the brain inhibit voiding.[42] The pontine micturition center in the brain controls and coordinates bladder detrusor and the urinary sphincters, dictating urine storage or excretion with fast bursts of activity, both during voiding and between voids.[42,53] It is speculated that a subset of pontine micturition center neurons signal to store urine through a distinct set of neurotransmitters.[54,55]

In humans, bladder pressure rises steadily during filling from 0 to 50 mL.[42] Bladder pressure increases more slowly during filling from 50–300 mL, then increases rapidly during filling beyond 400 mL.[42] As the bladder fills, feelings of fullness are transmitted to the spinal cord via the pelvic and hypogastric nerves.[56,57] The myelinated afferent axons in these nerves convey information about bladder filling while the unmyelinated afferent axons (C-fibers) respond to noxious stimuli.[56,58,59] The cell bodies for both axon types are located in the dorsal root ganglion of T11-L2 and S2-S4 and synapse with interneurons involved in spinal reflexes and spinal-tract neurons that transmit information to higher brain centers such as the pontine micturition center.

Many medical textbooks describe the bladder as a sentient balloon, passively filling until "stretch receptors" trigger voiding. The bladder is not sentient during the urine storage phase. Urine storage is the result of complex interplay of sensory feedback occurring during a dynamic state of the bladder. The detrusor contracts transiently during filling. Transient contractions (TCs) occur during bladder filling in multiple species,[60–63] have been observed in ex vivo preps,[64–67] and increase in frequency with bladder pathologies.[57,68–71] Studies have shown that afferent nerve activity occurs within these TCs[62,72–75] and a recent study suggests that TC-induced bursts of afferent nerve activity play an important role in conveying the sensation of bladder fullness and regulating micturition frequency.[76]

3.3 Urine excretion

Early life urine excretion in humans and dogs occurs through the micturition reflex. As humans and dogs age, they gain voluntary control over initiation and termination.[47] Newborn human infants void through the micturition reflex but do not require perineal stimulation to do so. When a puppy is born, urination is involuntary and occurs through reflex voiding following perineal stimulation. As both humans and dogs age, they receive information on when and where it is appropriate to urinate. Toddlers are potty trained and dogs house trained. Both humans and dogs can experience nocturia (having to urinate at night) leading to bedwetting and accidents. Stress can also impact continence in both humans and dogs.

In a human or dog with voluntary control over initiation and termination of urination, urination is the result of a coordinated relaxation of the internal and external urethral sphincters, contraction of the detrusor, combination of relaxation and contraction of urethral musculature that results in the widening of the urethral lumen and bearing down of the pelvic floor.

4. Top 5 human urologic diseases: Comparison to spontaneous disease in dog, past/current use of dogs in research, and recommendations for future studies using dogs as large animal models

4.1 Urinary tract infections (adult female)

Much like in human females, urinary tract infections (UTIs) are a prevalent and problematic urinary disease in female dogs. The risk of UTI increases with age in both species.[77–81] Circulating estrogens protect against UTI in both species and spayed female dogs and post-menopausal women are at increased risk for infections.[82] *Escherichia coli* is the most common organism cultured in human and canine patients with UTI.[80,81,83–86] There is zoonotic transmission of infectious organisms such that uropathogenic isolates detected in dog owners are often the same as those detected in their pet.[87] Although several studies compare bacterial characteristics between dogs and humans,[88,89] this seems to be the limit of veterinary and human physician collaborative UTI research. There has been an underwhelming use of dogs, despite the similarity of disease and ample opportunities to utilize dogs in clinical studies. This is an area of urologic research where collaborations between veterinarians and urologists could be extremely fruitful, especially the use of dogs in clinical trials for UTI preventatives and therapeutics.

4.2 Prostate cancer (adult male)

It is common practice to use veterinary patients as models of human disease for carcinogenesis and tumor cell biology research. A body of work is focused specifically on comparing prostate morphology and function between men and dogs and also comparing and contrasting prostatic cancer etiology, progression, and treatment.[90,91] Men and dogs have a similar prostate structure; they are among the few animals known to develop spontaneous prostatic neoplasia.[92]

Prostate cancer is common in men with most cases being microscopic, low-grade, and slow progressing.[90] It is thought that early life castration may protect against prostate cancer in men; however, studies of early life castrate men are rare. Studies of men with steroid 5α-reductase type 2 (*SRD5A2*) polymorphisms are more common but associations between *SDR5A2* polymorphisms and prostate cancer risk are highly controversial.[93–95] Prostate cancer is rare in dogs, more common in castrate males, and aggressive with a poor prognosis. In men, most prostatic carcinomas arise from the peripheral zone (glandular region furthest from the urethra, lying just under the capsule).[96] In dogs, most prostatic carcinomas are thought to arise from the periurethral region (glandular region immediately adjacent to the urethra[97]) or from the urethral epithelium itself. In men, histologic prostatic intra-epithelial neoplasia (PIN) is a well characterized lesion and an accurate predictor of future prostatic carcinoma.[98,99] In dogs, lesions similar to PIN are noted but evidence of predictive value has not been established.[100–102] In fact, the limited available data points to a lack of association between PIN and prostatic carcinoma in dogs.[90] Much is known about potential genetic influences on the development of prostatic carcinoma in men[103]; men often have multiple foci of prostatic carcinoma, and prostatic carcinoma comprises a very heterogeneous population of neoplasia. There is also an established grading system, The Gleason scoring system, which is a highly accurate predictor of disease progression.[104] In dogs, little is known about the etiology and pathogenesis of prostatic carcinomas; compared to men, lesions are fairly homogeneous with most having a gland-like or acinar structure; there is not a well-established scoring system with accurate predictive value for dogs like there is for men.

Despite the differences that exist between prostate cancer in men and dogs, dogs remain an excellent large animal model for niche areas in prostate cancer research. In both men and dogs, androgen resistant prostatic carcinomas are rapidly progressing, metastasize to bone, and are difficult to treat. It is within the niche of developing novel therapeutics for high-grade and rapidly progressing prostatic carcinoma that dogs are a highly effective and beneficial animal model.

4.3 Benign prostatic hyperplasia (adult male)

Canine prostate anatomically and histologically resembles human prostate (Fig. 2).[97,105] Dog benign prostatic hyperplasia research was the foundation for understanding human benign prostatic hyperplasia.[106–110] Dogs and men develop urinary, ejaculatory, excretory, systemic, and/or locomotive dysfunction as a secondary to benign prostatic hyperplasia.[111,112] Dogs and men also develop prostatic inflammation and collagen accumulation with age, likely contributing to urinary symptoms in both species.[97,105,113–115]

Despite the long list of features that make dogs ideal for benign prostatic hyperplasia research, their use has lost favor in recent years, likely due to ethical considerations around using research beagle colonies, and the higher per diems of a large animal model. For these reasons, along with the ease of genetic modification in rodent models, rodents have largely replaced dogs in benign prostatic hyperplasia in the past 10–15 years. However, prostate structure in rodents differs from man and associated physiologic changes remain difficult to characterize in rodent models.

The field of benign prostatic hyperplasia research could benefit from a resurgence of canine research and not just traditional canine research using beagle colonies. Clinical populations of intact male dogs that spontaneously develop benign prostatic hyperplasia make ideal research subjects for clinical trials. Canine prostates are readily available from necropsy services and provide high quality tissue for immunohistochemistry, in situ hybridization, and other studies using FFPE tissue.

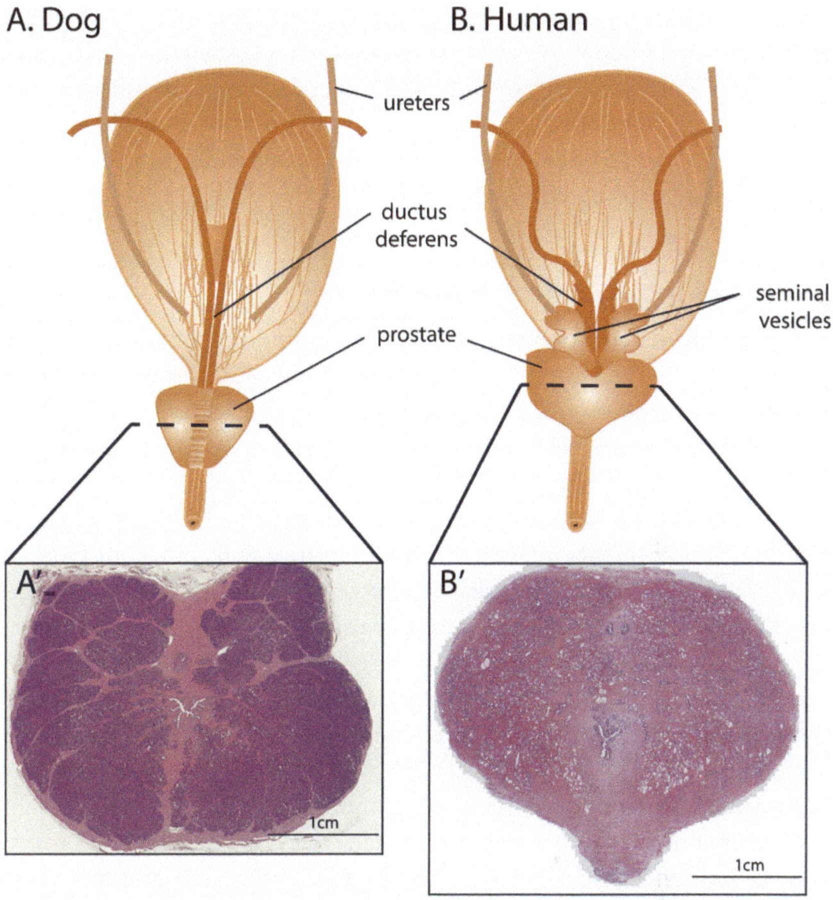

Fig. 2 Lower urinary tract anatomy and prostate histology in intact male dogs and men. (A) Diagram of intact male dog lower urinary tract. (A′) Cross-section of prostate stained with hematoxylin and eosin. (B) Diagram of male human lower urinary tract. (B′) Cross-section of prostate stained with hematoxylin and eosin.

4.4 Erectile dysfunction (adult male)

Spontaneous erectile dysfunction is rare/absent in dogs; however, dogs have still been utilized as a large animal model for investigating mechanisms and therapies. Unlike humans, dogs have two separate corpora cavernosa, and cAMP is not the predominate second messenger system for erectile physiology.[116] These two features, along with sizeable variations from dog to dog, make the dog less ideal as an animal model.[116] In this area of research, cats, which have a single corpora cavernosa and utilize cAMP as the predominate second messenger system for erectile physiology, are actually the more favorable animal model.[116] However, dogs have been established as the standard model for evaluating safety and toxicology of many pharmaceuticals.[116] Therefore they are used favorably in the study of drug pharmacokinetics and safety and toxicity evaluations in the development and approval of new therapeutics for erectile dysfunction.[117,118]

4.5 Urinary incontinence (adult female)

Urinary incontinence is prevalent in dogs and improving incontinence treatment has become increasingly important for the dog and owner's quality of life. This is because dogs have moved from being primarily outside/farm dogs to the now more common household member.

There are a range of causes for urinary incontinence in dogs which are not fully understood. Incontinence in dogs can have both a behavioral and a physiologic component, making a detailed history and thorough work-up necessary to determine potential etiology and optimal course of treatment.[119] Common physiologic causes of urinary incontinence in dogs are urethral sphincter mechanism incompetence, ectopic ureters, pelvic bladder syndrome, detrusor instability, or nerve damage.[119] The majority of urinary incontinence in dogs is attributed to urethral sphincter mechanism incompetence.[119–123]

In women, urinary incontinence is commonly divided into four groups: overflow incontinence, stress incontinence, urge incontinence, or mixed. Stress incontinence occurs when a stressor is placed on the urethral closure mechanism, such as a cough or sneeze, and urinary leaking occurs. Urge incontinence is a sudden urgency to urinate that cannot be delayed.[124] Another category of incontinence is overflow incontinence. The symptoms of overflow incontinence are similar to those associated with stress and urge incontinence but overflow incontinence is characterized by incomplete bladder emptying.[4] In women, stress incontinence is the most common, followed by mixed, and then urge incontinence.[125–128] Overflow incontinence usually occurs as a result of a systemic neurologic disorder, nerve damage, or urethral obstruction.

Diagnosis of urinary incontinence in dogs and women includes a detailed patient history, a physical exam, and urinalysis and urine culture.[4,119] The work-up in dogs also includes physician observance of a void. Additionally, canine diagnosis of urinary incontinence usually includes a complete blood cell count and serum chemistry to determine if there are any noticeable underlying causes. In atypical cases or those that do not respond to therapy, further diagnostics such as abdominal ultrasound to assess urinary tract for any abnormalities, including post-void residual, and urodynamics (if equipment is available) may be pursued. In women, urinary stress testing and assessment of post-void residual are more commonly performed.[4]

Treatment in both dogs and women is centered on controlling symptoms. This could include behavioral and lifestyle modifications, medications, and/or surgical procedures.

Common treatments for urethral sphincter mechanism incompetence in dogs include phenylpropanolamine and estriol. Estriol is commonly used in dogs which experience urethral sphincter mechanism incompetence following a spay surgery. Phenylpropanolamine as a sole intervention results in clinical resolution in 85% of dogs[129,130] and estriol in 80%.[131] Several surgical procedures also exist for dogs, although due to effectiveness of medical therapeutics and cost, they are rarely pursued. Surgical interventions include endoscopic injection of a bulking agent in the urethral submucosa[132–134] and placement of a hydraulic urethral occuluder.[135] Sling procedures have also been performed but are not as broadly accepted in dogs.[136–139] The treatment of choice for dogs with detrusor instability (the most commonly accepted etiology of urge incontinence) is anti-cholinergic drugs: oxybutynin, flavoxate hydrochloride, and dicyclomine.[119]

Almost all women begin with noninvasive therapies, including behavioral and lifestyle modifications and pelvic floor exercises.[4] Behavioral and lifestyle modifications include smoking cessation (if applicable), weight loss (if applicable), reducing excessive fluid consumption, reducing caffeine consumption, frequent intake of small volume fluids, and voiding at set intervals.[4] For stress incontinence, women are limited to lifestyle/behavioral modifications, pelvic floor exercises, and surgical interventions; there are no FDA approved drugs for stress incontinence. Surgical interventions are similar to dogs, including urethral bulking agent injection and urethral sling; however, urethroplexy is performed rather than a hydraulic urethral occulder.[4] In women, placing a urethral sling is the most common procedure. For urge incontinence, there are 6 FDA approved drugs in the primary medication class available for treatment: oxybutynin, darifenacin, fesoterodine, solifenacin, tolterodine, and trospium. All are anticholinergic medications. Mirabegron, a β3-adrenergic receptor agonist, is an alternative treatment option when anticholinergics are contraindicated. Unlike in dogs, where surgical interventions are not pursued for urge incontinence, there are 3 FDA approved interventions for women. They are percutaneous tibial nerve stimulation, onabotulinumtoxinA injection, and sacral neuromodulation.[4]

Dogs have been used minimally as animal models in studies of urinary incontinence, but presentation and treatment are nearly identical. Urethral anatomy is similar with the only two differences being that 1) in women, the external urethral sphincter striated muscles take the form of a two horse-shoes stacked on top of each other, whereas the muscle is circular in female dogs,[30] and 2) the individual striated muscle bundles are smaller in women than in female dogs.[30,38] There is much to be learned from collaborative urinary incontinence research between physicians and veterinarians. Dogs would make excellent candidates for clinical trials of both medical and surgical interventions.

5. Concluding thoughts

Despite the similarity of anatomy and physiology and the similarity of the etiology, diagnosis, and treatment of many urologic diseases between dogs and humans, there has been little evidence of collaboration of the human and veterinary fields to tackle these complex multifactorial diseases. There is abounding literature that demonstrates the benefit of "One Health" research, comparative medicine, and translational research.[140] Dogs as animal models are critical to the medical advancements that are being made today.[140] Collaborating with veterinarians and employing dogs as animal models for urologic diseases is critical to continue progressing forward in the field of urologic research.

References

1. Litwin M, Saigal C. *Urologic Diseases in America*. US Department of Health and Human Services, Public Health Service, National Institutes of Health, National Institute of Diabetes and Digestive and Kidney Diseases; 2012.
2. Abrams P, Kelleher CJ, Kerr LA, Rogers RG. Overactive bladder significantly affects quality of life. *Am J Manag Care*. 2000;6(11 Suppl):S580–S590.
3. Ricci JA, Baggish JS, Hunt TL, et al. Coping strategies and health care-seeking behavior in a US national sample of adults with symptoms suggestive of overactive bladder. *Clin Ther*. 2001;23(8):1245–1259.
4. Lukacz ES, Santiago-Lastra Y, Albo ME, Brubaker L. Urinary incontinence in women: a review. *JAMA*. 2017;318(16):1592–1604.
5. Coyne KS, Wein AJ, Tubaro A, et al. The burden of lower urinary tract symptoms: evaluating the effect of LUTS on health-related quality of life, anxiety and depression: EpiLUTS. *BJU Int*. 2009;103(Suppl 3):4–11.

6. Stewart WF, Van Rooyen JB, Cundiff GW, et al. Prevalence and burden of overactive bladder in the United States. *World J Urol.* 2003;20(6):327–336.
7. Dunphy C, Laor L, Te A, Kaplan S, Chughtai B. Relationship between depression and lower urinary tract symptoms secondary to benign prostatic hyperplasia. *Rev Urol.* 2015;17(2):51–57.
8. Evans HE, Miller ME. *Miller's Anatomy of the Dog.* 4th ed. St. Louis, Missouri: Elsevier; 2013.
9. Gray H. *Anatomy: Descriptive and Surgical.* 15th ed. London, UK: Bounty Books; 2012.
10. Becker GJ, Hewitson TD. Animal models of chronic kidney disease: useful but not perfect. *Nephrol Dial Transplant.* 2013;28(10):2432–2438.
11. Abel JJ, Rowntree LG, Turner BB. On the removal of diffusable substances from the circulating blood by means of dialysis. Transactions of the Association of American Physicians, 1913. *Transfus Sci.* 1990;11(2):164–165.
12. Yarger WE, Griffith LD. Intrarenal hemodynamics following chronic unilateral ureteral obstruction in the dog. *Am J Phys.* 1974;227(4):816–826.
13. Carvalho M, Lulich JP, Osborne CA, Nakagawa Y. Role of urinary inhibitors of crystallization in uric acid nephrolithiasis: Dalmatian dog model. *Urology.* 2003; 62(3):566–570.
14. Furrow E, Lees GE, Brown CA, Cianciolo RE. Glomerular Lesions in Proteinuric Miniature Schnauzer Dogs. *Vet Pathol.* 2017;54(3):484–489.
15. Furrow E, McCue ME, Lulich JP. Urinary metals in a spontaneous canine model of calcium oxalate urolithiasis. *PLoS One.* 2017;12(5):e0176595.
16. Aiello SE, Moses MA, Allen DG. *The Merck Veterinary Manual.* 11th ed. Kenilworth, NJ: Merck & Co., Inc.; 2016.
17. Pewowaruk R, Rutkowski D, Hernando D, Kumapayi BB, Bushman W, Roldán-Alzate A. A pilot study of bladder voiding with real-time MRI and computational fluid dynamics. *PLoS One.* 2020;15(11):e0238404.
18. Dorschner W, Stolzenburg JU. A new theory of micturition and urinary continence based on histomorphological studies. 3. The two parts of the musculus sphincter urethrae: physiological importance for continence in rest and stress. *Urol Int.* 1994;52(4):185–188.
19. Dorschner W, Stolzenburg JU, Leutert G. A new theory of micturition and urinary continence based on histomorphological studies. 1. The musculus detrusor vesicae: occlusive function or support of micturition? *Urol Int.* 1994;52(2):61–64.
20. Dorschner W, Stolzenburg JU, Dieterich F. A new theory of micturition and urinary continence based on histomorphological studies. 2. The musculus sphincter vesicae: continence or sexual function? *Urol Int.* 1994;52(3):154–158.
21. Dorschner W, Stolzenburg JU, Rassler J. A new theory of micturition and urinary continence based on histomorphological studies. 4. The musculus dilator urethrae: force of micturition. *Urol Int.* 1994;52(4):189–193.
22. Dorschner W, Stolzenburg JU. A new theory of micturition and urinary continence based on histomorphological studies. 5. The musculus ejaculatorius: a newly described structure responsible for seminal emission and ejaculation. *Urol Int.* 1994;53(1):34–37.
23. Stolzenburg JU, Schwalenberg T, Do M, et al. Is the male dog comparable to human? A histological study of the muscle systems of the lower urinary tract. *Anat Histol Embryol.* 2002;31(4):198–205.
24. Hutch JA. The internal urinary sphincter: a double-loop system. *J Urol.* 1971; 105(3):375–383.
25. Hutch JA. A new theory of the anatomy of the internal urinary sphincter and the physiology of micturition. *Investig Urol.* 1965;3:36–58.
26. Hutch JA, Rambo ON. A new theory of the anatomy of the internal urinary sphincter and the physiology of micturition. 3. Anatomy of the urethra. *J Urol.* 1967; 97(4):696–704.

27. Hutch JA. A new theory of the anatomy of the internal urinary sphincter and the physiology of micturition. IV. The urinary sphincteric mechanism. *J Urol.* 1967; 97(4):705–712.
28. Cullen WC, Fletcher TF, Bradley WE. Histology of the canine urethra II. Morphometry of the male pelvic urethra. *Anat Rec.* 1981;199(2):187–195.
29. Bruschini H, Schmidt RA, Tanagho EA. The male genitourinary sphincter mechanism in the dog. *Investig Urol.* 1978;15(4):284–287.
30. Stolzenburg JU, Dorschner W, Postenjak M, et al. Sphincteric musculature of female canine urethra in comparison to woman including 3D reconstruction. *Cells Tissues Organs.* 2002;170(2–3):151–161.
31. Medved M, Sammet S, Yousuf A, Oto A. MR imaging of the prostate and adjacent anatomic structures before, during, and after ejaculation: qualitative and quantitative evaluation. *Radiology.* 2014;271(2):452–460.
32. de Menezes HB, Sampaio FJB, de Resende JAD, et al. Anatomic study of verumontanum during endoscopic surgeries in patients with benign prostatic hyperplasia. *Int Braz J Urol.* 2021;47(2):308–321.
33. Kaye KW, Milne N, Creed K, van der Werf B. The 'urogenital diaphragm', external urethral sphincter and radical prostatectomy. *Aust N Z J Surg.* 1997;67(1):40–44.
34. Elbadawi A, Schenk EA. A new theory of the innervation of bladder musculature. 4. Innervation of the vesicourethral junction and external urethral sphincter. *J Urol.* 1974;111(5):613–615.
35. Augsburger HR, Cruz-Orive LM, Arnold S. Morphology and stereology of the female canine urethra correlated with the urethral pressure profile. *Acta Anat (Basel).* 1993;148(4):197–205.
36. Tanagho EA. Urodynamics of female urinary incontinence with emphasis on stress incontinence. *J Urol.* 1979;122(2):200–204.
37. Rud T, Andersson KE, Asmussen M, Hunting A, Ulmsten U. Factors maintaining the intraurethral pressure in women. *Investig Urol.* 1980;17(4):343–347.
38. Augsburger HR, Cruz-Orive LM. Morphological, histochemical and stereological analysis of the female canine M. urethralis. *Histochemistry.* 1994;102(5):373–382.
39. van Haarst EP, Heldeweg EA, Newling DW, Schlatmann TJ. The 24-h frequency-volume chart in adults reporting no voiding complaints: defining reference values and analysing variables. *BJU Int.* 2004;93(9):1257–1261.
40. Ridley JW. *Fundamentals of the Study of Urine and Body Fluids.* Cham: Springer International Publishing; 2018. Imprint: Springer http://link.springer.com/10.1007/978-3-319-78417-5.
41. Rizzi ET, Valenciano A, Bowles M, Cowell R, Tyler R, DeNicola BD. *Atlas of Canine and Feline Urinalysis.* 1st ed. Hoboken, NJ: John Wiley & Sons, Inc; 2017.
42. Boron WF, Boulpaep EL. *Medical Physiology.* 3rd ed. Philadelphia, PA: Elsevier; 2017.
43. Hoy WE, Douglas-Denton RN, Hughson MD, Cass A, Johnson K, Bertram JF. A stereological study of glomerular number and volume: preliminary findings in a multiracial study of kidneys at autopsy. *Kidney Int Suppl.* 2003;83:S31–S37.
44. Hughson M, Farris AB, Douglas-Denton R, Hoy WE, Bertram JF. Glomerular number and size in autopsy kidneys: the relationship to birth weight. *Kidney Int.* 2003;63 (6):2113–2122.
45. Nyengaard JR, Bendtsen TF. Glomerular number and size in relation to age, kidney weight, and body surface in normal man. *Anat Rec.* 1992;232(2):194–201.
46. Eisenbrandt DL, Phemister RD. Postnatal development of the canine kidney: quantitative and qualitative morphology. *Am J Anat.* 1979;154(2):179–193.
47. Tortora G, Derrickson B. *Principles of Anatomy and Physiology.* 13th ed. Hoboken, NJ: John Wiley and Sons, Inc; 2012.

48. Fenton A, Montgomery E, Nightingale P, et al. Glomerular filtration rate: new age- and gender- specific reference ranges and thresholds for living kidney donation. *BMC Nephrol.* 2018;19(1):336.
49. Von Hendy-Willson VE, Pressler BM. An overview of glomerular filtration rate testing in dogs and cats. *Vet J.* 2011;188(2):156–165.
50. Treuting PM, Dintzis SM, Montine KS. *Comparative Anatomy and Histology: A Mouse, Rat and Human Atlas.* 2nd ed. London: Academic Press, an imprint of Elsevier; 2018. http://www.sciencedirect.com/science/book/9780128029008.
51. Jamison RL. Short and long loop nephrons. *Kidney Int.* 1987;31(2):597–605.
52. Lang RJ, Davidson ME, Exintaris B. Pyeloureteral motility and ureteral peristalsis: essential role of sensory nerves and endogenous prostaglandins. *Exp Physiol.* 2002; 87(2):129–146.
53. Manohar A, Curtis AL, Zderic SA, Valentino RJ. Brainstem network dynamics underlying the encoding of bladder information. *elife.* 2017;6.
54. Malykhina AP. How the brain controls urination. *elife.* 2017;6.
55. Hou XH, Hyun M, Taranda J, et al. Central control circuit for context-dependent micturition. *Cell.* 2016;167(1):73–86.e12.
56. Jänig W, Morrison JF. Functional properties of spinal visceral afferents supplying abdominal and pelvic organs, with special emphasis on visceral nociception. *Prog Brain Res.* 1986;67:87–114.
57. Fowler CJ, Griffiths D, de Groat WC. The neural control of micturition. *Nat Rev Neurosci.* 2008;9(6):453–466.
58. Häbler HJ, Jänig W, Koltzenburg M. Activation of unmyelinated afferent fibres by mechanical stimuli and inflammation of the urinary bladder in the cat. *J Physiol.* 1990;425:545–562.
59. Fall M, Lindström S, Mazières L. A bladder-to-bladder cooling reflex in the cat. *J Physiol.* 1990;427:281–300.
60. Robertson AS. Behaviour of the human bladder during natural filling: the Newcastle experience of ambulatory monitoring and conventional artificial filling cystometry. *Scand J Urol Nephrol Suppl.* 1999;201:19–24.
61. Streng T, Hedlund P, Talo A, Andersson KE, Gillespie JI. Phasic non-micturition contractions in the bladder of the anaesthetized and awake rat. *BJU Int.* 2006;97(5): 1094–1101.
62. Zvara P, Wright AJ, Roach K, et al. A non-anesthetized mouse model for recording sensory urinary bladder activity. *Front Neurol.* 2010;1:127.
63. Biallosterski BT, van Koeveringe GA, van Kerrebroeck PE, Gillespie JI, de Wachter SG. Nonvoiding activity of the guinea pig bladder. *J Urol.* 2011;186(2):721–727.
64. Drake MJ, Hedlund P, Harvey IJ, Pandita RK, Andersson KE, Gillespie JI. Partial outlet obstruction enhances modular autonomous activity in the isolated rat bladder. *J Urol.* 2003;170(1):276–279.
65. Gillespie JI. Phosphodiesterase-linked inhibition of nonmicturition activity in the isolated bladder. *BJU Int.* 2004;93(9):1325–1332.
66. Parsons BA, Drake MJ, Gammie A, Fry CH, Vahabi B. The validation of a functional, isolated pig bladder model for physiological experimentation. *Front Pharmacol.* 2012;3:52.
67. Vahabi B, Drake MJ. Physiological and pathophysiological implications of micromotion activity in urinary bladder function. *Acta Physiol (Oxford).* 2015;213(2): 360–370.
68. Bristow SE, Neal DE. Ambulatory urodynamics. *Br J Urol.* 1996;77(3):333–338.
69. Brading AF. A myogenic basis for the overactive bladder. *Urology.* 1997;50 (6A Suppl):57–67; [discussion 68-73].

70. Gillespie JI, Palea S, Guilloteau V, Guerard M, Lluel P, Korstanje C. Modulation of non-voiding activity by the muscarinergic antagonist tolterodine and the β(3)-adrenoceptor agonist mirabegron in conscious rats with partial outflow obstruction. *BJU Int.* 2012;110(2 Pt 2):E132–E142.
71. Li P, Liao L, Chen G, Zhang F, Tian Y. Early low-frequency stimulation of the pudendal nerve can inhibit detrusor overactivity and delay progress of bladder fibrosis in dogs with spinal cord injuries. *Spinal Cord.* 2013;51(9):668–672.
72. Iijima K, Igawa Y, Wyndaele JJ, De Wachter S. Mechanosensitive primary bladder afferent activity in rats with and without spinal cord transection. *J Urol.* 2009;182(5):2504–2510.
73. McCarthy CJ, Zabbarova IV, Brumovsky PR, Roppolo JR, Gebhart GF, Kanai AJ. Spontaneous contractions evoke afferent nerve firing in mouse bladders with detrusor overactivity. *J Urol.* 2009;181(3):1459–1466.
74. Yu Y, de Groat WC. Effects of stimulation of muscarinic receptors on bladder afferent nerves in the in vitro bladder-pelvic afferent nerve preparation of the rat. *Brain Res.* 2010;1361:43–53.
75. Daly DM, Nocchi L, Liaskos M, McKay NG, Chapple C, Grundy D. Age-related changes in afferent pathways and urothelial function in the male mouse bladder. *J Physiol.* 2014;592(3):537–549.
76. Heppner TJ, Tykocki NR, Hill-Eubanks D, Nelson MT. Transient contractions of urinary bladder smooth muscle are drivers of afferent nerve activity during filling. *J Gen Physiol.* 2016;147(4):323–335.
77. Kivistö AK, Vasenius H, Sandholm M. Canine bacteruria. *J Small Anim Pract.* 1977;18(11):707–712.
78. Bush BM. The incidence of significant bacteriuria in the dog. *Tijdschr Diergeneeskd.* 1978;103(14):750–757.
79. Thomsen MK, Svane LC, Poulsen PH. Canine urinary tract infection. Detection, prevalence and therapeutic consequences of bacteriuria. *Nord Vet Med.* 1986;38(6):394–402.
80. Ling GV, Norris CR, Franti CE, et al. Interrelations of organism prevalence, specimen collection method, and host age, sex, and breed among 8,354 canine urinary tract infections (1969-1995). *J Vet Intern Med.* 2001;15(4):341–347.
81. Cohn LA, Gary AT, Fales WH, Madsen RW. Trends in fluoroquinolone resistance of bacteria isolated from canine urinary tracts. *J Vet Diagn Investig.* 2003;15(4):338–343.
82. Mulholland SG, Qureshi SM, Fritz RW, Silverman H. Effect of hormonal deprivation on the bladder defense mechanism. *J Urol.* 1982;127(5):1010–1013.
83. Krieger JN. Urinary tract infections: what's new? *J Urol.* 2002;168(6):2351–2358.
84. Forrester SD, Troy GC, Dalton MN, Huffman JW, Holtzman G. Retrospective evaluation of urinary tract infection in 42 dogs with hyperadrenocorticism or diabetes mellitus or both. *J Vet Intern Med.* 1999;13(6):557–560.
85. Norris CR, Williams BJ, Ling GV, Franti CE, Johnson RAL. Recurrent and persistent urinary tract infections in dogs: 383 cases (1969-1995). *J Am Anim Hosp Assoc.* 2000;36(6):484–492.
86. Seguin MA, Vaden SL, Altier C, Stone E, Levine JF. Persistent urinary tract infections and reinfections in 100 dogs (1989-1999). *J Vet Intern Med.* 2003;17(5):622–631.
87. Johnson JR, Clabots C. Sharing of virulent Escherichia coli clones among household members of a woman with acute cystitis. *Clin Infect Dis.* 2006;43(10):e101–e108.
88. Low DA, Braaten BA, Ling GV, Johnson DL, Ruby AL. Isolation and comparison of Escherichia coli strains from canine and human patients with urinary tract infections. *Infect Immun.* 1988;56(10):2601–2609.
89. Johnson JR, O'Bryan TT, Low DA, et al. Evidence of commonality between canine and human extraintestinal pathogenic Escherichia coli strains that express papG allele III. *Infect Immun.* 2000;68(6):3327–3336.

90. Leroy BE, Northrup N. Prostate cancer in dogs: comparative and clinical aspects. *Vet J.* 2009;180(2):149–162.
91. Argyle DJ. Prostate cancer in dogs and men: a unique opportunity to study the disease. *Vet J.* 2009;180(2):137–138.
92. Leav I, Ling GV. Adenocarcinoma of the canine prostate. *Cancer.* 1968;22 (6):1329–1345.
93. Li Q, Zhu Y, He J, et al. Steroid 5-alpha-reductase type 2 (SRD5A2) V89L and A49T polymorphisms and sporadic prostate cancer risk: a meta-analysis. *Mol Biol Rep.* 2013;40(5):3597–3608.
94. Li J, Coates RJ, Gwinn M, Khoury MJ. Steroid 5-{alpha}-reductase type 2 (SRD5a2) gene polymorphisms and risk of prostate cancer: a HuGE review. *Am J Epidemiol.* 2010;171(1):1–13.
95. Li X, Huang Y, Fu X, et al. Meta-analysis of three polymorphisms in the steroid-5-alpha-reductase, alpha polypeptide 2 gene (SRD5A2) and risk of prostate cancer. *Mutagenesis.* 2011;26(3):371–383.
96. McNeal JE. Normal histology of the prostate. *Am J Surg Pathol.* 1988;12(8):619–633.
97. Ruetten H, Wegner KA, Romero MF, et al. Prostatic collagen architecture in neutered and intact canines. *Prostate.* 2018.
98. Bostwick DG, Montironi R, Sesterhenn IA. Diagnosis of prostatic intraepithelial neoplasia: prostate working group/consensus report. *Scand J Urol Nephrol Suppl.* 2000;205:3–10.
99. Bostwick DG. Prostatic intraepithelial neoplasia. *Curr Urol Rep.* 2000;1(1):65–70.
100. Waters DJ, Bostwick DG. The canine prostate is a spontaneous model of intraepithelial neoplasia and prostate cancer progression. *Anticancer Res.* 1997;17(3A): 1467–1470.
101. Waters DJ, Hayden DW, Bell FW, Klausner JS, Qian J, Bostwick DG. Prostatic intraepithelial neoplasia in dogs with spontaneous prostate cancer. *Prostate.* 1997; 30(2):92–97.
102. Waters DJ, Bostwick DG. Prostatic intraepithelial neoplasia occurs spontaneously in the canine prostate. *J Urol.* 1997;157(2):713–716.
103. DeMarzo AM, Nelson WG, Isaacs WB, Epstein JI. Pathological and molecular aspects of prostate cancer. *Lancet.* 2003;361(9361):955–964.
104. Humphrey PA. Gleason grading and prognostic factors in carcinoma of the prostate. *Mod Pathol.* 2004;17(3):292–306.
105. Ruetten H, Cole C, Wehber M, et al. An immunohistochemical prostate cell identification key indicates that aging shifts procollagen 1A1 production from myofibroblasts to fibroblasts in dogs prone to prostate-related urinary dysfunction. *PLoS One.* 2020; 15(7):e0232564.
106. Coffey DS, Walsh PC. Clinical and experimental studies of benign prostatic hyperplasia. *Urol Clin North Am.* 1990;17(3):461–475.
107. Berry SJ, Coffey DS, Walsh PC, Ewing LL. The development of human benign prostatic hyperplasia with age. *J Urol.* 1984;132(3):474–479.
108. Ehrlichman RJ, Isaacs JT, Coffey DS. Differences in the effects of estradiol on dihydrotestosterone induced prostatic growth of the castrate dog and rat. *Investig Urol.* 1981; 18(8):466–470.
109. Isaacs JT, Coffey DS. Etiology and disease process of benign prostatic hyperplasia. *Prostate Suppl.* 1989;2:33–50.
110. Brendler CB, Berry SJ, Ewing LL, et al. Spontaneous benign prostatic hyperplasia in the beagle. Age-associated changes in serum hormone levels, and the morphology and secretory function of the canine prostate. *J Clin Invest.* 1983;71(5):1114–1123.
111. Zambelli D, Cunto M, Gentilini F. Validation of a model to develop a symptom index for benign prostatic hyperplasia in dogs. *Reprod Domest Anim.* 2012;47 (Suppl 6):229–231.

112. Ruetten H, Wehber M, Murphy M, et al. A retrospective review of canine benign prostatic hyperplasia with and without prostatitis. *Clin Theriogenol.* 2021;13:360–366.
113. Macoska JA, Uchtmann KS, Leverson GE, McVary KT, Ricke WA. Prostate transition zone fibrosis is associated with clinical progression in the MTOPS study. *J Urol.* 2019;202(6):1240–1247.
114. Ma J, Gharaee-Kermani M, Kunju L, et al. Prostatic fibrosis is associated with lower urinary tract symptoms. *J Urol.* 2012;188(4):1375–1381.
115. Rodriguez-Nieves JA, Macoska JA. Prostatic fibrosis, lower urinary tract symptoms, and BPH. *Nat Rev Urol.* 2013;10(9):546–550.
116. Hellstrom WJ. Functional measurements of penile erection in feline, canine and primate animal models. *Int J Impot Res.* 2001;13(3):149–150.
117. Juenemann KP, Lue TF, Abozeid M, Hellstrom WJ, Tanagho EA. Blood gas analysis in drug-induced penile erection. *Urol Int.* 1986;41(3):207–211.
118. Takahashi Y, Ishii N, Lue TF, Tanagho EA. Pharmacological effects of adenosine on canine penile erection. *Tohoku J Exp Med.* 1991;165(1):49–58.
119. Acierno MJ, Labato MA. Canine incontinence. *Vet Clin North Am Small Anim Pract.* 2019;49(2):125–140.
120. Holt PE. Urinary incontinence in dogs and cats. *Vet Rec.* 1990;127(14):347–350.
121. Angioletti A, De Francesco I, Vergottini M, Battocchio ML. Urinary incontinence after spaying in the bitch: incidence and oestrogen-therapy. *Vet Res Commun.* 2004;28(Suppl 1):153–155.
122. Stöcklin-Gautschi NM, Hässig M, Reichler IM, Hubler M, Arnold S. The relationship of urinary incontinence to early spaying in bitches. *J Reprod Fertil Suppl.* 2001; 57:233–236.
123. Forsee KM, Davis GJ, Mouat EE, Salmeri KR, Bastian RP. Evaluation of the prevalence of urinary incontinence in spayed female dogs: 566 cases (2003-2008). *J Am Vet Med Assoc.* 2013;242(7):959–962.
124. Abrams P, Andersson KE, Birder L, et al. Fourth international consultation on incontinence recommendations of the international scientific committee: evaluation and treatment of urinary incontinence, pelvic organ prolapse, and fecal incontinence. *Neurourol Urodyn.* 2010;29(1):213–240.
125. Minassian VA, Stewart WF, Wood GC. Urinary incontinence in women: variation in prevalence estimates and risk factors. *Obstet Gynecol.* 2008;111(2 Pt 1):324–331.
126. Hannestad YS, Rortveit G, Sandvik H, Hunskaar S, Nord-Trøndelag NEsEoIitCo. A community-based epidemiological survey of female urinary incontinence: the Norwegian EPINCONT study. Epidemiology of incontinence in the county of Nord-Trøndelag. *J Clin Epidemiol.* 2000;53(11):1150–1157.
127. Minassian VA, Drutz HP, Al-Badr A. Urinary incontinence as a worldwide problem. *Int J Gynaecol Obstet.* 2003;82(3):327–338.
128. Hunskaar S, Lose G, Sykes D, Voss S. The prevalence of urinary incontinence in women in four European countries. *BJU Int.* 2004;93(3):324–330.
129. Richter KP, Ling GV. Clinical response and urethral pressure profile changes after phenylpropanolamine in dogs with primary sphincter incompetence. *J Am Vet Med Assoc.* 1985;187(6):605–611.
130. Scott L, Leddy M, Bernay F, Davot JL. Evaluation of phenylpropanolamine in the treatment of urethral sphincter mechanism incompetence in the bitch. *J Small Anim Pract.* 2002;43(11):493–496.
131. Mandigers RJ, Nell T. Treatment of bitches with acquired urinary incontinence with oestriol. *Vet Rec.* 2001;149(25):764–767.
132. Barth A, Reichler IM, Hubler M, Hässig M, Arnold S. Evaluation of long-term effects of endoscopic injection of collagen into the urethral submucosa for treatment of urethral sphincter incompetence in female dogs: 40 cases (1993-2000). *J Am Vet Med Assoc.* 2005;226(1):73–76.

133. Wood JD, Simmons-Byrd A, Spievack AR, Badylak SF. Use of a particulate extracellular matrix bioscaffold for treatment of acquired urinary incontinence in dogs. *J Am Vet Med Assoc.* 2005;226(7):1095–1097.
134. Byron JK, Chew DJ, McLoughlin ML. Retrospective evaluation of urethral bovine cross-linked collagen implantation for treatment of urinary incontinence in female dogs. *J Vet Intern Med.* 2011;25(5):980–984.
135. Reeves L, Adin C, McLoughlin M, Ham K, Chew D. Outcome after placement of an artificial urethral sphincter in 27 dogs. *Vet Surg.* 2013;42(1):12–18.
136. Hamon M, Hamaide AJ, Noël SM, Claeys S. Long-term outcome of the transobturator vaginal tape inside out for the treatment of urethral sphincter mechanism incompetence in female dogs. *Vet Surg.* 2019;48(1):29–34.
137. Claeys S, Ruel H, de Leval J, Heimann M, Hamaide A. Transobturator vaginal tape inside out for treatment of urethral sphincter mechanism incompetence in female dogs: cadaveric study and preliminary study in continent female dogs. *Vet Surg.* 2010;39 (8):957–968.
138. Claeys S, de Leval J, Hamaide A. Transobturator vaginal tape inside out for treatment of urethral sphincter mechanism incompetence: preliminary results in 7 female dogs. *Vet Surg.* 2010;39(8):969–979.
139. Deschamps JY, Roux FA. Transobturator vaginal tape for treatment of urinary incontinence in spayed bitches. *J Am Anim Hosp Assoc.* 2015;51(2):85–96.
140. Stroud C, Dmitriev I, Kashentseva E, et al. A one health overview, facilitating advances in comparative medicine and translational research. *Clin Transl Med.* 2016;5(Suppl 1):26.

CHAPTER THREE

Canine models of human cancer: Bridging the gap to improve precision medicine

Rebecca L. Nance[a,b], Abdul Mohin Sajib[c], and Bruce F. Smith[a,b,*]
[a]Scott-Ritchey Research Center, Auburn University College of Veterinary Medicine, Auburn, AL, United States
[b]Department of Pathobiology, Auburn University College of Veterinary Medicine, Auburn, AL, United States
[c]Division of Cellular and Gene Therapies, Center for Biologics Evaluation and Research, US Food and Drug Administration, Silver Spring, MD, United States
*Corresponding author: e-mail address: smithbf@auburn.edu

Contents

Abstract

Dogs are remarkable, adaptable, and dependable creatures that have evolved alongside humans while contributing tremendously to our survival. Our canine companions share many similarities to human disease, particularly cancer. With the advancement of next-generation sequencing technology, we are beginning to unravel the complexity of cancer and the vast intra- and intertumoral heterogeneity that makes treatment difficult. Consequently, precision medicine has emerged as a therapeutic approach to improve patient survival by evaluating and classifying an individual tumor's molecular profile. Many canine and human cancers share striking similarities in terms of genotypic, phenotypic, clinical, and histological presentations. Dogs are superior to rodent models

Progress in Molecular Biology and Translational Science, Volume 189
ISSN 1877-1173
https://doi.org/10.1016/bs.pmbts.2021.12.003

of cancer because they are a naturally heterogeneous population in which tumors occur spontaneously, are exposed to similar environmental conditions, and show more similarities in key modulators of tumorigenesis and clinical response, including the immune system, drug metabolism, and gut microbiome. In this chapter, we will explore various canine models of human cancers and emphasize the dog's critical role in advancing precision medicine and improving the survival of both man and man's best friend.

1. Cancer and the precision medicine approach to treatment

Cancer is a highly complex and heterogeneous disease characterized by extreme genomic instability, sustained cell proliferation, suppressed cell death, induced angiogenesis, stimulated cell invasion, and evaded immune destruction.[1] Mutations in oncogenes and/or tumor suppressor genes contribute to initial aberrant cellular properties and can result in the accumulation of additional genetic changes that drive tumor formation and progression. Our understanding of tumorigenesis has grown exponentially in the past decade as the molecular processes continue to be elucidated, largely due to the efforts of translational research made possible by the reduced cost of next-generation sequencing and development of new technologies. In recent years, the study and treatment of cancer has evolved from an organ-centric approach to one that seeks to characterize tumors based on their molecular profile. Advances in the field of "omics" technologies have made it possible to integrate genomic, transcriptomic, and proteomic information to create a more comprehensive strategy to improve patient outcomes and lower treatment costs.[2] Extensive sequencing has exposed the overwhelming complexity of cancer and the overall genetic uniqueness of each individual tumor, regardless of type or stage, which often share common mutations in critical genes and dysregulated pathways. This observation gave rise to the concept of precision or personalized medicine, which seeks to overcome both intra- and intertumoral heterogeneity as well as interindividual variation in drug response.[3–5] By providing insight into druggable targets such as pathways, receptors, or genes, critical therapeutic decisions and development of novel therapies are guided by the molecular landscape of an individual tumor.

However, many obstacles compromise the adequate implementation of such treatment approaches, including the lack of appropriate infrastructure and communication among medical doctors and research scientists.[6] Additionally, the clinical validity of precision medicine faces some scrutiny due to the lack of appropriate or efficient standardization.[7] As such, the clinical utility is criticized, and additional evidence is needed to support the claim that "omics"-based personalized medicine generates an overall cost reduction.[8,9] There is a critical need for an infrastructure that supports the integration of a patient's genomic data with relevant clinical information such as age, diagnosis, family history, symptom severity, and other confounding variables such as lifestyle, diet, and drug interactions.[2]

In spite of these obstacles, many molecular-based treatments have been successfully implemented in humans. In fact, more than 1 out of 4 drugs approved by the U.S. Food and Drug Administration (FDA) since 2014 are precision medicine therapies.[10] According to an FDA news release, the first approved drug for cancer treatment based solely on molecular features rather than tumor type was released in May 2017. The drug, known as pembrolizumab, or more commonly as Keytruda, was granted accelerated approval for the treatment of unresectable or metastatic solid tumors with biomarkers for high microsatellite instability or deficient mismatch repair. Amid a global pandemic in the first half of 2020, the FDA approved a record total of 21 precision drugs for cancer treatment; of these, 7 were characterized by novel molecular entities and 14 were previously approved drugs with expanded molecular indications.[11] These include the tyrosine kinase inhibitor imatinib, which was approved for use in chronic myeloid leukemia tumors bearing the t(9,22)(q34;q11) translocation which generates the fusion gene BCR-ABL1.[12] Breast cancer treatment has improved substantially with the identification of molecular subtypes, including the application of the anti-HER2 kinase inhibitor neratinib, monoclonal antibody trastuzumab that targets the HER2 receptor, and PI3Kα inhibitor alpelisib in patients bearing HER2 amplification and/or overexpression and PIK3CA mutations, respectively.[13–16] These are just a few of the many successful applications of precision medicine in the treatment of cancer that continue to improve patient outcomes.

2. Limitations of rodent models

Despite significant progress in precision oncology research and treatment, development is hindered by the extremely low percentage of therapeutics that show similar outcomes in humans and preclinical rodent studies.[17,18] Indeed, the rate of effective translation from animal models to clinical trials is less than 8%.[19] The abundant failures of murine models are disappointing, time-consuming, and expensive. While rodent models remain fundamentally important in the study of cancer, they frequently do not accurately recapitulate the full complexity of human cancer, including tumor heterogeneity, microenvironment, accumulated genomic instability, immune influence, and metastatic disease, and therefore suffer significant translational limitations.[19–23]

The controlled environment of laboratory rodents is thought to contribute to the inaccurate representation of human disease. These inbred animals are typically homogeneous in terms of genetics, size, age, sex, and diet, and are housed in pathogen-free environments that are unrepresentative of natural environments.[24] The use of young mice is controversial given the age-related changes that occur to the immune system and the propensity of cancer occurrence in older humans.[25] Even patient-derived xenografts with implantation of humanized immune system components lack critical functions and responses that are representative of human malignancies.[24]

In the past decade, discovery of the dynamic interactions between the immune system and gut microbiome has uncovered a new field of research.[26] The gut microbiome has emerged as an important and influential factor affecting immune function, and consequently, tumorigenesis and response to treatment.[27] In a 2015 study, Sivan et al. uncovered significant differences in the growth rate of subcutaneous B16.SIY melanoma in genetically similar mice housed in two different facilities as a result of immune-mediated differences in tumor-specific T cells and intratumoral $CD8^+$ cell accumulation. After cohousing the two groups of mice prior to tumor implantation, the team found these differences in tumor growth and immune response disappeared, suggesting an environmental effect. They evaluated the gut microbiome in both groups, performed fecal transplantations, and discovered that *Bifidobacterium* conferred an antitumor immune response that was augmented with anti-PDL1 monoclonal antibody immunotherapy.[28] This data suggests that commensal gut microbiota can influence the immune response to tumor and immunotherapy effectiveness and therefore should be considered in the selection of disease models.

In addition to the molecular and physiological dissimilarities in rodent models, differences in drug metabolism are also likely to blame for the poor translational outcomes. Tumor response to drugs, particularly chemotherapy, depends on a plethora of elements, including tumor-specific factors such as histology, growth rate, and drug-resistance mechanisms, as well as individual-specific features such as physiology, absorption efficiency, and immunologic response.[29] A 1976 study found differences in the drug sensitivities of human and murine bone marrow upon exposure to chemotherapeutic agents, calling into question the applicability of mouse models in the evaluation of chemotherapy and combination therapy response.[30] Transgenic mice bearing alterations in metabolic enzymes, such as humanized cytochrome P450, and transplantation of human liver hepatocytes have been developed to better mimic *in vivo* drug response.[31,32] Despite these improvements, the implementation of these models does not generate improved translational benefits to human responses.

Simply put, rodent models do not faithfully mimic the inherent tumor evolution and drug response observed in humans; a natural disease model exhibiting spontaneous tumor development is likely to generate more promising results.[33] The success of precision medicine is critically dependent upon a reliable and appropriate model for monitoring tumor development, progression to metastases, immune influence, and therapeutic safety and effectiveness.

3. Emergence of the canine model

In 2003, the National Cancer Institute's Center for Cancer Research (CCR) recognized the value of spontaneously occurring tumors in pet animals as models of human disease and launched the Comparative Oncology Program (NCI-COP). The establishment of this program inspired many researchers to explore the genetics of companion dogs. Two years later, whole-genome shotgun sequencing (WGS) of a boxer named Tasha resulted in the first high quality draft sequence of the dog, covering ~99% of the genome (CanFam2.0).[34] After the completion of the canine genome project in 2005, researchers had the genomic information necessary to launch a gene expression database of normal canine tissue.[34,35] The Canine Normal Tissue Database established the foundation for a more robust assessment of the biological functions of specific genes and cross-species comparisons of gene expression. In fact, during the development of this database, Briggs et al. found remarkable similarities in tissue-specific expression profiles between dogs and humans, thus providing additional support for the dog's role in translational research.

3.1 Evolution of the canine genome

Man's best friend, *Canis lupus familiaris*, has evolved alongside humans for centuries. Diverging from the gray wolf, the earliest record of a confirmed domestic dog was discovered in a maxillary fragment discovered in the Kesserloch Cave in Switzerland and dated to 12.225 ± 45 bp or c.

14,100–14,600 BP from a 2012 study.[36] First discovered in 1928 and later confirmed in 1964, karyotype analysis reveals 38 pairs of autosomal acrocentric chromosomes and a pair of metacentric sex chromosomes in the dog.[37] Dogs were the first species to be domesticated, occurring more than 15,000 years ago when humans were hunter-gatherers.[38] Since then, selection for certain phenotypic traits such as temperament and function has resulted in over 400 breeds that can be differentiated by distinct genetic components and morphological variation.[39–41] Linkage disequilibrium in the dog is roughly 20–50 times more extensive within breed groups than in the human population, even for isolated populations, which allow genome-wide association studies (GWAS) to be accomplished with significantly fewer SNPs. For example, only about 10,000 canine SNPs are necessary as opposed to the ~500,000 SNPs needed in humans, providing an economical advantage to disease mapping.[42]

Conserved synteny analyses show that human and dog genomes are more closely related to each other in terms of nucleotide divergence and chromosomal aberrations than either one is to rodent genomes, including mice.[20,43] The average rate of nucleotide divergence in both humans and dogs is approximately 0.35 substitutions per site, a rate that is much more similar than that observed in the mouse.[42] Interchromosomal rearrangements occur less frequently in the canine genome than in rodents but significantly more than in humans. However, the rate of intrachromosomal reshuffling is more similar between humans and dogs than rodents.[42] Despite evolutionary divergence, the canine and human genomes share remarkable similarities that highlight the importance of the dog as an essential animal model.

3.2 Similarities in the canine and human immune system, drug metabolism, and microbiome

The relatedness between dogs and humans extends beyond genetics to include immune system development and function. Unlike rodents, both human and canine immune systems are fully developed before birth and maturation is believed to continue postnatally.[44] Studies of X-linked severe combined immunodeficiency (XSCID) caused by mutations in the common gamma chain, a component of at least six different cytokine receptors, have compared the outcomes on lymphocyte development and function in the human, dog, and mouse. These results have shown the effects on B cells, T cells, natural killer (NK) cells, and postnatal T cell development are almost identical in dogs and humans, while mice show distinctly different responses.[44] Similar findings of age-related changes in immune cell distribution have been found in humans and dogs. Age-related decline in the absolute numbers of B-, T-, and $CD4^+$ cells, but not in $CD8^+$ cells, in dogs resemble previously published data in humans.[45,46] The major immune effector cells have been characterized in dogs, including $CD4^+$, $CD8^+$, $CD90^+$, and dendritic cells, and show similar homology to humans.[47,48]

Differences in pharmacokinetics and pharmacodynamics has been observed in various dog breeds. Many breed-specific features affecting drug absorption and metabolism have been elucidated, including polymorphisms in cytochrome P450 enzymes and cyclooxygenase 2 inhibitors. Due to these differences that can lead to inconsistent results, caution should be used in studies employing multiple different breeds. All domestic dog breeds and other canids lack the *n*-acetyltransferase genes that detoxify many drugs and metabolites, including arylamine and hydrazine compounds.[49] While canine pharmacodynamics and metabolism are not a perfect model for humans, the similarities can be useful for drawing comparative conclusions.

The gut microbiome can influence drug pharmacokinetics and the immune system and therefore should be considered in experimental models. In a 2018 study cataloging the canine gut microbiome, researchers found more similarities to the human microbiome than to mouse or pig.[50] Furthermore, gut microbial changes associated with colorectal cancer, irritable bowel syndrome, and diabetes mellitus, have shown comparable results between human and dog.[51–53]

3.3 Cancer in the dog

Dogs develop spontaneous tumors that share similar biological and histopathological properties as human neoplasms at comparable rates.[54] The predicted cancer incidence is similar in dogs and humans, with an estimated 300 per 100,000 people or dogs per year at risk of developing a tumor.[43] In addition to the relative ease of sample acquisition, dogs also share similar environments and are exposed to comparable risk factors as humans.[8] Age, nutrition, sex, and environment, influence tumorigenesis in similar ways in both species.[55] While unfortunate for man's best friend, the condensed lifespan of dogs allows the accelerated study of tumors that would normally take decades to develop in humans to be performed in 2–3 years.[56] Veterinary clinical trials are less expensive and patients are also typically more homogeneous in terms of pretreatment and age than human clinical trials.[57] Owners can evaluate their dog's quality of life in response to treatment within their natural environment, a task that is not easily accomplished and often not considered in laboratory rodent studies. Additionally, with less stringent control on patient records and confidentiality, access to patient data is more easily available in dogs.

Some tumors have also shown a significant breed disposition, providing the opportunity to study the genetic components, such as germline mutations, that influence the risk of tumor development.[58] Genome mapping studies have revealed breeds with significantly increased risk for specific tumor types, including osteosarcoma in the Greyhound and Rottweiler, hemangiosarcoma in the Golden Retriever, lymphoma in the Golden Retriever and Boxer, melanoma in the Scottish Terrier, mast cell tumors in the Pug and Golden Retriever, transitional cell carcinoma in the Scottish Terrier, brain tumors in Boxers, and stomach cancer in Chows.[59,60] Breed-specific cancers show patterns that reflect the inherited mutations that predispose human children and families to an increased risk of specific cancers.[61] Genome-wide association studies (GWAS) in breed groups have successfully identified inherited risk loci for several tumor types, which will be discussed in the following sections. Due to the genetic homogeneity within a breed, GWAS analysis can be accomplished using a smaller number of canine patients compared with human studies, providing a more efficient opportunity to identify risk loci that may be translational to humans.[61]

Comparative gene expression studies in human and canine tumors have exposed the striking similarities in terms of genotypic and phenotypic properties, including specific germline and somatic mutations, chromosomal aberrations, physiological processes, enriched gene sets or pathways, and molecular markers in various cancers that are not naturally reflected in rodent models. In addition to the genetic and environmental aspects driving tumor formation, the immune system, gut microbiome, and drug metabolism play a critical role, especially for studying the effects of immunotherapy. Dogs and humans share more commonalities in their immune system development, function, microbiome composition, and metabolism than rodents. In addition to genomic and pathophysiological similarities, dogs share similar environmental influences that may impact cancer. For these reasons, canine models of cancer are superior to rodents in oncological research. In the following sections, we will discuss these similarities in depth for each tumor type. While dogs also serve as valuable models for numerous other human disorders, including inherited retinal degeneration, dilated cardiomyopathy, cystinuria, Duchenne muscular dystrophy, chronic obstructive pulmonary disorder (COPD), and rare human disorders such as Caffey, van den Dende-Gupta, and Raine syndromes, the focus of this chapter will be on cancer specifically.[43,62–64] Ultimately, our canine companions serve as promising models for uncovering the molecular dynamics of tumorigenesis, improving clinical outcomes, lowering treatment costs, evaluating drug safety and efficacy, and advancing the era of precision medicine in the treatment of cancer—for both man and man's best friend.[65]

4. Osteosarcoma

Osteosarcoma (OSA) is a highly aggressive bone tumor that typically originates in the appendicular skeleton of pediatric patients and large breed dogs. The canine model has been extensively studied and is perhaps the most well-known of cancer models due to its striking similarity to human OSA. In both humans and dogs, primary tumors and metastases occur in similar sites, show comparable clinical presentations, and share treatment approaches.[66] Metastases primarily occur in the lungs and drug resistance is exceedingly common. The standard of care for dogs involves amputation followed by chemotherapy, though the 1-year survival rate is <50%.[67]

OSA is characterized by complex karyotypes with significant chromosomal aberrations.[68] In fact, in terms of gene expression, OSA is one of the most heterogeneous tumor types.[69] Flow cytometry analysis of multiple sections from 25 untreated human OSA tumors indicated DNA aneuploidies present in 86% and multiple aneuploid DNA stemlines in 48% of highly malignant cases, rates that far exceed those from other tumors.[70] Contrary to the human disease, canine OSA shows a disposition for large breed dogs, suggesting a heritable component.[71] A genome-wide association study in Greyhounds, Rottweilers, and Irish Wolf Hounds identified 33 loci associated with OSA, accounting for 50–80% of the disease risk.[71] Notably, none of the genomic regions overlap between the three breeds. A noncoding regulatory element upstream of the *CDKNA2A/B* locus was identified as a main causal variant, causing subsequent dysregulation of *CDKN2A/ARF*. *CDKN2A/ARF* encodes multiple cyclin-dependent kinase inhibitors, including INK4a/b and ARF, which inactivate D-cyclins to promote G1 progression and control senescence via the RB (Retinoblastoma) and p53 pathways.[71] Germline variants in the regulatory regions of *CDKN2A* may explain the observed heritability and predisposition of OSA development. Investigation of the copy number aberrations via comparative genomic hybridization arrays in 12 Greyhounds and 10 Rottweilers revealed substantial conservation in regional DNA copy numbers between the two breeds. Furthermore, the results showed striking similarities in both the frequency and distribution of copy number aberrations (CNAs) between dogs and humans, including *MYC* gain (dog=60%, human=67%), *RB1* loss (dog=36%, human=33%), and *CDKN2A/B* loss (dog=73%, human=67%).[71]

Perhaps the most well documented genetic alteration in both human and canine OSA involves *TP53*. Mutations in *TP53* have been reported in 41% of primary canine OSA tumors with the majority consisting of point mutations (74%) and fewer being deletions (26%).[72] In contrast, approximately 20% of primary human OSA tumors have been reported to harbor *TP53* mutations, most of which include genomic deletions.[73]

At the gene expression level, canine and human OSA are nearly indistinguishable. A cross-species gene signature study using oligonucleotide arrays to examine a set of orthologous genes could not distinguish the human and canine diseases by hierarchical clustering.[74] The researchers identified two genes, interleukin-8 (IL-8) and solute carrier family I member 3 (SLCIA3), which were expressed in all dogs, but only a subset of pediatric patients. These genes were validated by protein expression using tissue microarray immunohistochemistry and were found to be associated with poor outcome in an independent population of pediatric OSA patients.[74] The identification of these progression-associated genes would likely not have been uncovered without the cross-species approach. In another study, Scott et al. evaluated the tumor transcriptional profiles using RNA sequencing of OSA tumors and cell lines derived from human, mouse, and dog. Despite the complex genomic instability characterizing OSA, their results showed conserved intertumoral expression and patterns of transcriptional variation that were distinct from other tumor types.[68]

In a 2011 study, genome-wide expression profiling revealed two molecular subtypes in 27 OSA tumors obtained from dogs prior to any clinical intervention. The cell lines derived from the tumors clustered into two distinct branches; Branch A, consisting of 16 samples, and Branch B, consisting of 10 samples, with two groups of reciprocally expressed groups, defined as "gene cluster 1" and "gene cluster 2." The pathways evident in gene cluster 1, which were overexpressed in the Branch A samples, consisted of genes responsible for mitosis, chromosome segregation, and mitotic spindle formation. The cluster 2 genes, overexpressed in Branch B dogs, were predicted to be associated with cancer cell-microenvironment interactions. Upon analysis of patient outcome, the researchers found that dogs from Branch A exhibited a worse survival than those from Branch B.[75] Scott et al. proposed that this may be due to the greater metastatic potential of more aggressive tumors, resulting in their ability to overcome limitations established by the tumor niche and rapidly divide and survive regardless of their environment.

Canine studies have demonstrated the practicality and effectiveness of precision medicine in the treatment of OSA. A multisite feasibility study evaluated the turn-around time for the molecular analysis of 20 dogs with OSA.[76] The team established a processing pipeline that generated a report of tumor expression within 5 days of sample receipt, demonstrating that the practical application of precision medicine in the clinical setting is reasonable.[76] Another canine case study was used to show the efficacy of personalized therapy for OSA. Davis et al. used a primary OSA tumor sample from a 7 year old Golden Retriever to establish a cell line, which was then screened for sensitivity to a panel of kinase inhibitors. The molecule with the highest activity was the Src and Abl inhibitor dasatinib, which had not been previously administered in dogs. After immunohistochemical staining to confirm overexpression of Src, the dog began receiving treatment with dasatinib and a dose of 0.75 mg/kg/day was established and well-tolerated. Twenty-four months after initial diagnosis and 8 months after completion of the precision medicine adjuvant therapy, the canine patient remained disease-free with no evidence of metastases. For comparison, the average survival rate after initial diagnosis is 33–65% for 12 months and 16–28% for 24 months.[77] While these results are based on only a single case, this 2013 study confirms the critical role dogs play in demonstrating the efficacy and feasibility of precision medicine therapy.

OSA is perhaps the most well-documented of tumors in canine comparative oncology. The shared characteristics with human OSA provides support for the canine model. In OSA, where treatment has remained essentially unchanged for 40 years, there is an urgent need to develop and improve therapies, and it is likely that dogs hold the key to benefitting patient survival.

5. B-cell lymphoma

Lymphoma, a disease of malignant lymphocytes, is one of the most common cancers occurring in the dog. As with many cancers, the exact etiology is unknown, but environmental and genetic factors are believed to play a role. In both humans and dogs, B-cell malignancies are more common than T-cell. Breeds with an increased risk of lymphoma include the boxer, bulldog, bull mastiff, Bernese mountain dog, Gordon setter, and Scottish terrier, suggesting heritable disease risk factors.[66] Canine lymphoma most closely resembles non-Hodgkin's lymphoma (NHL) in humans, which typically occurs in older adults, due to the similarities in clinical presentation, molecular components, and response to treatment. While no apparent sex predisposition has been established, intact female dogs have been reported to have a reduced risk.[78]

Diffuse large B-cell lymphoma (DLBCL) is the most common subtype of NHL in both humans and dogs and occurs at a frequency of 47% in dogs.[79] The importance of the canine model of lymphoma was reflected in a 1993 study that evaluated the xenotransplantation of human NHL tumors into severe combined immunodeficient mice and found that only 10 of 50 were successful at retaining the genotype and phenotype reflective of the primary tumor.[80] While 13 other tumors were successfully engrafted, they were composed of Epstein-Barr Virus (EBV)-transformed B cells, indicating that the prevalence of EBV in humans, which is suspected to drive instability due to proliferation, poses limitations to the establishment of xenograft models.[80] On the contrary, there are few reports of viruses driving lymphoma progression in the dog.[81]

Treatment in both dogs and humans involves CHOP chemotherapy (cyclophosphamide, doxorubicin, vincristine, and prednisolone). The median survival time for dogs is 10–14 months for B-cell neoplasms and 6–9 months for T-cell malignancies.[66] Drug resistance is suspected in cases that do not respond to chemotherapy, and is believed to be related to the upregulation of drug transporters, particularly in the ATP-Binding Cassette (ABC)-transporter family, including P-gp and BCRP.[81] Most canine lymphoma cases can be managed with chemotherapy for extended periods, though most will eventually develop drug-resistance.[81]

Exposure to environmental toxins, including household chemicals, has been suggested to play a role in lymphoma carcinogenesis.[81] Glutathione-S-transferase enzymes (GSTs) are involved in detoxification of environmental carcinogens, and defects in these genes, particularly *GSTT1*, are common in human lymphomas. In the dog, 27 *GSTT1* variants were identified, and one was found in 18% of all canine lymphomas ($n = 93$). This intronic variant is predicted to impair mRNA splicing and thus, enzyme detoxification.[82]

Common cytogenetic aberrations have been identified in human and canine lymphoma. A translocation of *MYC* resulting in its control by the immunoglobulin heavy-chain enhancer has been identified as a recurrent abnormality in canine B-cell lymphoma, causing overexpression of MYC.[83] This translocation is analogous to the t(8;14) rearrangement observed in Burkitt's lymphoma in humans, which occurs in approximately 80% of B-cell lymphomas.[83] Globally, genomic instability is lower in canine NHL compared to human NHL, with chromosomal copy number aberrations largely restricted to particular genomic regions. Comparative genomic hybridization has revealed abnormalities restricted to canine chromosomes 13 and 31 and human chromosomes 8 and 21.[84] While the gain of chromosome 13 is regularly reported in canine lymphoma, this chromosomal abnormality has been found in other tumors, including carcinomas and sarcomas, suggesting its involvement in pan-tumor progression rather than specific to lymphoma tumorigenesis.[81]

Comparative transcriptional analysis using microarray data has revealed similarities in canine and human DLBCL, including activation of the NF-κB/p65 canonical pathway.[85] Gene expression profiling was used to uncover two distinct subtypes of human DLBCL, with higher NF-κB activity in the more aggressive activated B-cell type.[86] In this subtype, enhanced NF-κB activity is attributed to CARD11-dependent chronic B-cell receptor signaling, preventing apoptosis.[87] Activation of the NF-κB pathway also results in upregulation of other antiapoptotic mechanisms, including Bcl-2 proteins, with have been implicated in chemotherapy resistance.[88] Microarray data supported by immunohistochemical staining and drug studies have demonstrated activation of the NF-κB pathway in both human and canine DLBCL samples.[85] Constitutive NF-κB expression and activity has been confirmed in a secondary study using primary canine CLBCL tissue, providing further support for its role as a model for activated B-cell DLBCL.[89]

Due to the vast heterogeneity in lymphoma as evidenced by the characterization of multiple subtypes, the canine model should be cautiously and rigorously evaluated at the morphologic, phenotypic, and molecular level to ensure appropriate similarity to the subset of human lymphoma being modeled.

6. Melanoma

Melanoma is the most aggressive, drug-resistant, and deadly of skin cancers. According to the *New England Journal of Medicine*, melanoma accounts for only 4% of all dermatologic cancers, but is responsible for 80% of deaths due to skin cancer.[90] Originating from malignant melanocytes, melanoma can be characterized as mucosal or cutaneous with differing etiologies. In dogs, melanoma primarily affects the oral cavity, lips, eyes, and footpads, with oral melanoma identified as the most aggressive, and the most frequently diagnosed type of canine melanoma.[91] The median survival time for oral melanoma is <36 months.[91] Surgical resection is the first line of treatment for melanoma followed by radiation and/or chemotherapy. Unfortunately, patients usually succumb to metastases, especially in oral melanoma cases.[91] Depending on prognostic indicators, post-treatment median survival time in dogs ranges from 5 months to 2 years.[92] Malignant melanoma occurs most frequently in the Cocker Spaniel, Poodle, Pekinese, Gordon Setter, Chow Chow, Golden Retriever, Dachshund, and mixed-breeds, compared to Boxers and German Shepherds where the incidence is much lower.[91]

Melanocytes are tissue-resident dendritic cells that migrate during embryogenesis and they have demonstrated aberrant expression of cell-cell adhesion genes, including N-cadherin, Mel-CAM1, ICAM1, and αβ integrins in malignant melanoma.[93] Molecular analysis of cutaneous melanomas has revealed persistent mutations in genes related to cell proliferation (*BRAF, NRAS, NF1*), metabolism (*PTEN, KIT*), apoptosis-resistance (*TP53*), cell cycle control (*CDKN2A*), and cell lifespan (*TERT*).[94] In 2002, Davies et al. uncovered the major role of the somatic point mutation V600E in the *BRAF* gene, a mutation that occurs in over 50% of human melanoma patients.[95] Approximately 70% of melanomas harbor mutations in the *CDKN2A* locus. Melanoma tumorigenesis sometimes involves transformation of normal melanocytes into a melanocytic nevus, and *BRAF, NRAS*, and *INK4a/ARF* are altered during, or as a result of, this process. Subsequent transformation of a nevus into a melanoma demonstrates mutation and/or dysregulation of the same genes (*BRAF, NRAS, INK4a/ARF*) with the addition of *PTEN, KIT, NEDD9*, and *MITF*.[91] Mutations in the proto-oncogene tyrosine kinase receptor *KIT* were found in 5 out of 49 dogs with malignant melanoma from various anatomical sites.[96] In dogs and humans, mutations in *KIT* are usually localized to exon 11.[96] Imatinib, a receptor tyrosine kinase inhibitor, has shown clinical efficacy at treating canine mast cell tumors bearing *KIT* mutations, and has also been shown to be beneficial in treating human melanoma patients with *KIT* mutations.[96] Interestingly, the genes affected by recurrent mutations are not highly correlated between human and canine mucosal malignant melanomas. Mutations in *NRAS* and *PTEN* have been reported in approximately 15% and 48% of human cases, while only occurring in 6% and 4% of dogs, respectively.[91]

In a 2014 study, Gillard et al. compared human and canine melanoma and found similar cytological and histopathological presentations. They studied the recurrent mutations in human melanomas, including *BRAF, KIT*,

GNAQ, *PTEN*, and *NRAS*, and out of 95 canine cases, no mutations were found in *BRAF*, *KIT*, *GNAQ*, and *CDK4* genes. Only three dogs had mutations in *NRAS* and two had a *PTEN* mutation. In the human cutaneous melanoma samples, *BRAF* and *NRAS* mutations were observed in 57% and 17% of cases, respectively.[97] The low frequency of *BRAF* mutations and anatomical location led the authors to conclude that canine oral melanoma can serve as a model for nonultraviolet-associated melanomas.

Melanoma is a particularly immunogenic tumor and was the first model to reveal $CD4^+$ and $CD8^+$ specificity against tumor-derived antigens gp100 and tyrosinase.[21] Many preclinical studies have been developed to evaluate the potential for an antitumor response using various vaccines with immunized mice or B16 melanoma cell lines, but the translational benefits are modest at best. Using dogs, more promising results have been uncovered. Oncept, the first xenogenic DNA-based canine cancer vaccine, was licensed by the United States Department of Agriculture in 2010 for use in dogs with stage II or III oral melanoma.[92] Oncept encodes human tyrosinase and has shown to improve survival times for dogs when used in combination with surgery and radiotherapy. Despite reports of conflicting efficacy, Oncept is widely used in clinics.[92] Studies using autologous dendritic cells (DCs) have been explored, though reports in the dog are scarce. In a 2005 study, Gyorffy et al. transduced bone-marrow derived DCs from canine melanoma patients with an adenovirus encoding the xenoantigen, human melanoma antigen gp100, to be used in combination with radiation therapy. One of the dogs showed antigen-specific cytotoxic lymphocyte activity in peripheral blood and remained disease-free for 48 months after initial vaccination. The other dog enrolled in the study did not show evidence of cytotoxic T lymphocyte activity and relapsed 22 months after initial DC injection.[98] Three years later, Tamura et al. published a study showing that autologous DCs pulsed with cell lysate from canine malignant melanoma cell line CMM2 elicit T cell recruitment when administered to dogs *in vivo*.[99] These results reinforce the concept of exploring dendritic cell immunotherapies using dogs. Lupeol has emerged as another promising novel therapy as it has shown to inhibit growth of melanoma cells *in vivo* and *in vitro* by inducing apoptosis.[100] Lupeol is a triterpene found in various fruits and vegetables and possesses robust anti-inflammatory, antimutagenic, and antioxidant activity.[100] In 2015, postoperative administration of subcutaneous lupeol was evaluated in 11 dogs with stage I, II, or III melanoma. No adverse effects were reported, and in 7 of the dogs, no local reoccurrence or distant metastasis was evident 180 days after surgical resection.[101]

Immune checkpoint inhibitors have been explored to restore anticancer immune responses in the treatment of melanoma. The expression of checkpoint molecules such as CTLA-4 and PD-1 has been confirmed in canine melanoma in both peripheral blood lymphocytes and tumor biopsies. This discovery has led to the development of a canine-chimeric monoclonal antibody against canine PD-L1 in dogs with advanced oral melanoma or undifferentiated sarcoma which underwent a clinical trial in 2017.[102] In humans, antibodies to PD-1 such as pembrolizumab and nivolumab have been approved by the FDA for treatment of BRAF inhibitor-resistant advanced melanoma bearing the common *BRAF* V600 mutation. Ipilimumab, an antibody against the cytotoxic T lymphocyte antigen (CTLA-4), has also shown to benefit patient survival in unresectable or metastatic melanoma.[103]

7. Canine mammary tumors/human breast cancer

Canine mammary tumors (CMT) are reflective of human breast cancer (HBC) in a number of ways. The most common histological type of HBC is invasive carcinoma. Malignant CMTs are classified as sarcomas or carcinomas, though the overwhelming majority are carcinomas.[104] In both species, larger tumor size, evidence of local lymph node metastasis, and advanced clinical stage are associated with a poor prognosis.[105,106]

There is considerable evidence for the importance of hormonal influence in both CMT and HBC. However, when evaluating shared hormonal etiology, it is important to remember that dogs do not undergo menopause; therefore, considerations should be made for premenopausal women who undergo cyclic hormonal changes similar to the dog.[107] In both humans and dogs, males are rarely affected. In terms of CMT risk, reproductively intact female dogs have a fourfold higher risk compared to dogs ovariohysterectomized (spayed) before the age of two.[108]

HBC and CMT share four distinct subtypes based on the molecular profile of the two receptors, ER and HER2: luminal A ($ER^+/HER2^-$), luminal B ($ER^+/HER2^+$), basal-like ($ER^-/HER2^-$), and $HER2^-$overexpressing ($ER^-/HER2^+$). The overlap in these phenotypic subtypes between the two species indicates that the molecular heterogeneity is maintained in CMT, thus lending additional support for canine mammary tumors as models of human breast cancer.

The levels of estrogen receptors (ER) have been useful in predicting response to endocrine therapy in humans. A study evaluating ER expression in CMT found that 61% of epithelial neoplastic mammary tumors displayed positive ER expression, as defined by 10 fmol/mg cytosol protein.[109] Typically, ERs are more actively expressed on benign rather than malignant neoplasms.[110] Expression of one of the two ER isoforms, ERβ, was found to be related to increased survival in ERα-negative patients who were treated with tamoxifen.[111] Knowledge of ER status is important for informing endocrine therapy, such as selective ER modulators (SERMs).[112] These drugs act as antagonists for ERα and agonists for ERβ, and therefore may be useful in the treatment of some canine mammary tumors.

Other genes have shown similar alterations, such as the frequency of *TP53* mutations (20% in both humans and dogs), which is similar in both CMT and HBC.[107] Tumors involving *TP53* mutations have been reported in more aggressive carcinomas, including HBC. The organization of the *TP53* gene, including exons and gene products, is similar between dogs and humans. In fact, Veldhoen et al. discovered a *TP53* point mutation on Ala125Val in CMT, which corresponded to the same mutation occurring on the human gene.[113] Similarities have also been found in the *BRCA* genes, including *BRCA1*, *BRCA2*, and *RAD51*, which are involved in maintaining genomic stability. In humans, low expression of the BRCA1/BRCA2/RAD51 complex has been documented in more aggressive HBC tumors. In dogs with CMT, however, these three genes display more variation and inconsistencies in their expression. The results of a 2009 study using laser microdissection and RT-qPCR (reverse transcriptase quantitative polymerase chain reaction) in 10 dogs with mammary adenocarcinomas and regional lymph node metastases suggested that RAD51 is upregulated in the majority (80%) of CMT lymph node metastases compared to normal canine mammary epithelium.[114]

8. Urothelial carcinoma/bladder cancer

Urothelial carcinoma (UC), also known as transitional cell carcinoma (TCC), is the most common subtype of bladder cancer, accounting for 90% of all bladder cancers in humans. It is also a common tumor in dogs, as there are more than 60,000 estimated new cases in United States each year.[115] Canine UC shares significant histological, biological, molecular, immunological, and clinical phenotypes with human UC and can act as a suitable animal model to provide improved predictive value for bladder cancer assessment and treatment for humans.

Pathological studies in dogs with tissue biopsies showed that more than 90% of bladder cancers consist of intermediate to high-grade UC. Knapp et al. reported a high level of similarity between human and canine UC at the microscopic level.[116] Canine UC showed similar patterns of local invasiveness, including metastatic frequency and site, such as bone metastases, to human UC. Canine studies have reported distant metastases in approximately 14% of dogs at the time of diagnosis and 49% of dogs at the time of death.[116] Maintaining control of the primary tumor is of critical importance; death due to urinary tract obstruction from an aggressive tumor often occurs in these dogs, prior to metastatic disease.

Gene expression analyses show similar molecular subtypes and features including mutational signatures and epigenetic events between canine and human UC. Several studies reported that both human and canine UC consist primarily of luminal and basal subtypes, and additional subgroupings within these subtypes arise that still need further validation in dogs.[117–120] These studies, utilizing The Cancer Genome Atlas (TCGA) project, reported a significant level of mutations in key cellular regulatory pathways using whole-exome sequencing of 131 muscle-invasive, high grade human bladder tumors. It has been reported that 50% of human UC tumors harbor alterations in the p53 pathway, the EGF-EGFR signaling pathway, and demonstrate epithelial-to-mesenchymal transition (EMT).[121–124] Several studies have reported aberrant expression of p63 (a homolog of p53), p53, the p53-inducible gene 14-3-3σ protein, and vimentin (responsible for EMT transition leading to metastasis) in canine UC tumors when compared to normal bladder epithelium, as is also the case for human UC.[125–128]

Dhawan et al. conducted microarray analysis with 18,000 canine mRNA transcripts and over 20,000 nonredundant predicted genes and found a significant level of similarities in gene expression pattern between human and canine UC. They reported that the canine UC samples showed enriched expression of genes indicative of luminal subtypes (*PPARG, TBX2, ERBB2, ERBB3*), basal subtypes (*MMP9, EGFR, JUN* and *MYC*), and EMT markers (*ZEB1, ZEB2, VIM, CDH1* and *CLDN3*), similar to that observed in human UC. The authors also reported the defective expression of genes involved in the p53 pathway and EGF-EGFR signaling pathway. They found that 73% of primary canine UC tumors displayed *EGFR* overexpression.[129] These findings are similar to those obtained for human UC by Liedberg et al., who found EGFR overexpression in 79% of human UC by immunohistochemistry.[129,130]

Decker et al. conducted complete transcriptome sequencing for multiple canine UC tumors. The authors found a somatic (as confirmed by DNA genotyping) *BRAF*^{V600E} mutation in all the tested canine tumors, as well as in 87% of 62 additional canine UC tumors by using Sanger sequencing. *BRAF*^{V600E} is a recurring mutation across many different tumor types, occurring in approximately 8% of all human cancers.[131] The *BRAF*^{V600E} mutation is rare in human UC, occurring in <1% of cases, although a significant number of bladder tumors have mutations resulting in over-activity of the RTK/RAS/RAF pathway, including downstream activation of BRAF.[131]

To further characterize canine UC at the molecular level, Maeda et al. conducted detailed gene expression analysis using RNA sequencing. Results from a total of 11 dogs with UC (five intact female, five neutered female and 1 intact male) and 5 healthy dogs showed similar gene expression patterns between human and canine UCs. The authors performed differential expression analysis and found a total of 2532 differentially expressed genes (DEGs) in canine UC when compared to normal bladder tissue. Among them, 1007 and 1524 DEGs were reported as upregulated and downregulated genes, respectively, with a fold-change >2. Furthermore, ingenuity pathway analysis (IPA) using the top 500 up- and downregulated DEGs revealed *PTGER2* and *ERBB2* as the most activated upstream regulators and Irg1, *TP53*, let-7 (micro-RNA), and *ZFP36* as those predicted to be inhibited downregulators in canine UC compared to normal bladder tissue. In addition, upregulated expression of several transcription factors (TFs) such as *CCND1*, *FOXM1*, *E2F3*, *FOXO1*, and *JUN* were also reported in canine UC.[132] The authors concluded that highly expressed genes in canine UC were associated with tumorigenesis and leukocyte migration, while downregulated genes were associated with muscle function.[132]

The expression of ERBB2, also known as HER2, was also evaluated using immunohistochemistry. Tsuboi et al. reported that 61% of canine UC samples were positive for HER2 at increased levels compared to that of healthy bladders. Immunoreactivity was observed to be varied based on tumor grade, where grade 2 showed higher immunoreactivity than grade 3. In addition, a significant level of intratumoral heterogeneity was also observed in high-grade UC samples, indicating the need to develop new treatment approaches for anti-HER2 therapy-resistant HER2-negative UC.[133,134]

In addition to the similar molecular and other features, there are similarities in terms of treatment options and responses between dogs and humans with UC. The standard treatment for both human and canine UC includes surgery. Both species showed effective response to platinum and cisplatin agents. The remission rate after cisplatin-based therapy was reported as 12–20% and 17–34% for dog and human, respectively.[135,136] The canine model showed enhanced response after treatment with Cox inhibitors. Cyclooxyegnease-2 (Cox-2) has also been reported to be overexpressed in both canine and human bladder cancer and Cox inhibitors have been implicated as key drugs for bladder cancer treatment.[137,138] Several studies reported that Cox inhibitors enhance the activity of chemotherapy drugs such as cisplatin or vinblastine.[139–141] The remission rate of cisplatin combined with a Cox inhibitor ranged from 50% to 70%, whereas cisplatin alone showed <20% remission rate. The encouraging findings for Cox inhibitors in the dog have been translated into human UC and similar remission in humans has also been observed.[141]

Naturally occurring canine bladder cancer has excellent potential as a model for studying human UC. Due to the similarities with human UC with respect to molecular pathogenesis and response to treatment, canine UC can model the origin, progression, and response to treatment for the human cancer.

9. Soft tissue sarcomas

Sarcoma accounts for a diverse group of cancers that originates in the bones (nonsoft tissue sarcoma) as well as in connective tissue. The latter, known collectively as soft tissue sarcomas (STS), are derived from the tissues that connect, support, and surround other body structures, and accounts for 80% of all human sarcomas.[142] According to National Cancer Institute (NCI), it is estimated that there will be 13,460 new STS cases diagnosed in adults and children, resulting in an estimated 5350 deaths in the United States in 2021.[143] STS includes many subtypes, including hemangiosarcoma, fibrosarcoma, peripheral nerve sheath tumors, histiocytic sarcoma, myxosarcoma, liposarcoma, rhabdomyosarcoma, leiomyosarcoma, synovial cell sarcoma, and lymphangiosarcoma.[144,145]

Hemangiosarcoma (HSA) represents the most common, most aggressive, and most highly metastatic form of STS in dogs. HSA occurs mainly in skin, spleen, heart, liver and subcutis.[146,147] An elevated incidence of HSA has been reported in the German shepherd, boxer, Labrador retriever, and golden retriever.[144,148] In affected dogs, the clinical signs first present as vague illness, but tumor rupture and massive blood loss often results in acute death.[144,148] Several studies have shown that canine HSA shares similarities with human angiosarcoma (AS, a subgroup of STSs), including cells of origin, anatomical location, and immunophenotypic characteristics. Both human AS and canine HSA show similar anatomical distribution, with tumors arising in cutaneous, subcutaneous, and visceral sites.[149,150] Gorden et al. conducted genome-wide expression profiling of 24 canine HSA tumor samples by employing microarray analysis and RNA sequencing and showed that, similar to human AS, this tumor originates from multipotent progenitors and differentiates into three different lineages. These three distinct molecular subtypes are associated with angiogenesis or endothelial cell development/function, inflammation and myeloid differentiation/function, and adipogenesis and lipid transport pathways.[151] In canine HSA, immunohistochemical studies, reverse transcriptase polymerase chain reaction quantification (qRT-PCR), and *in situ* hybridization assays (ISH), have reported upregulated expression of clinically significant vascular endothelial markers, genes, or proteins related to signal transduction or angiogenesis, such as CD31, CD34, Factor VIII-related Antigen (F8RA), CD117 (also known as c-KIT), FGFR1, *VEGFR 1–3, TIE1, TIMP1, TNFa,* and *PECAM1.*[152,153] Thomas et al. collected 75 canine HSA tumor samples from five highly predisposed dog breeds and evaluated somatic DNA CNAs through the use of an array-based comparative genomic hybridization assay. The authors observed aberrant expression in oncogene and tumor suppressor genes, including *PDGFRB*, *TP53*, *MYC*, and *PTEN*. Most importantly, they reported gain of function (GOF) mutations in canine chromosomes 13, 24, and 31, including *PDGFRA*, *KIT*, *VEGFR2*, *ANGPT1*, and *ANGPT4*, and loss of chromosome 16, including the genes *ANGPT2* and *CDKN2AIP*. The authors utilized the Genomic Identification of Significant Targets in Cancer (GISTIC) algorithm to differentiate between driver and passenger mutations and reported most of the gene mutations as passenger mutations. However, the authors identified *CDKN2A*, *VEGFA*, and *SKI* oncogene mutations as drivers leading to canine HSA tumor development.[154]

CNAs and inconsistent expression of onco- and tumor-suppressor genes have also been observed in human AS. Leduc et al. assessed 10 human AS samples via immunohistochemical staining and a whole-genome, single-nucleotide polymorphism-based microarray platform called OncoScan, and reported trisomy for several chromosomes, including Chromosomes

4, 8, 11, 17, 20 as well 1q+. Homozygous deletion of *CDKN2* was also identified.[155] Furthermore, while specific amplification of the *MYC* gene was not observed, 8 of the 10 samples harbored an additional copy of *MYC* as a result of trisomy 8. Similar findings were reported in a 2017 study utilizing whole exome sequencing of 20 canine HSA samples, including somatic mutations in potential therapeutic target genes largely responsible for controlling the PI3K pathway, such as PIK3CA, TP53, PTEN, FOXO3 and PLCG1. The mutation in *PLCG1* was found to be identical to a mutation observed previously in human AS. Based on these findings, the authors reported activation of the PI3K pathway as the driver promoting canine HSA progression, similar to human cancers.[156]

In addition to defects in cancer-associated genes, abnormal expression of microRNAs (miRNA) associated with angiogenesis and regulation of PI3K has also been observed in canine HSA. Grimes et al. conducted miRNA analysis in 10 dog samples (5 with HSA and 5 with nodular hyperplasia) and found 22 differentially expressed miRNAs in HSA samples compared to nodular hyperplasia and/or normal spleen. These miRNAs include mir-26a, mir-126, mir-139, mir-140, mir-150, mir-203, mir-424, mir-503, mir-505, mir-542, mir-30e, mir-33b, mir-365, mir-758, mir-22, and mir-452. A previous study showed high prevalence of mir-126 in the heart and liver, as well as lung and endothelial precursor cells, which are the cells-of-origin of HSA. Additionally, mir-126 is involved in regulating angiogenesis, specifically through the PI3K pathway, by promoting overexpression of VEGF, as well as G protein signaling pathways. The authors reported upregulation of mir-126 and mir-452 (3.58- and 13-fold, respectively) and downregulation of mir-150 and mir-203 (2.8- and 2.56-fold, respectively).[157] Collectively, these results suggest similarities at the molecular level between human AS and canine HSA.

Like HSA, other canine STSs show similarities with various other human STSs at both the pathological and molecular levels. For example, fibrosarcoma and histiocytic sarcoma in the dog have been reported to contain CNAs in canine chromosomes 4, 11, 27, and 30 as well as mutations in *TP53*, *CDKN2A/B*, *MTAP*, *PTEN*, *KIT*, and $BRAF^{V600E}$.[158–161]

STSs are relatively common in both human and canine cohorts and share similarities in terms of immune-histology, pathology, dysregulated molecular pathways, and gene expression profile. Thus, the study for therapeutic intervention, including disease detection mechanisms, subtyping, as well as novel targeted therapies, if developed in dogs, can be both applicable and beneficial for human AS patients.

10. Summary and conclusion

Cancer is a complex disease that can arise from somatic and/or germline mutations resulting in a phenotype favorable for tumorigenesis. The consequential effects include evasion of host responses, such as DNA damage checkpoints and immune regulatory functions, aberrant proliferation and/or antiapoptotic mechanisms promoting uncontrolled cell division, and accumulated genomic instability. The past two decades have been filled with significant breakthroughs in cancer research, including diagnosis and treatment that have extended and improved the lives of many. As sequencing costs have fallen, the era of precision medicine has risen. Personalized or precision medicine uses a patient's genes, proteins, and/or environmental factors to guide drug selection based on the individual tumor and its potential sensitivity to specific therapies.

While many successful therapies have been approved by the United States FDA for precision oncology, many therapies fail to translate from rodent models to humans because the promising results seen in laboratory mice are rarely reflected accurately in humans. Rodent models are severely limited in their ability to (1) generate spontaneous tumors, (2) recapitulate the human immune response in its entirety, (3) share similar environments with humans and thus face comparable environmental exposures, (4) mimic human-related drug metabolism, (5) undergo natural tumoral progression and evolution, and (6) accurately reflect tumorigenesis. These weaknesses hinder the full potential of precision medicine concepts.

Humans and dogs have evolved together, and incidentally, share tumors in similar locations that are nearly indistinguishable from each other. Due to the relatively restricted gene pool within the canine breed structure, the genetic heterogeneity within dogs is reduced, therefore allowing reduction of noise in genome-wide association studies to more easily identify associated risk factors. Thus, fewer patient samples are needed to reach comparable conclusions to human studies. Not only do dogs share our environmental exposures such as tobacco smoke or cleaning chemicals, they also develop spontaneous, inherently heterogeneous tumors that undergo natural immune-crosstalk and interaction with the microenvironment. Breed disposition has revealed an inherited component to some tumors, similar to that observed in human families. Canine studies have many benefits, including a shortened lifespan that facilitates the study of natural tumor progression, ability to obtain patient samples and clinical information more easily, maintenance of normal home activities within the patient's own environment, and ability to control food, water, and drug intake. Most importantly, pet owners are more eager than ever to explore treatment options for their beloved furry family members. In addition, our canine companions and their tumors may hold the key to improving our survival as well.

In this chapter, we discussed the origin of the canine model, its genomic evolution, and several tumors that are excellent models for the human disease, including osteosarcoma, B-cell lymphoma, melanoma, canine mammary tumor/human breast cancer, urothelial carcinoma/bladder cancer, and soft tissue sarcomas. Many of these tumors share multiple features between humans and dogs, including common mutational signatures, gene expression patterns, clinical and histological presentation, and treatment/response. The cancers discussed herein are in no way exhaustive, and as additional tumors are explored, sequenced, and compared, it is likely that additional cancers will be added to the list as suitable canine models. While rodent models are highly valuable and will always have a place in advancing oncology research, their value in predicting drug response and recapitulating the complex intra- and intertumoral heterogeneity observed in human tumors is rather poor. The dog is a bridge between the lab and the clinic; they represent a patient sample pool from which tumor data can be generated, drugs can be tested *in vivo* for safety and efficacy, and most importantly, precision medicine concepts can converge and be tested and refined in a biologically relevant and logistically feasible context.

References

1. Hanahan D, Weinberg RA. The hallmarks of cancer. *Cell.* 2000;100(1):57–70. https://doi.org/10.1016/S0092-8674(00)81683-9.
2. Morse BL, Kim RB. Is personalized medicine a dream or a reality? *Crit Rev Clin Lab Sci.* 2015;52(1):1–11. https://doi.org/10.3109/10408363.2014.950407.
3. LeBlanc VG, Marra MA. Next-generation sequencing approaches in cancer: Where have they brought us and where will they take us? *Cancer.* 2015;7(3):1925–1958. https://doi.org/10.3390/cancers7030869.
4. Meyerson M, Gabriel S, Getz G. Advances in understanding cancer genomes through second-generation sequencing. *Nat Rev Genet.* 2010;11(10):685–696. https://doi.org/10.1038/nrg2841.
5. Rao SR, Somarelli JA, Altunel E, et al. From the clinic to the bench and back again in one dog year: How a cross-species pipeline to identify new treatments for sarcoma illuminates the path forward in precision medicine. *Front Oncol.* 2020;10. https://doi.org/10.3389/fonc.2020.00117.
6. Misra SC, Bisui S. Feasibility of large scale implementation of personalized medicine in the current scenario. *Int J E-Health Med Commun IJEHMC.* 2016;7(2):30–49. https://doi.org/10.4018/IJEHMC.2016040103.

7. Brunak S, Collin CB, Cathaoir KEÓ, et al. Towards standardization guidelines for in silico approaches in personalized medicine. *J Integr Bioinform*. 2020;17(2–3). https://doi.org/10.1515/jib-2020-0006.
8. Król M, Motyl T. Exploiting cancer genomics in pet animals to gain advantage for personalized medicine decisions. *J Appl Genet*. 2014;55(3):337–341. https://doi.org/10.1007/s13353-014-0206-0.
9. Vizirianakis IS, Fatouros DG. Personalized nanomedicine: Paving the way to the practical clinical utility of genomics and nanotechnology advancements. *Adv Drug Deliv Rev*. 2012;64(13):1359–1362. https://doi.org/10.1016/j.addr.2012.09.034.
10. Kent J. FDA approvals advance precision medicine, genomics treatments. *Health IT Anal*. 2020;25. Published online February https://healthitanalytics.com/news/fda-approvals-advance-precision-medicine-genomics-treatments.
11. Ray T. FDA approves record number of precision oncology drugs in H1. *Precis Oncol News*. 2020;21:2020. Published online July https://www.precisiononcologynews.com/cancer/fda-approves-record-number-precision-oncology-drugs-h1-2020#.YQnWcI43m71.
12. Soverini S, Mancini M, Bavaro L, Cavo M, Martinelli G. Chronic myeloid leukemia: The paradigm of targeting oncogenic tyrosine kinase signaling and counteracting resistance for successful cancer therapy. *Mol Cancer*. 2018;17(1):49. https://doi.org/10.1186/s12943-018-0780-6.
13. André F, Ciruelos E, Rubovszky G, et al. Alpelisib for PIK3CA-mutated, hormone receptor–positive advanced breast cancer. *N Engl J Med*. 2019;380(20):1929–1940. https://doi.org/10.1056/NEJMoa1813904.
14. Hyman DM, Piha-Paul SA, Won H, et al. HER kinase inhibition in patients with HER2- and HER3-mutant cancers. *Nature*. 2018;554(7691):189–194. https://doi.org/10.1038/nature25475.
15. Smyth LM, Piha-Paul SA, Won HH, et al. Efficacy and determinants of response to HER kinase inhibition in HER2-mutant metastatic breast cancer. *Cancer Discov*. 2020;10(2):198–213. https://doi.org/10.1158/2159-8290.CD-19-0966.
16. Gambardella V, Tarazona N, Cejalvo JM, et al. Personalized medicine: Recent progress in cancer therapy. *Cancer*. 2020;12(4):1009. https://doi.org/10.3390/cancers12041009.
17. de Jong M, Maina T. Of mice and humans: Are they the same?—Implications in cancer translational research. *J Nucl Med*. 2010;51(4):501–504. https://doi.org/10.2967/jnumed.109.065706.
18. Day C-P, Merlino G, Van Dyke T. Preclinical mouse cancer models: A maze of opportunities and challenges. *Cell*. 2015;163(1):39–53. https://doi.org/10.1016/j.cell.2015.08.068.
19. Mak IW, Evaniew N, Ghert M. Lost in translation: Animal models and clinical trials in cancer treatment. *Am J Transl Res*. 2014;6(2):114–118.
20. Ranieri G, Gadaleta CD, Patruno R, et al. A model of study for human cancer: Spontaneous occurring tumors in dogs. Biological features and translation for new anticancer therapies. *Crit Rev Oncol Hematol*. 2013;88(1):187–197. https://doi.org/10.1016/j.critrevonc.2013.03.005.
21. Barutello G, Rolih V, Arigoni M, et al. Strengths and weaknesses of pre-clinical models for human melanoma treatment: Dawn of dogs' revolution for immunotherapy. *Int J Mol Sci*. 2018;19(3):799. https://doi.org/10.3390/ijms19030799.
22. Riccardo F, Aurisicchio L, Impellizeri JA, Cavallo F. The importance of comparative oncology in translational medicine. *Cancer Immunol Immunother*. 2015;64(2):137–148. https://doi.org/10.1007/s00262-014-1645-5.
23. Talmadge JE, Singh RK, Fidler IJ, Raz A. Murine models to evaluate novel and conventional therapeutic strategies for cancer. *Am J Pathol*. 2007;170(3):793–804. https://doi.org/10.2353/ajpath.2007.060929.

24. Park JS, Withers SS, Modiano JF, et al. Canine cancer immunotherapy studies: Linking mouse and human. *J Immunother Cancer.* 2016;4(1):97. https://doi.org/10.1186/s40425-016-0200-7.
25. Weyand CM, Goronzy JJ. Aging of the immune system. Mechanisms and therapeutic targets. *Ann Am Thorac Soc.* 2016;13(Suppl 5):S422–S428. https://doi.org/10.1513/AnnalsATS.201602-095AW.
26. Hooper LV, Littman DR, Macpherson AJ. Interactions between the microbiota and the immune system. *Science.* 2012;336(6086):1268–1273. https://doi.org/10.1126/science.1223490.
27. Ivanov II, Honda K. Intestinal commensal microbes as immune modulators. *Cell Host Microbe.* 2012;12(4):496–508. https://doi.org/10.1016/j.chom.2012.09.009.
28. Sivan A, Corrales L, Hubert N, et al. Commensal Bifidobacterium promotes antitumor immunity and facilitates anti–PD-L1 efficacy. *Science.* 2015;350(6264):1084–1089. https://doi.org/10.1126/science.aac4255.
29. Lawrence J, Cameron D, Argyle D. Species differences in tumour responses to cancer chemotherapy. *Philos Trans R Soc B Biol Sci.* 2015;370(1673):20140233. https://doi.org/10.1098/rstb.2014.0233.
30. Gordon MY, Blackett NM. The sensitivities of human and murine hemopoietic cells exposed to cytotoxic drugs in an in vivo culture system. *Cancer Res.* 1976;36(8):2822–2826.
31. Sanoh S, Ohta S. Chimeric mice transplanted with human hepatocytes as a model for prediction of human drug metabolism and pharmacokinetics. *Biopharm Drug Dispos.* 2014;35(2):71–86. https://doi.org/10.1002/bdd.1864.
32. Cheung C, Gonzalez FJ. Humanized mouse lines and their application for prediction of human drug metabolism and toxicological risk assessment. *J Pharmacol Exp Ther.* 2008;327(2):288–299. https://doi.org/10.1124/jpet.108.141242.
33. Tarone L, Barutello G, Iussich S, et al. Naturally occurring cancers in pet dogs as pre-clinical models for cancer immunotherapy. *Cancer Immunol Immunother.* 2019;68(11):1839–1853. https://doi.org/10.1007/s00262-019-02360-6.
34. Lindblad-Toh K, Wade CM, Mikkelsen TS, et al. Genome sequence, comparative analysis and haplotype structure of the domestic dog. *Nature.* 2005;438(7069):803–819. https://doi.org/10.1038/nature04338.
35. Briggs J, Paoloni M, Chen Q-R, Wen X, Khan J, Khanna C. A compendium of canine normal tissue gene expression. *PLoS One.* 2011;6(5), e17107. https://doi.org/10.1371/journal.pone.0017107.
36. Napierala H, Uerpmann H-P. A 'new' palaeolithic dog from Central Europe. *Int J Osteoarchaeol.* 2012;22(2):127–137. https://doi.org/10.1002/oa.1182.
37. Breen M, Thomas R. Karyotype and chromosomal organization. In: Ostrander EA, Giger U, Lindblad-Toh K, eds. *The Dog and its Genome.* Cold Spring Harbor Laboratory Press; 2005.
38. Price TD, Gebauer AB. *Last Hunters - First Farmers : New Perspectives on the Prehistory Transition to Agriculture.* School of American Research Press; 1996.
39. Parker HG, Kim LV, Sutter NB, et al. Genetic structure of the purebred domestic dog. *Science.* 2004;304(5674):1160–1164. https://doi.org/10.1126/science.1097406.
40. Ostrander E, Dreger D, Evans J. Canine cancer genomics: Lessons for canine and human health. *Annu Rev Anim Biosci.* 2019;7(1):449–472.
41. vonHoldt BM, Pollinger JP, Lohmueller KE, et al. Genome-wide SNP and haplotype analyses reveal a rich history underlying dog domestication. *Nature.* 2010;464(7290):898–902. https://doi.org/10.1038/nature08837.
42. Wade CM, Karlsson EK, Mikkelsen TS, Zody MC, Lindblad-Toh K. The dog genome: Sequence, evolution, and haplotype structure. In: *The Dog and its Genome.* Cold Spring Harbor Laboratory Press; 2006.

43. Ostrander E, Urs G, Lindblad-Toh K, eds. *The Dog and its Genome*. Cold Spring Harbor Laboratory Press; 2006.
44. Felsburg PJ. Overview of immune system development in the dog: Comparison with humans. *Hum Exp Toxicol*. 2002;21(9–10):487–492. https://doi.org/10.1191/0960327102ht286oa.
45. Greeley EH, Kealy RD, Ballam JM, Lawler DF, Segre M. The influence of age on the canine immune system. *Vet Immunol Immunopathol*. 1996;55(1):1–10. https://doi.org/10.1016/S0165-2427(96)05563-8.
46. Sansoni P, Cossarizza A, Brianti V, et al. Lymphocyte subsets and natural killer cell activity in healthy old people and centenarians. *Blood*. 1993;82(9):2767–2773. https://doi.org/10.1182/blood.V82.9.2767.2767.
47. Cobbold S, Metcalfe S. Monoclonal antibodies that define canine homologues of human CD antigens: Summary of the first international canine leukocyte antigen workshop (CLAW). *Tissue Antigens*. 1994;43(3):137–154. https://doi.org/10.1111/j.1399-0039.1994.tb02315.x.
48. Isotani M, Katsuma K, Tamura K, et al. Efficient generation of canine bone marrow-derived dendritic cells. *J Vet Med Sci*. 2006;68(8):809–814. https://doi.org/10.1292/jvms.68.809.
49. Trepanier LA, Ray K, Winand NJ, Spielberg SP, Cribb AE. Cytosolic arylamine n-acetyltransferase (NAT) deficiency in the dog and other canids due to an absence of NAT genes. *Biochem Pharmacol*. 1997;54(1):73–80. https://doi.org/10.1016/S0006-2952(97)00140-8.
50. Coelho LP, Kultima JR, Costea PI, et al. Similarity of the dog and human gut microbiomes in gene content and response to diet. *Microbiome*. 2018;6(1):72. https://doi.org/10.1186/s40168-018-0450-3.
51. Jergens AE, Guard BC, Redfern A, et al. Microbiota-related changes in unconjugated fecal bile acids are associated with naturally occurring, insulin-dependent diabetes mellitus in dogs. *Front Vet Sci*. 2019;6. https://doi.org/10.3389/fvets.2019.00199, 199.
52. Vázquez-Baeza Y, Hyde ER, Suchodolski JS, Knight R. Dog and human inflammatory bowel disease rely on overlapping yet distinct dysbiosis networks. *Nat Microbiol*. 2016;1(12):1–5. https://doi.org/10.1038/nmicrobiol.2016.177.
53. Herstad KMV, Moen AEF, Gaby JC, Moe L, Skancke E. Characterization of the fecal and mucosa-associated microbiota in dogs with colorectal epithelial tumors. *PLoS One*. 2018;13(5), e0198342. https://doi.org/10.1371/journal.pone.0198342.
54. MacEwen EG. Spontaneous tumors in dogs and cats: Models for the study of cancer biology and treatment. *Cancer Metastasis Rev*. 1990;9(2):125–136. https://doi.org/10.1007/BF00046339.
55. Paoloni M, Khanna C. Translation of new cancer treatments from pet dogs to humans. *Nat Rev Cancer*. 2008;8(2):147–156. https://doi.org/10.1038/nrc2273.
56. Parker HG, Shearin AL, Ostrander EA. Man's best friend becomes biology's best in show: Genome analyses in the domestic dog. *Annu Rev Genet*. 2010;44:309–336. https://doi.org/10.1146/annurev-genet-102808-115200.
57. Fowles JS, Brown KC, Hess AM, Duval DL, Gustafson DL. Intra- and interspecies gene expression models for predicting drug response in canine osteosarcoma. *BMC Bioinformatics*. 2016;17(1):93. https://doi.org/10.1186/s12859-016-0942-8.
58. Gardner HL, Fenger JM, London CA. Dogs as a model for cancer. *Annu Rev Anim Biosci*. 2016;4:199–222. https://doi.org/10.1146/annurev-animal-022114-110911.
59. Olson PN. Using the canine genome to cure cancer and other diseases. *Theriogenology*. 2007;68(3):378–381. https://doi.org/10.1016/j.theriogenology.2007.04.016.
60. Waters DJ, Wildasin K. Cancer clues from pet dogs. *Sci Am*. 2006;295(6):94–101.
61. Schiffman JD, Breen M. Comparative oncology: What dogs and other species can teach us about humans with cancer. *Philos Trans R Soc B Biol Sci*. 2015;370(1673):20140231. https://doi.org/10.1098/rstb.2014.0231.

62. Hytönen MK, Lohi H. Canine models of human rare disorders. *Rare Dis*. 2016;4(1), e1241362. https://doi.org/10.1080/21675511.2016.1241362.
63. Kornegay JN, Bogan JR, Bogan DJ, et al. Canine models of Duchenne muscular dystrophy and their use in therapeutic strategies. *Mamm Genome*. 2012;23(1):85–108. https://doi.org/10.1007/s00335-011-9382-y.
64. Chapman RW. Canine models of asthma and COPD. *Pulm Pharmacol Ther*. 2008;21 (5):731–742. https://doi.org/10.1016/j.pupt.2008.01.003.
65. Rowell JL, McCarthy DO, Alvarez CE. Dog models of naturally occurring cancer. *Trends Mol Med*. 2011;17(7):380–388. https://doi.org/10.1016/j.molmed.2011.02.004.
66. Vail DM, Page RL, Withrow SJ. Tumors of the skeletal system. In: Withrow and MacEwen's Small Animal Clinical Oncology. 5th ed; 2013:463–503.
67. Simpson S, Dunning MD, de Brot S, Grau-Roma L, Mongan NP, Rutland CS. Comparative review of human and canine osteosarcoma: Morphology, epidemiology, prognosis, treatment and genetics. *Acta Vet Scand*. 2017;59(1):71. https://doi.org/10.1186/s13028-017-0341-9.
68. Scott MC, Temiz NA, Sarver AE, et al. Comparative transcriptome analysis quantifies immune cell transcript levels, metastatic progression, and survival in osteosarcoma. *Cancer Res*. 2018;78(2):326–337. https://doi.org/10.1158/0008-5472.CAN-17-0576.
69. Heymann M-F, Lézot F, Heymann D. The contribution of immune infiltrates and the local microenvironment in the pathogenesis of osteosarcoma. *Cell Immunol*. 2019;343, 103711. https://doi.org/10.1016/j.cellimm.2017.10.011.
70. Hiddemann W, Roessner A, Wörmann B, et al. Tumor heterogeneity in osteosarcoma as identified by flow cytometry. *Cancer*. 1987;59(2):324–328. https://doi.org/10.1002/1097-0142(19870115)59:2<324::aid-cncr2820590226>3.0.co;2-9.
71. Karlsson EK, Sigurdsson S, Ivansson E, et al. Genome-wide analyses implicate 33 loci in heritable dog osteosarcoma, including regulatory variants near CDKN2A/B. *Genome Biol*. 2013;14(12):R132. https://doi.org/10.1186/gb-2013-14-12-r132.
72. Kirpensteijn J, Kik M, Teske E, Rutteman GR. TP53 gene mutations in canine osteosarcoma. *Vet Surg*. 2008;37(5):454–460. https://doi.org/10.1111/j.1532-950X.2008.00407.x.
73. Wunder JS, Gokgoz N, Parkes R, et al. TP53 mutations and outcome in osteosarcoma: A prospective, multicenter study. *J Clin Oncol*. 2016;23(7):1483–1490. https://doi.org/10.1200/JCO.2005.04.074.
74. Paoloni M, Davis S, Lana S, et al. Canine tumor cross-species genomics uncovers targets linked to osteosarcoma progression. *BMC Genomics*. 2009;10:625. https://doi.org/10.1186/1471-2164-10-625.
75. Scott MC, Sarver AL, Gavin KJ, et al. Molecular subtypes of osteosarcoma identified by reducing tumor heterogeneity through an interspecies comparative approach. *Bone*. 2011;49(3):356–367. https://doi.org/10.1016/j.bone.2011.05.008.
76. Monks NR, Cherba DM, Kamerling SG, et al. A multi-site feasibility study for personalized medicine in canines with osteosarcoma. *J Transl Med*. 2013;11 (1):158. https://doi.org/10.1186/1479-5876-11-158.
77. Davis LE, Hofmann NE, Li G, et al. A case study of personalized therapy for osteosarcoma. *Pediatr Blood Cancer*. 2013;60(8):1313–1319. https://doi.org/10.1002/pbc.24512.
78. Villamil JA, Henry CJ, Hahn AW, Bryan JN, Tyler JW, Caldwell CW. Hormonal and sex impact on the epidemiology of canine lymphoma. *J Cancer Epidemiol*. 2009;2009, 591753. https://doi.org/10.1155/2009/591753.
79. Marconato L, Gelain ME, Comazzi S. The dog as a possible animal model for human non-Hodgkin lymphoma: A review. *Hematol Oncol*. 2013;31(1):1–9. https://doi.org/10.1002/hon.2017.

80. Itoh T, Shiota M, Takanashi M, et al. Engraftment of human non-Hodgkin lymphomas in mice with severe combined immunodeficiency. *Cancer.* 1993;72(9):2686–2694. https://doi.org/10.1002/1097-0142(19931101)72:9<2686::aid-cncr2820720927>3.0.co;2-t.
81. Zandvliet M. Canine lymphoma: A review. *Vet Q.* 2016;36(2):76–104. https://doi.org/10.1080/01652176.2016.1152633.
82. Ginn J, Sacco J, Wong YY, Motsinger-Reif A, Chun R, Trepanier LA. Positive association between a glutathione- S-transferase polymorphism and lymphoma in dogs. *Vet Comp Oncol.* 2014;12(3):227–236. https://doi.org/10.1111/vco.12000.
83. Breen M, Modiano JF. Evolutionarily conserved cytogenetic changes in hematological malignancies of dogs and humans Y man and his best friend share more than companionship. *Chromosome Res.* 2008;2008(16):145–154.
84. Thomas R, Seiser EL, Motsinger-Reif A, et al. Refining tumor-associated aneuploidy through 'genomic recoding' of recurrent DNA copy number aberrations in 150 canine non-Hodgkin lymphomas. *Leuk Lymphoma.* 2011;52(7):1321–1335. https://doi.org/10.3109/10428194.2011.559802.
85. Mudaliar MAV, Haggart RD, Miele G, et al. Comparative gene expression profiling identifies common molecular signatures of NF-κB activation in canine and human diffuse large B cell lymphoma (DLBCL). *PLoS One.* 2013;8(9), e72591. https://doi.org/10.1371/journal.pone.0072591.
86. Bavi P, Uddin S, Bu R, et al. The biological and clinical impact of inhibition of NF-κB-initiated apoptosis in diffuse large B cell lymphoma (DLBCL). *J Pathol.* 2011;224(3):355–366. https://doi.org/10.1002/path.2864.
87. Davis RE, Ngo VN, Lenz G, et al. Chronic active B cell receptor signaling in diffuse large B cell lymphoma. *Nature.* 2010;463(7277):88–92. https://doi.org/10.1038/nature08638.
88. Brien G, Trescol-Biemont M-C, Bonnefoy-Bérard N. Downregulation of Bfl-1 protein expression sensitizes malignant B cells to apoptosis. *Oncogene.* 2007;26(39):5828–5832. https://doi.org/10.1038/sj.onc.1210363.
89. Gaurnier-Hausser A, Patel R, Baldwin AS, May MJ, Mason NJ. NEMO binding domain peptide inhibits constitutive NF-κB activity and reduces tumor burden in a canine model of relapsed, refractory diffuse large B-cell lymphoma. *Clin Cancer Res.* 2011;17(14):4661–4671. https://doi.org/10.1158/1078-0432.CCR-10-3310.
90. Miller AJ, Mihm MC. Melanoma. *N Engl J Med.* 2006;355(1):51–65. https://doi.org/10.1056/NEJMra052166.
91. Nishiya AT, Massoco CO, Felizzola CR, et al. Comparative aspects of canine melanoma. *Vet Sci.* 2016;3(1):7. https://doi.org/10.3390/vetsci3010007.
92. Turek M, LaDue T, Looper J, et al. Multimodality treatment including ONCEPT for canine oral melanoma: A retrospective analysis of 131 dogs. *Vet Radiol Ultrasound.* 2020;61(4):471–480. https://doi.org/10.1111/vru.12860.
93. Li G, Herlyn M. Dynamics of intercellular communication during melanoma development. *Mol Med Today.* 2000;6(4):163–169. https://doi.org/10.1016/s1357-4310(00)01692-0.
94. Shain AH, Yeh I, Kovalyshyn I, et al. The genetic evolution of melanoma from precursor lesions. *N Engl J Med.* 2015;373:1926–1936. https://doi.org/10.1056/NEJMoa1502583. https://doi.org/10.1056/NEJMoa1502583.
95. Davies H, Bignell GR, Cox C, et al. Mutations of the BRAF gene in human cancer. *Nature.* 2002;417(6892):949–954. https://doi.org/10.1038/nature00766.
96. Chu P-Y, Pan S-L, Liu C-H, Lee J, Yeh L-S, Liao AT. KIT gene exon 11 mutations in canine malignant melanoma. *Vet J.* 2013;196(2):226–230. https://doi.org/10.1016/j.tvjl.2012.09.005.
97. Gillard M, Cadieu E, De Brito C, et al. Naturally occurring melanomas in dogs as models for non- UV pathways of human melanomas. *Pigment Cell Melanoma Res.* 2014;27(1):90–102. https://doi.org/10.1111/pcmr.12170.

98. Gyorffy S, Rodriguez-Lecompte JC, Woods JP, et al. Bone marrow-derived dendritic cell vaccination of dogs with naturally occurring melanoma by using human gp100 antigen. *J Vet Intern Med*. 2005;19(1):56–63. https://doi.org/10.1111/j.1939-1676.2005.tb02659.x.
99. Tamura K, Yamada M, Isotani M, et al. Induction of dendritic cell-mediated immune responses against canine malignant melanoma cells. *Vet J*. 2008;175(1):126–129. https://doi.org/10.1016/j.tvjl.2006.11.005.
100. Saleem M, Maddodi N, Zaid MA, et al. Lupeol inhibits growth of highly aggressive human metastatic melanoma cells in vitro and in vivo by inducing apoptosis. *Clin Cancer Res*. 2008;14(7):2119–2127. https://doi.org/10.1158/1078-0432.CCR-07-4413.
101. Yokoe I, Azuma K, Hata K, et al. Clinical systemic lupeol administration for canine oral malignant melanoma. *Mol Clin Oncol*. 2015;3(1):89–92. https://doi.org/10.3892/mco.2014.450.
102. Maekawa N, Konnai S, Takagi S, et al. A canine chimeric monoclonal antibody targeting PD-L1 and its clinical efficacy in canine oral malignant melanoma or undifferentiated sarcoma. *Sci Rep*. 2017;7:8951. https://doi.org/10.1038/s41598-017-09444-2.
103. Schadendorf D, Hodi FS, Robert C, et al. Pooled analysis of long-term survival data from phase II and phase III trials of ipilimumab in unresectable or metastatic melanoma. *J Clin Oncol*. 2015;33(17):1889–1894. https://doi.org/10.1200/JCO.2014.56.2736.
104. Tavasoly A, Golshahi H, Rezaie A, Farhadi M. Classification and grading of canine malignant mammary tumors. *Vet Res Forum Int Q J*. 2013;4(1):25–30.
105. Hellmén E, Bergström R, Holmberg L, Spångberg IB, Hansson K, Lindgren A. Prognostic factors in canine mammary tumors: A multivariate study of 202 consecutive cases. *Vet Pathol*. 1993;30(1):20–27. https://doi.org/10.1177/030098589303000103.
106. Sarli G, Preziosi R, Benazzi C, Castellani G, Marcato PS. Prognostic value of histologic stage and proliferative activity in canine malignant mammary tumors. *J Vet Diagn Investig Off Publ Am Assoc Vet Lab Diagn Inc*. 2002;14(1):25–34. https://doi.org/10.1177/104063870201400106.
107. Queiroga FL, Raposo T, Carvalho MI, Prada J, Pires I. Canine mammary tumours as a model to study human breast cancer: Most recent findings. *In Vivo*. 2011;25(3):455–465.
108. Schneider R. Comparison of age, sex, and incidence rates in human and canine breast cancer. *Cancer*. 1970;26(2):419–426. https://doi.org/10.1002/1097-0142(197008)26:2<419::aid-cncr2820260225>3.0.co;2-u.
109. MacEwen EG, Patnaik AK, Harvey HJ, Panko WB. Estrogen receptors in canine mammary tumors. *Cancer Res*. 1982;42(6):2255–2259.
110. Rutteman GR, Misdorp W, Blankenstein MA, van den Brom WE. Oestrogen (ER) and progestin receptors (PR) in mammary tissue of the female dog: Different receptor profile in non-malignant and malignant states. *Br J Cancer*. 1988;58(5):594–599.
111. Gruvberger-Saal SK, Bendahl P-O, Saal LH, et al. Estrogen receptor beta expression is associated with tamoxifen response in ERalpha-negative breast carcinoma. *Clin Cancer Res*. 2007;13(7):1987–1994. https://doi.org/10.1158/1078-0432.CCR-06-1823.
112. Omoto Y, Iwase H. Clinical significance of estrogen receptor β in breast and prostate cancer from biological aspects. *Cancer Sci*. 2015;106(4):337–343. https://doi.org/10.1111/cas.12613.
113. Veldhoen N, Watterson J, Brash M, Milner J. Identification of tumour-associated and germ line p53 mutations in canine mammary cancer. *Br J Cancer*. 1999;81(3):409–415. https://doi.org/10.1038/sj.bjc.6690709.
114. Klopfleisch R, Gruber AD. Increased expression of BRCA2 and RAD51 in lymph node metastases of canine mammary adenocarcinomas. *Vet Pathol*. 2009;46(3):416–422. https://doi.org/10.1354/vp.08-VP-0212-K-FL.

115. Knapp DW, Dhawan D, Ramos-Vara JA, et al. Naturally-occurring invasive urothelial carcinoma in dogs, a unique model to drive advances in managing muscle invasive bladder cancer in humans. *Front Oncol.* 2020;9:1493. https://doi.org/10.3389/fonc.2019.01493.
116. Knapp DW, Glickman NW, DeNicola DB, Bonney PL, Lin TL, Glickman LT. Naturally-occurring canine transitional cell carcinoma of the urinary bladder. A relevant model of human invasive bladder cancer. *Urol Oncol Semin Orig Investig.* 2000;5(2):47–59. https://doi.org/10.1016/S1078-1439(99)00006-X.
117. Robertson AG, Kim J, Al-Ahmadie H, et al. Comprehensive molecular characterization of muscle invasive bladder cancer. *Cell.* 2017;171(3):540–556.e25. https://doi.org/10.1016/j.cell.2017.09.007.
118. McConkey DJ, Choi W. Molecular subtypes of bladder cancer. *Curr Oncol Rep.* 2018;20(10):77. https://doi.org/10.1007/s11912-018-0727-5.
119. Dhawan D, Hahn NM, Ramos-Vara JA, Knapp DW. Naturally-occurring canine invasive urothelial carcinoma harbors luminal and basal transcriptional subtypes found in human muscle invasive bladder cancer. *PLoS Genet.* 2018;14(8), e1007571. https://doi.org/10.1371/journal.pgen.1007571.
120. Choi W, Porten S, Kim S, et al. Identification of distinct basal and luminal subtypes of muscle-invasive bladder cancer with different sensitivities to frontline chemotherapy. *Cancer Cell.* 2014;25(2):152–165. https://doi.org/10.1016/j.ccr.2014.01.009.
121. Castillo-Martin M, Domingo-Domenech J, Karni-Schmidt O, Matos T, Cordon-Cardo C. Molecular pathways of urothelial development and bladder tumorigenesis. *Urol Oncol.* 2010;28(4):401–408. https://doi.org/10.1016/j.urolonc.2009.04.019.
122. Cancer Genome Atlas. Comprehensive molecular characterization of urothelial bladder carcinoma. *Nature.* 2014;507:315–322.
123. Watanabe R, Tomita Y, Nishiyama T, Tanikawa T, Sato S. Correlation of p53 protein expression in human urothelial transitional cell cancers with malignant potential and patient survival. *Int J Urol.* 1994;1(1):43–48. https://doi.org/10.1111/j.1442-2042.1994.tb00007.x.
124. Chatterjee SJ, Datar R, Youssefzadeh D, et al. Combined effects of p53, p21, and pRb expression in the progression of bladder transitional cell carcinoma. *J Clin Oncol.* 2004;22(6):1007–1013. https://doi.org/10.1200/JCO.2004.05.174.
125. Hanazono K, Nishimori T, Fukumoto S, et al. Immunohistochemical expression of p63, Ki67 and β-catenin in canine transitional cell carcinoma and polypoid cystitis of the urinary bladder. *Vet Comp Oncol.* 2016;14(3):263–269. https://doi.org/10.1111/vco.12095.
126. Suárez-Bonnet A, Herráez P, Aguirre M, et al. Expression of cell cycle regulators, 14-3-3σ and p53 proteins, and vimentin in canine transitional cell carcinoma of the urinary bladder. *Urol Oncol.* 2015;33(7). 332.e1-7. https://doi.org/10.1016/j.urolonc.2015.04.006.
127. Dhawan D, Ramos-Vara JA, Stewart JC, Zheng R, Knapp DW. Canine invasive transitional cell carcinoma cell lines: In vitro tools to complement a relevant animal model of invasive urinary bladder cancer. *Urol Oncol Semin Orig Investig.* 2009;27(3):284–292. https://doi.org/10.1016/j.urolonc.2008.02.015.
128. Baumgart E, Cohen MS, Silva Neto B, et al. Identification and prognostic significance of an epithelial-mesenchymal transition expression profile in human bladder tumors. *Clin Cancer Res Off J Am Assoc Cancer Res.* 2007;13(6):1685–1694. https://doi.org/10.1158/1078-0432.CCR-06-2330.
129. Dhawan D, Paoloni M, Shukradas S, et al. Comparative gene expression analyses identify luminal and basal subtypes of canine invasive urothelial carcinoma that mimic patterns in human invasive bladder cancer. *PLoS One.* 2015;10(9), e0136688. https://doi.org/10.1371/journal.pone.0136688.

130. Liedberg F, Anderson H, Chebil G, et al. Tissue microarray based analysis of prognostic markers in invasive bladder cancer: Much effort to no avail? *Urol Oncol.* 2008;26 (1):17–24. https://doi.org/10.1016/j.urolonc.2006.08.021.
131. Decker B, Parker HG, Dhawan D, et al. Homologous mutation to human BRAF V600E is common in naturally occurring canine bladder cancer—Evidence for a relevant model system and urine-based diagnostic test. *Mol Cancer Res MCR.* 2015;13 (6):993–1002. https://doi.org/10.1158/1541-7786.MCR-14-0689.
132. Maeda S, Tomiyasu H, Tsuboi M, et al. Comprehensive gene expression analysis of canine invasive urothelial bladder carcinoma by RNA-seq. *BMC Cancer.* 2018;18 (1):472. https://doi.org/10.1186/s12885-018-4409-3.
133. Tsuboi M, Sakai K, Maeda S, et al. Assessment of HER2 expression in canine urothelial carcinoma of the urinary bladder. *Vet Pathol.* 2019;56(3):369–376. https://doi.org/10.1177/0300985818817024.
134. Millanta F, Impellizeri J, McSherry L, Rocchigiani G, Aurisicchio L, Lubas G. Overexpression of HER-2 via immunohistochemistry in canine urinary bladder transitional cell carcinoma - a marker of malignancy and possible therapeutic target. *Vet Comp Oncol.* 2018;16(2):297–300. https://doi.org/10.1111/vco.12345.
135. Lei AQ, Cheng L, Pan C. Current treatment of metastatic bladder cancer and future directions. *Expert Rev Anticancer Ther.* 2011;11(12):1851–1862. https://doi.org/10.1586/era.11.181.
136. Ismaili N, Amzerin M, Flechon A. Chemotherapy in advanced bladder cancer: Current status and future. *J Hematol OncolJ Hematol Oncol.* 2011;4:35. https://doi.org/10.1186/1756-8722-4-35.
137. Knapp DW, Ramos-Vara JA, Moore GE, Dhawan D, Bonney PL, Young KE. Urinary bladder cancer in dogs, a naturally occurring model for cancer biology and drug development. *ILAR J.* 2014;55(1):100–118. https://doi.org/10.1093/ilar/ilu018.
138. Cekanova M, Uddin MJ, Bartges JW, et al. Molecular imaging of cyclooxygenase-2 in canine transitional cell carcinomas in vitro and in vivo. *Cancer Prev Res (Phila).* 2013;6 (5):466–476. https://doi.org/10.1158/1940-6207.CAPR-12-0358.
139. Mohammed SI, Craig BA, Mutsaers AJ, et al. Effects of the cyclooxygenase inhibitor, piroxicam, in combination with chemotherapy on tumor response, apoptosis, and angiogenesis in a canine model of human invasive urinary bladder cancer. *Mol Cancer Ther.* 2003;2(2):183–188.
140. Knapp DW, Henry CJ, Widmer WR, et al. Randomized trial of cisplatin versus firocoxib versus cisplatin/firocoxib in dogs with transitional cell carcinoma of the urinary bladder. *J Vet Intern Med.* 2013;27(1):126–133. https://doi.org/10.1111/jvim.12013.
141. Dhawan D, Craig BA, Cheng L, et al. Effects of short-term celecoxib treatment in patients with invasive transitional cell carcinoma of the urinary bladder. *Mol Cancer Ther.* 2010;9(5):1371–1377. https://doi.org/10.1158/1535-7163.MCT-10-0049.
142. Merlo DF, Rossi L, Pellegrino C, et al. Cancer incidence in pet dogs: Findings of the animal tumor registry of Genoa, Italy. *J Vet Intern Med.* 2008;22(4):976–984. https://doi.org/10.1111/j.1939-1676.2008.0133.x.
143. Cancer of soft tissue including heart. *Cancer Stat Facts*; 2021. SEER. Accessed October 27 https://seer.cancer.gov/statfacts/html/soft.html.
144. Grüntzig K, Graf R, Boo G, et al. Swiss canine cancer registry 1955-2008: Occurrence of the most common tumour diagnoses and influence of age, breed, body size, sex and neutering status on tumour development. *J Comp Pathol.* 2016;155(2–3):156–170. https://doi.org/10.1016/j.jcpa.2016.05.011.
145. Gustafson DL, Duval DL, Regan DP, Thamm DH. Canine sarcomas as a surrogate for the human disease. *Pharmacol Ther.* 2018;188:80–96. https://doi.org/10.1016/j.pharmthera.2018.01.012.

146. Hargis AM, Ihrke PJ, Spangler WL, Stannard AA. A retrospective clinicopathologic study of 212 dogs with cutaneous hemangiomas and hemangiosarcomas. *Vet Pathol.* 1992;29(4):316–328. https://doi.org/10.1177/030098589202900406.
147. Schultheiss PC. A retrospective study of visceral and nonvisceral hemangiosarcoma and hemangiomas in domestic animals. *J Vet Diagn Investig Off Publ Am Assoc Vet Lab Diagn Inc.* 2004;16(6):522–526. https://doi.org/10.1177/104063870401600606.
148. Srebernik N, Appleby EC. Breed prevalence and sites of haemangioma and haemangiosarcoma in dogs. *Vet Rec.* 1991;129(18):408–409. https://doi.org/10.1136/vr.129.18.408.
149. Antonescu C. Malignant vascular tumors- -an update. *Mod Pathol Off J U S Can Acad Pathol Inc.* 2014;27(Suppl 1):S30–S38. https://doi.org/10.1038/modpathol.2013.176.
150. Fosmire SP, Dickerson EB, Scott AM, et al. Canine malignant hemangiosarcoma as a model of primitive angiogenic endothelium. *Lab Investig J Tech Methods Pathol.* 2004;84(5):562–572. https://doi.org/10.1038/labinvest.3700080.
151. Gorden BH, Kim J-H, Sarver AL, et al. Identification of three molecular and functional subtypes in canine hemangiosarcoma through gene expression profiling and progenitor cell characterization. *Am J Pathol.* 2014;184(4):985–995. https://doi.org/10.1016/j.ajpath.2013.12.025.
152. Sabattini S, Bettini G. An immunohistochemical analysis of canine haemangioma and haemangiosarcoma. *J Comp Pathol.* 2009;140(2–3):158–168. https://doi.org/10.1016/j.jcpa.2008.10.006.
153. Yonemaru K, Sakai H, Murakami M, Yanai T, Masegi T. Expression of vascular endothelial growth factor, basic fibroblast growth factor, and their receptors (flt-1, flk-1, and flg-1) in canine vascular tumors. *Vet Pathol.* 2006;43(6):971–980. https://doi.org/10.1354/vp.43-6-971.
154. Thomas R, Borst L, Rotroff D, et al. Genomic profiling reveals extensive heterogeneity in somatic DNA copy number aberrations of canine hemangiosarcoma. *Chromosome Res Int J Mol Supramol Evol Asp Chromosome Biol.* 2014;22(3):305–319. https://doi.org/10.1007/s10577-014-9406-z.
155. Leduc C, Jenkins SM, Sukov WR, Rustin JG, Maleszewski JJ. Cardiac angiosarcoma: Histopathologic, immunohistochemical, and cytogenetic analysis of 10 cases. *Hum Pathol.* 2017;60:199–207. https://doi.org/10.1016/j.humpath.2016.10.014.
156. Wang G, Wu M, Maloneyhuss MA, et al. Actionable mutations in canine hemangiosarcoma. *PLoS One.* 2017;12(11), e0188667. https://doi.org/10.1371/journal.pone.0188667.
157. Grimes JA, Prasad N, Levy S, et al. A comparison of microRNA expression profiles from splenic hemangiosarcoma, splenic nodular hyperplasia, and normal spleens of dogs. *BMC Vet Res.* 2016;12:272. https://doi.org/10.1186/s12917-016-0903-5.
158. Aguirre-Hernández J, Milne BS, Queen C, et al. Disruption of chromosome 11 in canine fibrosarcomas highlights an unusual variability of CDKN2Bin dogs. *BMC Vet Res.* 2009;5(1):27. https://doi.org/10.1186/1746-6148-5-27.
159. Go H, Jeon YK, Huh J, et al. Frequent detection of BRAF(V600E) mutations in histiocytic and dendritic cell neoplasms. *Histopathology.* 2014;65(2):261–272. https://doi.org/10.1111/his.12416.
160. Liu Q, Tomaszewicz K, Hutchinson L, Hornick JL, Woda B, Yu H. Somatic mutations in histiocytic sarcoma identified by next generation sequencing. *Virchows Arch Int J Pathol.* 2016;469(2):233–241. https://doi.org/10.1007/s00428-016-1965-2.
161. Mochizuki H, Kennedy K, Shapiro SG, Breen M. BRAF mutations in canine cancers. *PLoS One.* 2015;10(6). https://doi.org/10.1371/journal.pone.0129534, e0129534.

CHAPTER FOUR

Bovine models for human ovarian diseases

John F. Roberts[a,*] and Chen-Che Jeff Huang[b]

[a]Department of Comparative, Diagnostic & Population Medicine, College of Veterinary Medicine, University of Florida, Gainesville, FL, United States
[b]Department of Anatomy, Physiology and Pharmacology, College of Veterinary Medicine, Auburn University, Auburn, AL, United States
*Corresponding author: e-mail address: john.roberts1@ufl.edu

Contents

Abstract

During early embryonic development, late fetal growth, puberty, adult reproductive years, and advanced aging, bovine and human ovaries closely share molecular pathways and hormonal signaling mechanisms. Other similarities between these species include the size of ovaries, length of gestation, ovarian follicular and luteal dynamics, and pathophysiology of ovarian diseases. As an economically important agriculture species, cattle are a foundational species in fertility research with decades of groundwork using physiologic, genetic, and therapeutic experimental techniques. Many technologies used in modern reproductive medicine, such as ovulation induction using hormonal therapy, were first used in cows before human trials. Human ovarian diseases

Progress in Molecular Biology and Translational Science, Volume 189
ISSN 1877-1173
https://doi.org/10.1016/bs.pmbts.2022.02.001

with naturally occurring bovine correlates include premature ovary insufficiency (POI), polycystic ovarian syndrome (PCOS), and sex-cord stromal tumors (SCSTs). This article presents an overview of bovine ovary research related to causes of infertility, ovarian diseases, diagnostics, and therapeutics, emphasizing where the bovine model can offer advantages over other lab animals for translational applications.

Abbreviations

AMH	anti-Müllerian hormone
ART	assisted reproductive techniques
BFB	blood-follicular barrier
BMP	bone morphometric protein
CL	corpus luteum
COC	cumulus–oocyte complex
ECM	extracellular matrix
ESHRE	European Society of Human Reproduction and Embryology
FSH	follicle-stimulating hormone
GREL	gonadal ridge epithelial-like
GnRH	gonadotropin-releasing hormone
GW	gestational week
HPG axis	hypothalamic pituitary gonadal axis
IACUC	Institutional Animal Care and Use Committee
LH	luteinizing hormone
OFD	ovarian follicular dysplasia
PCOS	polycystic ovarian syndrome
POI	premature ovary insufficiency
PGC	primordial germ cell
ROS	reactive oxygen species
S-tGCT	Sertoli-type granulosa cell tumor
SCST	sex-cord stromal tumors
SOX	SRY-Box Transcription Factor
THI	temperature-humidity index
VEGF	vascular endothelial growth factor

1. Introduction

In general, phenotype-genotype correlations are difficult to confirm in humans especially when functional *in vivo* tests are not easy to coordinate for significant segments of the population. When rudimentary searches were conducted in PubMed for mammals using species + ovary, the top three species were rats, mice, and bovine with approximately 22,500, 18,600, and 9800 citations recorded for each species, respectively. Verifying that cattle are a well-established species in basic ovary research with ovarian physiologic, endocrine, and molecular process well conserved with humans. Moreover, the bovine genome of approximately 3 Gb (3 billion base pairs) contains approximately 22,000 genes that are common to multiple mammalian species.[1–3] The presence of large cohorts of tagged heifers and cows within agriculture production allows for targeted studies performed over extended time periods. Identifying genetic markers that are predicative of reproductive success and translatable to humans is possible given the ability to track reproduction records in individual cattle.[4]

Basic bovine ovary regulatory function is well studied in animal production science programs across the world. Some economically important bovine diseases, such as bovine follicular cysts and heat stress, have also been widely investigated.[5] Common human ovarian diseases including premature ovarian insufficiency (POI), polycystic ovarian syndrome (PCOS), follicular dysplasia, and ovary neoplasia, have clinical correlates with pathophysiologic overlap in cattle.[6–9] The terminology "premature ovarian insufficiency", "primary ovary insufficiency" and "premature ovarian failure" are all used interchangeably throughout human literature. This is confusing since two of these terms have the same synonym "POI". In this review article, we are using the term "premature ovarian insufficiency" (POI) for both humans and cattle, as defined by the European Society of Human Reproduction and Embryology (ESHRE).[10]

Other ovary diseases that share commonality between humans and cattle may be defined by morphology (luteinized anovulatory follicles), metabolic status (stress induced irregularity in ovary cycling),[11] and genetics (toxin induced oocyte damage).[12] Although commonly observed from slaughter house sampling, other disease such as bovine ovarian neoplasia[13] and ovarian follicular dysplasia[14] have received little attention in the literature but likely have well conserved pathways in oncogenesis that could be beneficial to human health.[15,16] This represents an opportunity for cancer researchers as physiologic and molecular pathways of bovine oogenesis and folliculogenesis are well studied yet oncogenetics in cattle have barely been explored. The high incidence of ovarian sex cord stromal tumors (SCSTs) in aged cows, as much as 7% by one estimate, could provide abundant research material that could be accessed at an abotoir.[13,15,17,18] Compared to rodent models, the whole body aging in cattle and humans shares a more closely related senescence homology.[19] Ovaries in humans have been deemed the "pacemaker of the female body" and also drive aging in other organ systems including vasculature, muscular skeletal,[20] and nervous systems, providing opportunities for systemic research on ovarian disease late in life.[21] We will limit our discussion to physiologic process occurring prior to fertilization and will provide an overview of ovarian physiology and pathophysiology, emphasizing studies of human and cattle, and then discuss overlapping areas where translational application exist for human ovarian diseases.

2. Practical aspects of using the bovine ovary model

The sizes of ovaries and follicular and corpus luteum (CL) stages are similar in both humans and cattle.[7] Studies requiring duplication of mechanical properties are benefited when consistency is maintained between weight, density, extracellular component, and cellular composition in tissues of compared species.[22] Along with morphologic parameters, cattle and humans share similar timeline during embryologic and fetal development, nearly congruent hormone cycling and monovular folliculogenesis.[11] The number of follicular waves during the reproductive cycle, the selection of dominant follicle, and the ovulation of a single follicle during the final follicular wave, are common between the two species.[23–31] The ability to access bovine ovaries transrectally and visualize with ultrasound allows for detailed synchronization of ovary sampling correlated with specific time points in follicular wave dynamics.[32] Ultrasound can also be used to assess ovarian stroma and the large size of bovine ovaries also makes them good candidates for visualization of mesenchymal and vascular structures with routine and Doppler ultrasound.[33–35] Other radiologic modalities such as magnetic resonance imaging (MRI) are morphologically comparable, making human *in vivo* vs bovine *ex situ* studies feasible.[36,37]

The ability to match events in the follicular cycle with circulating endocrine profiles and gene expression allows for diagnostic biomarker comparisons for reproducitve diseases.[38] Bovine anatomy allows for direct ovary access and *in vivo* transvaginal collection of follicular fluid used for hormone and extracellular vesicle assays, lipid analysis, quantitation of biproducts from systemic bacterial infections and quantitation of inflammatory cytokines.[39–42] Bovine ovarian biopsies have been utilized to correlate morphology and gene expression of the CL via immunohistochemistry with circulating lipopolysaccharide bacterial products, a procedure that would be difficult to perform in small ovaries of laboratory rodents.[43] Cow ovaries have also been used to investigate comparative inflammatory processes associated with PCOS.[44]

When established medical, surgical, or infertility treatments involving the ovary are used clinically in both human and veterinary medicine, therefore, comparative clinical efficacy studies are possible that can benefit both species. Assisted Reproductive Techniques (ART) and *in vitro* fertilization procedure, along with oocyte morphology, are similar between human and cattle. ART utilizes the transvaginal route for ultrasound-guided oocyte pickup and intraovarian injection of platelet-rich plasma.[45,46] During human IVF, immature oocytes are harvested from antral ovarian follicles and may be matured *in vitro* in enriched culture media. This technique is also being researched using bovine oocytes.[47,48] The large blood volume and ease of sampling in the cow further allows for serial endocrine profiling and correlation of ovarian parameters for circulating molecular, metabolic and protein biomarkers.[27,49] In cattle with known neoplastic ovarian disease, serial blood sampling has been used to test various biomarker regimes.[50] When hourly intervals of blood sampling are required, cattle can be fitted with indwelling venous catheters that can be maintained for longer periods compared to small laboratory rodents. The relationship between post-partum uterine infections and pregnancy status related to ovarian function is also more feasible to research in cows with well-maintained herd health records.[51,52]

The ability to sample cattle at processing capitalizes on an unused resource and poses minimal ethical dilemmas for Institutional Animal Care and Use Committees (IACUC). Bovine oocytes, granulosa cells, and theca cells extracted from ovaries at slaughter can be fertilized or manipulated *in vitro*, allowing for genetic studies on toxins, antioxidants, nutritional effects and stress factors.[12,53–55] Sampling of whole ovaries at slaughter can also facilitate large scale histopathology disease screening for ovarian disease or confirmation of physiologic processes.[56–59] In countries like Japan, where traceability of individual animals is required for processing, date of birth on each animal makes ovary aging studies possible without obtaining records for each individual farm.[60] Ethical and social considerations make sequential sampling of human fetal ovaries unfeasible in many countries but this can also be accomplished using bovine fetuses obtained at slaughter. Sampling at processing also presents the ability to simultaneously collect brain, pituitary gland, adrenal gland, blood, reproductive tract, placenta, visceral organs, and fetal tissues, when integrated studies of the hypothalamic pituitary gonadal axis (HPG axis) and the hypothalamic-pituitary- (adrenal) gonadal axis [HP(A)G axis], or ovary-uterine-fetal relationships are desired.[61] The large size of bovine embryos and fetuses also facilitates sequential developmental analysis of genetic expression and establishment of harvested cell lines from ovaries correlated with specific stages of gestation.[8,62]

Early puberty of heifers makes possible comparative studies of ovary aging in sets of daughters and even granddaughters of specific genetic lines of cattle.[30] Ability to detect functioning corpus luteum (CL) in heifers with ultrasound allows for necropsy at selected time points before and after onset of puberty, making genetic and physiologic studies of ovary maturation possible.[63]Theoretically, the relatively short interval of sexual maturity of the heifers, at approximately 2 years, allows for collection of umbilical cord origin stem cells at birth that can be utilized for experimental procedures in ovaries when these same females reaches sexual maturity. Stem cells from multiple anatomical locations including umbilical cord, bone marrow, adipose tissue, could be used in human regenerative cell-based therapies to treat female infertility.[64,65] *In vitro*, follicular granulosa cells also have mesenchymal stem cell-like properties with the ability to differentiate into multiple tissue types including osteoblast, neurons, and chondrocytes, thus also making them applicable to next generation tissue engineering.[65] The large size of the bovine ovary may also be applicable to the emerging field of tissue engineering, where *in vivo* follicle transplantation with or without accompanying mesenchymal scaffold or the creation of "artificial ovaries" may be applicable.[66]

Animal models with longer life span and similar reproductive life spans provide unique insight into aging-related diseases.[67] IACUC guidelines are increasingly less supportive of protocols using large number of animals in terminal studies for single use applications. Compared to small laboratory rodents, cattle offer an opportunity to examine ovarian diseases in a clinical setting that more closely mimics a natural model of occurring disease in mammals, providing opportunities in long term serial sampling of tissues and fluids without sacrifice of the animal. Techniques already perfected *in vitro* using bovine tissues make cattle suitable candidates for *in vivo* continuation of these techniques. Cattle reproductive studies with non-toxic agents provide the opportunity to combine reproductive studies with other teaching applications, such as meat science, reproductive techniques, and gross anatomy laboratories. Moreover, the wide distribution of cattle throughout the world in temperate and tropical environments makes them excellent candidates for monitoring environmental toxins that alter ovarian physiology.[68–70]

3. Ovarian development during gestation and before puberty

The use of bovine fetuses has been fundamental to understanding ovary embryogenesis in humans. Acquired and inherited genetic mutations and altered fetal gene expression during gestation may have effects on ovarian pathophysiology into adulthood.[9,71] The longer gestation of cattle compared to laboratory rodents allows for better temporal separation of gene expression linked to developmental processes and morphologic features.[71] When gestational week (GW) matched studies of gene expression between cattle and humans were possible, close correlation was observed during the 1st, 2nd, and 3rd trimesters of pregnancy.[71] Fundamental cellular pathways occurring along this timeline are key to understanding mechanisms of fertility and overlapping bovine pathophysiology found in diseases such as POI, PCOS, and ovarian neoplasia.

Differentiation of the gonad into the female phenotype (XX) in mammals is considered the default physiologic process in the absence of sex-determining region Y (*SRY*) gene located on the Y chromosome.[72] The functional unit of the ovary include oocytes, follicles composed of granulosa cells and theca cells and CLs that arise from dominant follicles. Absence of *SRY* and failure to activate SRY-Box Transcription Factor (*SOX*)*9* in XX mammalian ovaries leads to transcription of factors influencing the β-catenin signaling pathway and upregulated expression of Forkhead boxL2 (*FOXL2*) gene.[73] *FOXL2* is involved early in sexual differentiation and then granulosa cell differentiation, follicle activation, and folliculogenesis. Mutations and variation in expression in *FOXL2* have been described in pathogenesis of POI[74] and ovarian cancer.[75]

During early embryonic development, the precursors to oocytes in human and bovine ovaries are primordial germ cells (PGCs) originating from the proximal yolk sac. Surface epithelium and stroma originate from the mesonephros and immune cells originate from hematopoietic stem cells.[8] At gastrulation, the differentiation of PGC is dependent on the activity of factors in the bone morphometric protein (BMP) family.[76] In both humans and cattle, the viability of PGCs at early GWs also relies on Notch, Nodal/Activin, and BMP paracrine signaling pathways.[77] In mammals, early BMP signaling is regulated by chorionic gonadotropin (CG) produced in placental trophoblast. BMPs as part of the transforming growth-factor β (TGFβ) superfamily have roles in the pathogenesis of infertility disorders including POI, PCOS, and ovarian cancer.[78]

Initially, epithelial cells on the surface of the embryonic mesonephros form a cluster of pluripotent cells at the genital ridge location. Unlike the remaining epithelial cells on the surface of the mesonephros, gonadal ridge epithelial-like (GREL) cells lose a defined basal lamina. At about GW 5 in the bovine and human, clusters of GREL cells are populated by PGCs migrating from the proximal yolk sac.[79,80] Stroma from the underlying mesonephros then grows into this cluster of GREL, forming the hilum of the ovary. As the stroma progresses into GREL, it branches, segmenting the GREL and forming ovigerous cords containing PGCs. Within these cords, PGCs continue to undergo mitosis, form nest, but remain interconnected by intracellular bridges due to incomplete cytokinesis.[62] When the branching stroma cords reaches near the surface of the GREL at around GW11 in the bovine fetus, the stroma proliferates laterally enclosing the ovigerous cords and separating them from the most peripheral surface GREL.[81] GREL excluded from the ovigerous cords becomes the surface (coelomic) epithelium in mature ovaries.[9]

The origin of human epithelial origin ovarian neoplasia is of current interest.[82] The early embryologic differentiation of ovarian epithelium may hold clues to the pathogenesis of human ovarian carcinoma, perhaps explaining why cattle are largely resistant to epithelial origin ovarian cancer and humans have high susceptibility to this deadly disease. As the stroma expands, it compresses clefts of surface epithelium and has the potential to form inclusion cyst, which on rare occasions may persist in adult ovaries and have also been implicated in oncogenesis of ovarian high-grade serous carcinoma.[83] The presence of renewable ovarian germline stem cells (GSC) originating from the surface epithelium have been explored to offer treatment for some types of human infertility.[84] The lateral progressing stroma originating from the mesonephros forms the tunica albuginea and cortical stroma of the adult ovary.[8] As the cortical stroma proliferates, genes that regulate the extracellular matrix (ECM) collagen type I alpha 2 chain (COL1A2) are strongly expressed in the bovine fetal ovary.[85]

Embryonic stroma proliferation may be altered during development and is a focus of PCOS research in humans.[9] In humans, ovigerous cords are surrounded by a basement membrane during nest formation and PGCs differentiate into oogonia at around GW 6-8.[77,84] During migration and after the PGCs reach the gonadal ridge, PGCs are undergoing mitosis and mitotic divisions that continue after they become oogonia. Oogonia continue to undergo mitosis until approximately GW 9-22 in humans.[86] Oogonia initiate the first stage of meiosis to become oocytes at approximately GW 13 in bovine and GW 10 in humans.[87–89] In humans, meiotic entry into the oocyte pool is asynchronous and occurs with frequency toward GW 20.[90] During fetal meiosis, oocytes become arrested at the diplotene stage of prophase I and do not reenter the cell cycle until puberty.[81] The state of arrest is maintained by high concentrations of intra-oocyte cyclic adenosine 3′,5′-monophosphate (cAMP)-dependent protein kinase A (PKA).[91] Mechanisms which alter the function of these pathways may result in premature depletion of oocytes.

There is scientific intrigue and skepticism in some camps regarding the ability of ovarian pluripotent epithelium to form oocytes after birth.[92] Recent experiments in cattle using stimulated dermal origin fibroblast have achieved some pluripotency and created cells with oocyte characterisitics.[93] However, from an isolation on over 24,000 single-cells from the human ovarian cortex, a recent study was unable to find any evidence for any germline stem cells, thus supporting the long held belief that the ovarian reserve is fixed in humans.[94]

During fetal development, oocytes are surrounded by a single layer of flattened pre-granulosa cells enclosed by a basement membrane to form primordial follicles, a process called follicular assembly. Expression of *FOXL2* is correlated with primordial follicle formation.[95] The number of identified genes and proteins involved in follicle assembly, development, and ovulation is immense and has considerable overlap between humans and cattle. One meta-analysis of PubMed citations in 2015, identified 788 papers corresponding to 556 proteins involved in ovary follicle assembly and that number is undoubtedly growing.[96] Primordial follicle assembly is reported to start from GW 16-19 in humans and at approximately GW 10.5 in cattle.[81,97] Dysregulation of granulosa cell growth is key to understanding several ovarian diseases including POI, PCOS, Ovarian Follicular Dysplasia (OFD), and ovarian SCSTs.[73] In cattle, initially primordial follicles are grouped in nest but eventually these nests breakdown and during that time approximately two thirds of primordial follicles undergo apoptosis and autophagy, a process peaking at GW 25-26.[62,98,99] In humans, it is estimated that fetal ovaries contain 6–7 million oocytes at mid-gestation and this number decreases through apoptosis and autophagy to around 2 million at the time of birth. The number of reserve oocytes then further decreases to about 400,000 by puberty and then progressively declines to approximately 1,000 at the time of menopause.[100,101]

Primordial follicles that survive to puberty are deemed the "ovarian reserve" and lie dormant until they are committed to enter the growth phase during primordial growth activation.[102] Recruitment of primordial follicles to primary follicles initiates in the bovine fetus at approximately GW 13[81,97] and continues through adulthood until folliculogenesis ceases from age related factors. When primary follicle formation is complete, granulosa cells are more cuboidal and the oocyte is encompassed in a zona pellucida.[65]

Primary follicles start to develop into secondary follicles, at approximately GW 16-22 in cattle when granulosa cells proliferate forming three to six layers.[97] During the transition from primary to secondary follicles, stroma abutting granulosa cells begins to form the theca cell layer.[65] In cattle, BMP4 regulates endocrine signaling between granulosa cell layer and the newly formed theca interna.[103] It is also at this time that genes regulating collagen in the extracellular matrix increase over two fold in expression [*e.g.*, collagen type IV alpha 1 chain (*COL4A1*), metallopeptidase 2 (*MMP2*), and tissue inhibitor of metalloproteinases 2 (*TIMP2*)].[85] This is an interrelated process with other factors such as connective tissue growth factor (CTGF) and gremlin (GREM) regulating BMP4 responses in granulosa cells.[104]

The basal lamina between the granulosa cells and anastomosing capillaries of the theca interna form a "blood follicular barrier" (BFB) from tight junctions in endothelial cells and adjacent basement membranes of vessels and granulosa cells. Regulated by nitric oxide, the BFB restricts movement of some macromolecules to developing follicles.[105] The BFB is important in the pathophysiology of ovary toxin diffusion and the delivery route of therapeutics to the follicle.[106]

Early stages of follicular development are gonadotropin independent and are regulated by members of the transforming growth-factor β (TGFβ) superfamily, locally secreted ovarian factors such as KIT ligand (KITLG), nerve growth factor (NGF),[107]anti-Müllerian hormone (AMH), BMP, and platelet-derived growth factors (PDGF).[108] Late-stage secondary follicles develop receptors for anterior pituitary origin gonadotropins follicle stimulating hormone (FSH) and luteinizing hormone (LH) with upregulation of Growth Differentiation Factor 9 (GDF9) and Bone Morphogenetic Protein 15 (BMP15) during antrum formation.[109] These growth regulators have greater functionality in mono-ovulation species such as humans and cattle compared to poly-ovulating rodents.[108] Receptors for AMH [AMH receptor type 1 (AMHR1) and AMH receptor type 2 (AMHR2)] are more numerous on small bovine follicles than larger follicles.[110] Dominant follicles destined to ovulate have greater concentration of AMH receptors, indicating that AMH promotes this subtype of follicle.[111] This process may carry on through adult life. AMH measured by ELISA in the blood stream is an important clinical biomarker in human and bovine medicine used to judge the population of viable oocytes remaining in the reserve pool.[112]

Over-stimulation by FSH can induce autophagy of granulosa cells with the upregulation of autophagy related 7 (ATG7) gene, which encodes an enzyme involved in p53-dependent cell cycle pathways.[113] BMP-4, although expressed in PGCs,[76,114] has greater expression in cattle during primary to secondary follicle progression.[115] In Graafian follicles, granulosa cells that directly surround the oocyte differentiate into cumulus cells while those surrounding the periphery of the antrum are termed mural granulosa cells. The inner most layer of granulosa cells around the oocyte are termed the corona radiata and the layer immediately outside this is the cumulus oophorus. Cells of the corona radiata and cumulus oophorus have gap junctions crucial for paracrine signaling that drive further granulosa cell proliferation and maintenance of oocyte viability.[65]

There are transzonal cytoplasmic projections that interact with the zona pellucida matrix and this association is termed the cumulus–oocyte complex (COC).[65] Genetic expression of the developing oocyte and cumulus cells must be properly synchronized to allow viability of the oocyte.[116] The COC has a unique transcriptome compared to granulosa cells and the dynamic expression of genes is termed the folliculome.[117] Expression patterns of granulosa cells are also utilized to identify patterns of aging in human and bovine ovaries.[60] Gap junctions proteins, specifically, the gap junction α-1 protein (known as connexin 43, encoded by *GJA1*) become impaired during aging processes, decreasing the viability of oocytes in older cows.[118] Disruption of processes maintaining the gap junction integrity could decrease the fertility in cattle and humans.[119,120] Cattle are a common experimental animal used to determine genetic regulation in the COC.[121] Early antral follicles with COC form in late gestation in cattle and can be visualized by ultrasound at the time of birth.[122,123]

MicroRNAs (miRNAs) are a class of small noncoding RNAs (ncRNAs), well conserved between cattle and humans, regulating post-translational gene signaling in the majority of protein coding genes.[124] MiRNAs are generated when ncRNAs are processed by RNA polymerase II and RNase III family (DROSHA and Endoribonuclease dicer (DICER)) and then transported by exportin 5 (XPO5). MiRNAs are involved in processes of normal oogenesis, including cellular proliferation, differentiation, and apoptosis of granulosa cells in both human and cattle, and pathogenesis of PCOS and ovarian SCSTs.[125–127] Mouse study showed that miR-21 regulates oocyte-secreted factors (GDF-9 and BMP-15) by action on TGFβ superfamily signaling through DICER and the upregulation of miR-21 prevents apoptosis via the PI3K/Akt signaling.[128] In bovine experiments, miR-21 in cumulus cells is correlated with blastocyst survival after fertilization.[129] MiR-21 is also upregulated in human ovarian cancer and functions as a circulating biomarker,[130,131] while conversely low serum levels of miR-21 are observed in POI.[132]

4. Puberty and the ovarian cycle

When puberty initiates, cyclic folliculogenesis results in the first ovulation and the formation of CL.[133] Lowering the age of puberty onset is of economic interest to cattle industries to increase production efficiency and thus has received some research attention.[133] Initiation of folliculogenesis at puberty requires an inter-coordinated cycle involving catabolic differentiation and atresia of oocytes, granulosa cells, and thecal cells.[73] Folliculogenesis requires the synchronized expression of over 13,000 genes.[117] Oocytes remain in a suspended state of meiosis until onset of puberty and then become eligible to enter folliculogenesis throughout years of reproductive fertility. Human follicular recruitment peaks at puberty and then declines progressively throughout life. In humans, serum AMH level are rising at puberty and then peak at about 25 years of age, corresponding to the period of maximum fertility in a woman's life.[134] Dormant oocytes are suppressed by cyclic GMP (cGMP) produced by granulosa cells, maintaining the oocyte in meiotic prophase arrest.[135] During initial follicular recruitment, influence of cGMP is removed via binding of LH to receptors on granulosa cells. The removal of cGMP influence occurs through inactivation of NPR2 and activation of the PDE5 phosphodiesterase, allowing the oocyte to re-enter meiosis at the diplotene stage.[135,136] Maintenance of primordial follicles and follicle recruitment is also regulated by the PI3K/PTEN/Akt and TSC/mTOR pathways, through the regulation of antiapoptotic pathways and the control of autophagy.[137–139] AMH produced by growing follicles suppresses the development of the primordial follicles retaining them in a state of dormancy.[140]

"Germinal vesicle" is another term for the oocyte nucleus and is used in all species. Both human[141] and cattle oocytes[136] undergo protein synthesis to progress to germinal vesicle breakdown, at which time there is cessation of transcriptional processes. Oocyte viability and maturation is dependent on post-transcriptional gene regulation involving miRNAs. The oocyte maturation process occurs during a relatively short period of time, 24-40 h in humans and 20-24 hours in cattle.[12] Oocytes from the active pool are recruited from a pool of dormant primordial follicles, and once committed, go on to develop into antral follicles capable of undergoing atresia. During the estrus cycle, only one out of the many antral follicles go on to become a dominant follicle proceeding to ovulation.

During follicle progression, the oocyte and surrounding somatic cells (granulosa cells) are receptive to FSH and LH, as well as estradiol, testosterone and progesterone, sex hormones secreted by the adrenal cortex and ovary. Gonadotropin-releasing hormone (GnRH) from the hypothalamus controls the release of FSH and LH from gonadotrophs in the anterior pituitary. GnRH in turn is controlled by kisspeptin released from neurons located in the arcuate nucleus or preoptic areas of the hypothalamus. Circulating estrogen produced by follicles is thought to provide a feedback mechanism controlling kisspeptin release.[142] Ovulation induced by LH can be initiated by intramuscular injection of kisspeptin in both *Bos indicus* and *Bos taurus* species of bovids.[143] Members of the TGFβ family involved in various aspects of folliculogenesis are well conserved in humans and cattle, and include growth differentiation factor (GDF)-9, BMP15, BMP2, BMP4, BMP7, AMH, and Follistatin (FST).[73,115] When follicles enter atresia, AMH levels decrease.[144] During the initial waves, circulating LH is suppressed by progesterone produced by the CL.[145] Granulosa cells of the dominant follicle acquire LH receptors and aromatase to produce estradiol.[146] In bovines, when the formation of new corpus luteum begins, theca derived BMP-4 inhibits the progesterone production through the interaction with steroidogenic acute regulatory protein (StAR). This inhibition limits luteal progression to only dominant follicles.[147] Inflammatory cytokines such as interleukin 18 (IL18) may stimulate the proliferation and dysplasia of theca cells through upregulation of androgen production.[44] Increased androgen is well accepted as a factor leading to PCOS.[148]

Initial studies of ovarian folliculogenesis were based on palpation and histologic evaluation of surgical excised ovaries correlated with the endocrine profiles.[149] In the late 1970s to1980s, ultrasound imaging revolutionized the understanding of ovarian follicles and the luteal status during estrus cycle both in both human and cattle.[150–153] Internal ultrasound scan (*e.g.* transrectal or transvaginal ultrasonography) to verify the follicular and luteal dynamics were first conducted in cows.[150] Although initial studies only monitored populations of follicles of different size categories rather than individual follicles, a method of real-time characterization of follicular dynamics was quickly adapted.[153–156] By using daily ultrasonography with detailed sketches of follicular and luteal structures in each ovary, the detailed characterization of the temporal associations of ovarian follicular waves in bovine were established.[157,158]

This method of serially monitoring individual follicles by ultrasonography was initially developed using the bovine model and then was applied tohuman folliculogenesis.[159,160] The follicular wave is a synchronous growth of a cohort of follicles during the ovarian cycle, followed by a continued growing of one single follicle which becomes dominant, while other follicles undergo atresia.[6] In women post puberty, greater than 99% of antral follicles become atretic, whereas in normal cycling women, two to three follicle waves are observed during the follicular phase of the menstrual cycle,[6,161] similar to what is found in cattle.[162] As follicles grow, they are dependent on vascularization and vascular endothelial growth factor (VEGF), which increases in follicular fluid as follicles increase in size.[54]

Formation and regression of the CL is required to continue cyclic folliculogenesis. Progesterone production by the CL stops during regression. Regression of the CL occurs through an integrated mechanisms including autophagy, apoptosis and lysosome activity.[163]

Bovine and human ovaries have similar coupled mechanisms influencing follicle and luteal dynamics that are coordinated between right and left ovaries. This relationship selects the lateral preference, dominant follicles and CLs. This communitive relationship between right and left ovaries is termed intraovarianism and has variable levels of interdependence, depending on species. Humans and cattle have greater intraovarianism synchronization compared to horses.[32] During ovulation, the capsule and ovary cortex are expanding and tearing at the site of follicle rupture. Increased ovulation rate has clinical implications in women and cattle undergoing *in vitro* fertilization, creating increased ovulation sites over a shorter length of time compared to normal ovaries. Increased tissue repair may predispose to initiation of ovary cancer pathogenesis.[45,84]

5. Mesenchymal influence in ovarian function

The term "stroma" has broad definition in relation to ovarian microanatomy and includes all ovarian tissues excluding follicles and CLs. The dynamic stroma reconstruction is intrinsically linked to growth and viability of follicles but also undergoes continuous architectural remodeling during fetal development, calf hood, puberty, adult life and aging but also ovarian diseases such as POI and PCOS. Stroma of the ovary include surface epithelium, nerves, blood, lymph vessels, immune cell populations, connective tissue with ECM, and the mesonephros derived rete ovarii.[164] Most of these tissues are of mesenchymal origin, with the exception of surface epithelium and epithelium of the rete ovarii, which may be reactive for cytokeratin intermediate filaments.[165] ECM produced by mesenchymal cells is a scaffold of macromolecules such as collagen, elastin, glycoproteins, proteoglycans, laminins, and glycosaminoglycans (hyaluronan) that are required for cell support, direct and indirect signaling.[22,164,166] Hyaluronan provides the stability to the ECM through tissue hydration, and facilitation of cell signaling.[22] When ECM becomes dysregulated, detrimental effects to ovarian function such as fibrosis ensue and this may accompany other disease such as POI and PCOS.[167] Transcription factors, including early growth response-1 (EGR1) and connective tissue growth factor (CTGF) or cellular communication network factor 2 (CCN2), are promoters of fibrosis,[168] with CCN2 associated with aging in bovine ovaries.[60]

Aberrant development and maintenance of the ovarian stroma, including increased thickness of the tunica albuginea, cortical and subcortical stroma, has been increasingly recognized in PCOS.[8] Accumulation of collagen and loss of hyaluronan may result in ovary stiffness and mechanical stress contributing to primordial follicle dormancy.[22] It has been validated that both humans and cattle have similar molecular expression in pathways involving fibrillin and TGFβ that regulate stroma.[169]

Blood provides oxygenation, hormone delivery routes, nutrient transport, and waste removal for the parenchyma and stroma components of the ovary.[164] Pathways regulating arteriogenesis are important in follicular development and pathogenesis of ovarian diseases. The role of VEGFs are critical to vascular remodeling constantly occurring in the ovary.[170] VEGFA-164, VEGFA-164b and VEGFR2 receptors are expressed in both granulosa cell and theca cell layers of follicles and are upregulated in cystic ovaries of cows.[171] Vascular endothelial growth factor receptor (VEGF-R3) is required for formation of lymphatic vessels and necessary for follicle viability. Circulating lipopolysaccharides from uterine or other infections may affect angiogenesis in the corpora luteum.[51]

6. Age related changes in ovaries

Cattle make a comparable clinical model for ovarian aging based on ultrasound observations.[27] Decreased pool of primordial follicles by atresia and viability of remaining oocytes influence the fertility of both cattle and humans.[172] Decreased number of antral follicles observed with ultrasound directly correlates with decreased number of primary follicles observed by histology in women.[172] As antral follicles decrease, less AMH is secreted and the suppression of primordial follicle recruitment ceases, resulting in diminished primordial follicle reserve.[140] Decreased oocyte competency for remaining follicles is a well-known occurrence in women and cows.[30,101,173] Decrease in estrogen (hypoestrogenism) associated with the loss of follicles also has systemic effect, contributing to heart disease, osteoporosis, cancer obesity, mental decline, and POI in humans.[108,174] These complications of aging in women are well documented but have not been explored in the bovine.

Since ovarian hormone output may partially drive general whole-body senescence, it is interesting to know whether the slowing of ovarian senescence can improve general health for geriatric patients. For women, AMH and antral follicle counts are currently best markers for evaluating ovarian reserve.[175] Women entering menopause and the sensation of ovulation initiates at an age averaging 47 to 54 years (median 51 years) when the total number of primordial follicles in the ovary drops below a few thousand.[176,177] Comparatively, cows have a decreases in fertility (approximately 55%) at around the age of 13 years, although some cows may remain fertile the entirety of their lifespan which can be over 20 years.[178] In a more resent study, Nelore cows, in Brazil began to experience ovarian senescence at approximately 17 years of age.[179] The millions of aged unproductive cows slaughtered each year could provide abundant data for ovarian aging research.

Decreased antral follicle numbers experienced during early senescence appear to correlate with decreased viability of embryos produced during *in vitro* fertilization .[179] As humans and cattle age, the mean circulating FSH increases yet fewer follicles with lower average competency are recruited.[27,30] Circulating estrogen also increases even though the number of follicles entering the wave are fewer.[27] These observed changes mimic what occurs in women approaching menopause, as elevated FSH can be used to detect decreased fertility in women.[180] Concurrently with the rise in FSH, circulating GnRH increases in older cows and women, resulting in increased recruitment of follicles and depletion of the ovarian reserve.[27] At this time, CLs get smaller and there is a decrease in circulating progesterone in women and cows.[27,181] Gene expressions correlated with aging in bovine ovaries includes Coiled-Coil Domain Containing 80 (CCDC80), Claudin-11 (CLDN11), collagen type I alpha 1 chain (COL1A1), connective tissue growth factor (CTGF), DnaJ heat shock protein family (Hsp40) member B1 (DNAJB1), galectin 9 (LGALS9), Lactotransferrin (LTF), Lumican (LUM), Metallothionein 2A (MT2A), Myosin Heavy Chain 11 (MYH11), Neuromedin B (NMB), SOX4, (TIMP Metallopeptidase Inhibitors) TIMPs.[60]

Super stimulation fertility techniques used in both human and bovine fertility treatments increase the number of ovulation fossa on the ovarian surface and increase post ovulation tissue remodeling. This may increase mesenchymal associated aging changes in the cortex. Oocyte competence and follicular growth characteristics are also altered with increased age. When cattle are super-ovulated, the number of follicles in waves and size of antral follicles incrementally decrease.[30] Ovarian stimulation and oocyte retrieval for fertility procedures may induce increased chromosome damage and is thus a complicating factor in the administration of these procedures.[173,181,182] Pre-implantation diagnostics for aneuploidy (the incorrect number of chromosomes in a cell) are still being perfected for humans, thus allowing for cattle to serve as a model for this type of testing.[183] Loss of oocyte competency can also be attributed to DNA damage, reactive oxygen species, telomere damage, and mitochondria damage.[101] A bovine model is being utilized to increase oocyte competency using cysteamine, a free radical scavenger, *in vitro*, to improve successful fertilization of flushed oocytes.[183]

7. Nutrition and stress effects on ovaries

Both humans and cattle are subjected to nutritional and metabolic factors that, when in combination with increased circulating cortisol, have negative influences on ovarian development, oocyte viability, and folliculogenesis.[184] Ovarian cyclicity is completely interdependent on the hypothalamic-pituitary-ovarian axis.[185] In humans with anorexia nervosa or dairy cattle with post-partum negative energy balance, the insufficient uptake of dietary nutrients will reduce secretion of GnRH from the hypothalamus and cause ovulation failure.[186] In humans, multiple stressors and suboptimal nutrition can also lead to metabolic disorders including diabetes, obesity, and hyperlipemia that adversely affect oocyte health.[187] Gene expression associated with lipid dysregulation has been studied in ovaries of dairy cattle, which could be correlated with the obesity-associated infertility in humans.[188] Dysregulation of potentially clinically significant pathways have been identified in genes related to aberrant stress. In dairy cattle, crowding, slick flooring, improper housing designs, labor issues, and heat stress, disrupt estrus expression and reduce success of artificial insemination programs.[189]

Oxygenation status of ovaries is also an important mediator in the production of reactive oxygen species (ROS) that may induce autophagy in both oocytes and granulosa cells.[99] Even during normal folliculogenesis, ROS levels are elevated and antioxidant levels fall in response to the preovulatory gonadotropins.[190] When this becomes dysregulated, hypoxia induces oocyte degradation which could be a factor in the development of POI or infertility in cattle. Anti-oxidant therapy has been implemented experimentally to achieve higher fertility in Holstein cows.[191]

Environmental stressors that suppress GnRH also reduce LH and have downstream negative effects on estrogen production, ovulation, and menstruation.[185] GnRH pulsation can be negatively affected by excessive exercise, eating disorders, anorexia, and psychological stress.[185] This is recognized clinically as functional hypothalamic amenorrhea. GnRH function has been linked to the dysregulation of several genes, including fibroblast growth factor receptor 1 (*FGFR*), prokineticin receptor 2 (*PROKR2*), GnRH receptor (*GNRHR*) and the Kallmann syndrome 1 (*KAL1*).[192] Immune stress associated with systemic inflammation or infection can also disrupt pulsation of GnRH through the production of pro-inflammatory cytokines such as interleukin-1β (IL-1β), IL-6, and tumor necrosis factor-α (TNFα). Other mediators associated with infection such as lymphokines and bacteria also induce LH suppression and downstream disruption of GnRH pulsation.

Fertility treatments are mainly reliant on ovarian stimulation via hormones. Repeated ovarian stimulation alters gene expression in older cows, causing the increased expression of *GRP78*, a gene involved in ER stress, and variable expression of genes involved in mitochondrial maintenance. Increased expression of thioredoxin 2 (*TXN2)* and transcription factor A, mitochondrial (*TFAM*) occurs in cumulus cells and decreased expression of genes involved in apoptosis [*TXN2*, peroxiredoxin 3 (*PRDX3*), and BCL2 associated X, apoptosis regulator (*BAX*)] also occur at this time.[193]

Excessive heat is a well-studied stress parameter in cattle that could increases in human as well, with increases in the temperature-humidity index (THI) occurring environmentally.[33,194,195] Heat stress affects mitochondria-associated genes in the oocyte participating in ATP production and DNA transcription and replication.[196] Increases in reactive oxygen species (ROS) have also been documented in oocytes exposed to heat. Diameter of preovulatory follicle, CL vascularity and progesterone can be decreased as a result of increased THI.[33] When COC were subjected to heat stress, CLDN11 was the most differentially expressed gene in a *Bos indicus* mixed breed and antioxidant metallothionein 1E (MT1E) was the most differentially expressed gene in a pure *Bos taurus* breed.[197]

8. Effects of toxins on ovaries

Decreased oocyte competence has been linked with the exposure to numerous natural, synthetic, and pharmaceutical compounds existing in water, food, and the environment. These compounds may activate endocrine pathways by mimicking hormones or blocking receptors.[12] Toxins inducing ROS are also more toxic to mammalian oocytes than oocytes of lower phyla.[99] Widespread distribution of cattle worldwide, often utilizing the same water supply as humans, also allows an opportunity to monitor for ovary toxic environment compounds in both populations. The use of bovine oocytes to screen toxins *in vitro* yields results directly translatable to potential effects these toxins could have on human oocytes *in vivo*.[12] The list of potentially ovary and ootoxic compounds is large with varied complexity.

Minerals such as cadmium and cobalt can be screened using inductively coupled plasma atomic emission spectroscopy analysis of liver and kidney from processed cattle.[197] Cadmium is a widespread environmental pollutant that induces follicular abnormalities and apoptosis in granulosa cells.[198] Evidence is emerging that cadmium may be associated with PCOS in humans.[199] The bovine oocytes in culture have been used to evaluate therapeutic option to counteract chronic cadmium exposure, but *in vivo* experiments have yet to be conducted.[200] Other metals such as nanoparticulate titanium oxide (nano TiO2) has also been reported to be associated with POI.[139]

In areas of the world experiencing rising temperatures, human and animal food supplies are increasingly exposed to mycotoxins such as zearalenones,[201,202] solanidines,[203] and cytochalasin B.[204] Mycotoxins have been demonstrated to decrease oocyte quality in cattle *in vitro*. Epidemiology studies have established associations between polycystic ovary-like syndrome in humans and mycotoxin exposure.[205]

Endocrine disruptors, specifically compounds that mimic estrogen and interact with estrogen receptors on follicles, could alter ovarian development. Plant-derived phytoestrogens, such as isoflavones, have been shown to affect the onset of puberty in cattle and have been linked to the development of cystic ovaries and irregular estrus.[206,207] Experimentally in cattle, endocrine disrupting toxins could be administered and monitored by real time via ultrasound, blood collection, and follicular fluid draw.[12,208,209]

Coumestrol, a product of leguminous plant, has been demonstrated to impair CL function, which could affect the ability to maintain pregnancy.[210]

Phthalates are diesters of o-phthalic acid used as plasticizers to create a variety of products including cosmetics, plastics in toys, food packaging, medical devices, and vinyl flooring. Di(2-ethylhexyl) phthalate (DEHP) and its closely related monoester metabolite, mono(2-ethylhexyl) phthalate (MEHP), decrease estradiol levels by suppressing aromatase mRNA and protein expression in granulosa cells.[211] Both DEHP and MEHP increase ROS in follicles that experimentally can be suppressed with antioxidants.[190]

Plasticizers, Bisphenol A (BPA) and bisphenol S (BPS), are detectable in the urine and follicular fluid of humans and cattle throughout the world.[212,213] These compounds are ambiguous in the environment and may equally affect both cattle and humans.[214] Increasing serum levels of BPA correlate to hyperandrogenism and PCOS in women.[208] BPA and BPS disrupt maturation and cause chromosome abnormalities at low does *in vitro*.[57] Gap junction disruption has also been implied as a possible mechanism for infertility induced by BPA and BPS.[214]

A class of oleophilic hydrophobic chemicals produced in fabric, nonstick cookware, food packaging, and firefighting foams, have been shown to bioaccumulate in environmental water sources. Of these, the shorter-chain perfluoroalkyl substances (PFAS) have been shown to have lower accumulation potential in cattle compared to the long-chain compounds in this class like perfluoroctane sulfonate (PFOS).[215] Human accumulation of PFOS, which may have a half-life of up to 5 years in living tissue,[216] have been shown to decrease estrogen.[217] Although studies have been done to confirm half-life of PFAS in cattle, no studies are found that assess ovary function.[218]

Several classes of pesticides (herbicides and insecticides) are also ovotoxic. Methoxychlor is an organochlorine insecticide, capable of causing atresia of antral follicles but not of smaller follicles.[219] Glyphosate has been demonstrated to increase proliferation of granulosa and theca cells in fetus of lambs, research that may also be directly applicable to cattle.[220] This experiment also demonstrated changes in fetal FSHR and GDF9 gene expression after glyphosate exposure.[220]

Cancer therapy with chemotherapeutics is the most common predisposing condition in women with POI and predisposes to entering menopause early.[221–223] Primordial follicles are destroyed by alkylating chemotherapeutic agents through a mechanism of DNA double-strand breaks in the oocyte, leading to apoptosis.[224] Ovotoxic chemotherapeutic agents in the alkylating class include cyclophosphamide, busulphan, and dacarbazine.[225] Other chemotherapeutic agents including platinum complexes and taxanes are also toxic to ovaries.[225] Most of these drugs decrease ovarian function by inducing apoptosis of quiescent primordial follicles.[225] Chemotherapeutics may also induce mass recruitment of primordial follicles through activation of the phosphoinositide 3-kinase/protein kinase B/Forkhead box protein O3a (PI3K/AKT/FOXO3a) pathway resulting in "ovary burnout."[140] Decreased estrogen associated with chemotherapeutic induced follicular damage may also have systemic effects.[223] Damaged ovarian vessels and ovarian cortical fibrosis are other side effects of chemotherapeutics.[226]

Although bovine models may be suitable to further investigate pathophysiology of ovotoxic chemotherapeutics, little work has currently been published in this area. Comparable patterns of ovarian and whole-body senescence post make the cow a plausible model for repeated administration of chemotherapeutics and evaluation of long term affects. Follicular fluids may be a useful tool to assess viability of oocytes.[227] Loss of viable oocyte is a serious side effect of chemotherapeutic treatments used for cancer treatment. Models to evaluate the effect of alkylating class chemotherapeutics via serum AMH corollate population of reserve pool oocytes would be plausible in cattle.[228]

9. Premature ovarian insufficiency (POI)

The ESHRE definition of POI includes (1) diminished primordial follicle reserve acquired in fetal life, (2) accelerated follicular atresia, and (3) blocked follicle maturation following puberty and proceeding menarche.[229] Although considered multifactorial in both humans and cattle, differentiation between inherited genetic defects and acquired variation in gene expression from fetal through adult life makes the etiologic diagnosis of POI challenging in both humans and cattle. The prevalence of POI in humans has been estimated at 1% of the population.[230] Karyotype investigation of Tunisian women with POI was able to detect chromosome abnormalities in up to 10.8% patients.[231] Other general etiologies for human POI include autoimmune disorders, impaired metabolism, infectious diseases, environmental toxicity, and iatrogenic causes.[230] The cause of POI is unknown in approximately 50% of human cases.[108]

A review paper identifies 107 genes associated with POI in mammals.[232] Thirty-four of these genes are associated with syndromes in humans while the other non-syndromic genes are associated with pathways involving ovarian development, meiosis, DNA repair, and metabolic function.[232] Furthermore, genes can be divided into subcategories based on the highest expression in various stages of follicular development within gonadotropin-independent and dependent stages of folliculogenesis.[232] In humans, genetic counselling is increasingly available to determine inheritability of POI.[233,234] In women, POI has been associated with increased rates of apoptosis in granulosa cells and oocytes.[235] Within the human population, POI has been increasing in the United States and is defined by decreases in viable oocytes and follicles prior to the age of 40, concurrent with low circulating estrogen, increased FSH, and absence of menstruation (amenorrhea).[236,237] Decreased circulating estrogen results in early-onset menopause.

Clinically, women with primary inherited genetic defects that result in POI may present with streak ovaries devoid of follicles and low or undetectable levels of AMH.[233] In women, POI has been correlated with congenital genetic factors, autoimmune disorders, chemotherapy, and radiation therapy.[238] In a study of 18 women with POI utilizing genetic analysis, autoimmunity and inherited defect in BMPs were the most commonly represented etiologies.[239] POI in cattle represents a failure of the ovary to undergo assembly in late gestation and disruption in mechanisms controlling programed cell death of oocytes.[240] However, systemic clinical manifestation of POI in cattle are not well described.

Inherited etiologies for POI are well known in humans but remain to be studied extensively in cattle. These genetic defects are often accompanied by other systemic clinical parameters typical of the syndrome.[232] Genetic copy number variations have been associated with defect in humans and cattle leading to infertility.[4,241] Genetic variation with specific alleles regulation mitochondrial DNA copy number for *COQ9* gene have been linked to decreased follicles numbers and lower AMH concentration in Holstein cows.[242]

Specific genes identified with inherited POI include *BMP15,* holin-like protein; bacteriophage (*BHLB*), DNA meiotic recombinase 1 (*DMC1*), eukaryotic translation initiation factor 2 subunit beta (*EIF2S2*), eukaryotic translation initiation factor 4E nuclear import factor 1 (*EIF4ENIF1*), folliculogenesis-specific bHLH transcription factor (*FIGLA*), FMRP translational regulator 1 (*FMR1*), *FOXL2,* follicle stimulating hormone subunit beta (*FSHB*), *FSHR,* forkhead box O3A (*FOXO3A*), estrogen receptor 1 (*ESR1*), *GDF9,* G protein-coupled receptor 3 (*GPR3*), helicase for meiosis 1 (*HFM1*), luteinizing hormone subunit beta (*LHB*), LIM homeobox 8 (*LHX8*), mutS homolog 5 (*MSH5*), NOBOX oogenesis homeobox

(*NOBOX*), nuclear receptor subfamily 5 group A member 1 (*NR5A1*), progesterone receptor membrane component 1 (*PGRMC1*), stromal antigen 3 (*STAG3*), and X-prolyl aminopeptidase 2 (*XPNPEP2*).[234,243] Of these genes, *FIGLA*[244], *FOXL2*[8,245,246], *FSHR*[247,248], *GDF9*[249], *LHX8*[250], *NOBOX*[251], *NR5A1*[8,252], *PGRMC1*[253–255], and *FSHB*[9,256] are studied in relation to bovine oogenesis. Inherited "knockout" of *BMP15* could lead to POI, which has been reported in both cattle and humans.[10] Cattle do represent a better model for *BMP15* expression compared to poly-ovulating mice, because *Bmp15* is not required for folliculogenesis in mice.[257] Polymorphisms of *BMP15* in bulls has been correlated to lower sperm quality.[258] *FMR1* is a primary gene associated with POI in humans. Although *FMR1* is linked to male infertility in cattle,[259] it has not yet been verified to be involved in POI in cows. *EIF4ENIF1* has been identified by our group (unpublished data) as one of themost discriminating genes correlated with bovine OFD. and its expression is also highly correlated with POI in humans.[244] In induced POI mouse model, VEGF is significantly downregulated in affected ovaries.[64] A study using murine model showed that the neutralization of endogenous VEGF with antibodies designed to block interactions with VEGF receptors depletes primordial follicles in the mouse ovary,[260] confirming the requirement of VEGF in maintaining primary follicles.

10. Polycystic ovary syndrome (PCOS)

PCOS is the greatest single cause of anovulatory disorder in women and is one of the most widely studied ovarian diseases.[123] Depending on the base population used in literature, PCOS is affecting 1.4–23% of reproductive age women worldwide.[260–264] Although it is named PCOS, polycystic morphology of ovaries is not a consensus criterion listed on commonly used guidelines of PCOS diagnosis. The widely accepted diagnostic criteria of PCOS is the combination of hyperandrogenism and anovulation.[265] For studying human PCOS, various animal models have been developed. Since the excess of androgens is considered the main culprit of PCOS, the induced hyperandrogenism through direct hormonal interventions or gene modification becomes the most commonly used approaches when developing PCOS animal models.[266,267] However, many of these androgenized PCOS models only develop PCOS when excess androgen is given during specific windows of development (e.g., folliculogenesis, ovigerous cord formation, and germ cell meiosis),[268] suggesting the complexity of the role of androgen in

PCOS. Indeed, the excess androgen can be the consequence of PCOS rather than the primary cause.[269] Other than the induced hyperandrogenism, the PCOS-like phenotype could also be developed in animal models with metabolic dysfunctions.[270] Models that are made with mutations in key molecules in the reproductive endocrine system, including estrogen receptor-α (ERα) knockout,[271] aromatase knockout,[272] and the LH over-secretion mice (the bLHβ-CTP mice),[273] are also used to decipher the origins and sequelae of impaired levels of variety of hormones other than androgens.[274]

Although the contribution of genetic factors has been linked to PCOS, it is widely agreed that PCOS is a multifactorial and a heterogenetic disease. Although no single gene mutation has been identified in studies conducted on PCOS patients from multiple families,[275] genome-wide studies have linked several genetic factors to the development of PCOS. These factors include many hormone receptors such as FSHR, insulin receptor (INSR), and luteinizing hormone/chorionic gonadotrophin receptor (LHCGR).[276] Variable expression of multiple bone morphometric proteins (BMP4, BMP6 and BMP7) as well as bone morphometric protein receptors (BMPR1A, BMPR1B and BMPR2) in granulosa cells from human ovaries with PCOS indicate a possible role of BMP signaling in PCOS.[277]

In addition, many diet-, environmentally-, or chemically-induced animal models also develop PCOS-like phenotypes,[278–280] further reflecting the complex etiology of PCOS and the interplay of multiple risk factors. For example, hyperinsulinemia may cause or contribute to the hyperandrogenism in women and leads to PCOS.[281] Oxidative stress has also been identified as a key player in the pathogenesis of PCOS.[282] Increase of ROS and oxidative stress makers such as serum xanthine oxidase are found in PCOS women.[283,284] Mouse model showed that injection of D-galactose in mice causes excessive formation of ROS and induces irregular estrous cycle with increased androgen levels, leading to polycystic ovaries in about 30% of the treated mice.[285] Although a variety of animal models are available for understanding the mechanisms involved in PCOS, most models are artificially developed. Naturally occurring models of PCOS are limited. Advances in oocyte culture allow for early genetic analysis of oocytes, and transcriptome analysis allows for genetic profiling for disease such as PCOS.[286] Decreased expression of *GJA1* (connexin) gene occurs in women with PCOS.[287]

The cystic ovarian disease (COD) or cystic ovarian follicles (COF) is commonly diagnosed in dairy cattle. The prevalence rate of COD/COF in dairy cattle is from 6% to 23%, which is similar to PCOS in human.[288] As early as in the 1830s, COD/COF in cow was reported and was later associated with the naturally occurring nymphomania. The affected animals have an absence or reduced LH surge and an excess androgen. Cystic follicles could be found on one or both ovaries in the absence of active luteal tissues. The large polycystic ovaries are usually found at the pre-ovulatory stage and the disorder has been linked to the X chromosome.[289,290] Researchers soon then noticed that mechanisms leading to COD/COF in cow can be complicated. Metabolic abnormalities with poor energy balance could be the reason that leads to low insulin-like growth factor-I (IGF-I) in the serum and then influence LH pulsatility, leading to ovarian cysts.[291] Cysts in affected ovaries are similar to atretic follicles morphologically. However, the gene expression signatures in cysts in COD/COF are different from atretic follicles.[292]

Many factors, both genetic and environmental, now have been linked to the increased risk, leading to COD/COF in cows.[293,294] Interestingly, many of these risk factors are also identified as contributors to human PCOS. For example, at least 7 genetic loci identified as risk genes for human PCOS are differentially expressed in bovine COD/COF (Table 1).[9,276] In water buffalos, spontaneous COD/COF has been reported with increased serum ROS, suggesting the role of oxidative stress in the pathogenesis of COD/COF.[295] Because oxidative stress has been linked to PCOS, and the elevated ROS is also found in the D-galactose-induced PCOS mouse model, the increased serum ROS in water buffalos could be an excellent spontaneous model to decipher the mechanism of the formation of ovarian cysts.

Table 1 Common genes in bovine COD/COF and human PCOS.

Gene	Human reference	Bovine reference
DENND1A (DENN domain containing 1A)	258,296	9
ERBB4 (erb-b2 receptor tyrosine kinase 4)	297	9,298
FSHB (follicle stimulating hormone subunit β)	276	256
GATA4 (GATA binding protein 4)	296	8
IRF1 (interferon regulatory factor 1)	296,299	300
THADA (thyroid adenoma associated)	301	9
YAP1 (Yes-associated protein 1)	296,302	303

Table 2 Bovine COD/COF models.

Breeds	**Ovarian cysts**	**Menstrual cycles**	**Blood androgen level**	**Notes**
N/R	Yes	Abnormal	N/R	Induced nymphomania[289]
Swedish red	Yes	Abnormal (continuous heat)	N/R	Nymphomania[290]
Holstein-Friesian	Yes	Abnormal	Normal	Spontaneous[306]
Argentinean Holstein	Yes	N/R	Normal	Spontaneous[307]
Mixed	Yes	Abnormal	N/R (reduced in follicular fluid)	Spontaneous[308]
Holstein-Friesian	Yes	Abnormal	N/R	Spontaneous[291]
Holstein	Yes	Abnormal	N/R	Spontaneous[309]

Although many reports showed that bovine COD/COF is naturally occurred with symptoms of irregular estrous cycle and cystic ovaries, similar to PCOS in women, only a few studies mentioned the androgen level in bovine COD/COF (Table 2).[267] A research group at University of Nebraska reported identifying a subpopulation of cows with extreme high mean androstenedione concentration.[304] They found that this subpopulation not only has increased androgen production but also displays altered gene expressions of theca cell and oocyte mRNAs, similar to what was seen in PCOS women. This naturally occurring subpopulation of cows with hyperandrogenism could refine the use of bovine as a model for PCOS studies, especially it has been reported that many candidate genes for human PCOS are expressed in bovine fetal ovaries and their expressions are correlated with gestational age.[71,305] The connection between androgen level, gene expression profiles in bovine fetal ovaries, and PCOS development further suggests that bovine could serve as a model to study how the intra-uterine environment programs the PCOS candidate genes.

Insulin has been known to be an important mediator of follicular development. Hyperinsulinemia could lead to increased testosterone synthesis in theca cells. This connection highlights the role of insulin in the etiology of PCOS, especially theca cells from PCOS ovaries are more responsive to insulin than those from non-PCOS ovaries.[310] Although insulin resistance affects 10–25% of the general population, 50–70% of PCOS patients show variable degrees of insulin resistance.[311] However, in COD/COF cows, insulin insufficiency has been observed rather than insulin resistance.[306,312]

11. Ovarian follicular dysplasia

Ovarian follicular dysplasia (OFD) has been historically reported in cows with variable terminology and consists of non-functional static follicles and disrupted stroma. The etiology of OFD has not been determined. Late stage atretic follicles (type IV atretic follicles) have been described in beef cattle ovaries in the USA and Japan.[313,314] The term OFD was first used in 2002 in a small herd of infertile dairy cows in Estonia[17] and we choose the term OFD to describe dense clusters of dysplastic follicles rather than "late stage atretic follicles" or "type IV atretic follicles" used by other investigators. Clusters of dysplastic follicles in OFD can become proliferative and this is contrary to what would be expected from an atretic follicle. OFD is under reported in cattle because histologic examination of ovaries is required to make the diagnosis, and histologic descriptions of infertile bovine ovaries are scarce in the literature.

Affected OFD cows start with a slowly progressive bilateral ovarian follicle dysplasia. Some cases progress to bilateral ovarian SCSTs (Fig. 1). These ovarian SCSTs have some characteristic with Sertoli-type GCT as described in 1969.[313] Call–Exner bodies in these tumors are also arranged in a bilateral pattern consistent with the sex cord tumor with annular tubules observed in Peutz-Jeghers syndrome, a genetic disease predisposing to the development of infertility. When the reserve population in the cortex are examined, progressive stages of OFD shows similarity with oocyte loss in POI. Loss of primordial and primary follicles are observed in the early stage of OFD and this loss is accompanied by increase in secondary follicles.[315] Minimal genetic profiling or immunohistochemistry has been conducted on OFD or Sertoli-type GCT. Further study of OFD and Sertoli-type GCT should include screening for *FOXL2* and *DICER1* mutations[316] as well as genetic expession patterns associated with Peutz-Jeghers syndrome.[317]

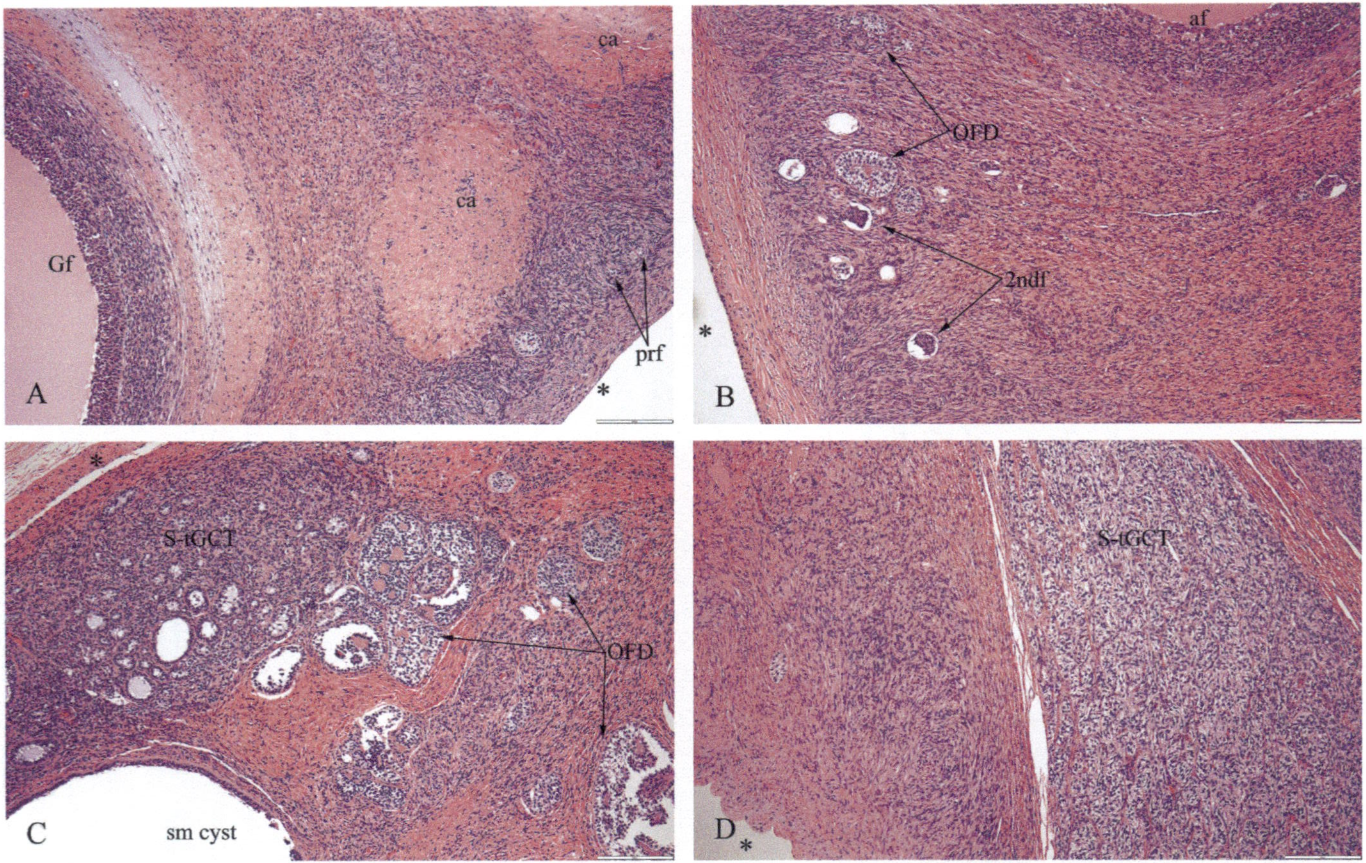

Fig. 1 See figure legend on opposite page.

12. Ovarian cancer

In humans, approximately 90% of malignant ovarian cancers originate from the ovarian surface epithelium, whereas the ovarian sex cord-stromal tumors (SCSTs) represent approximately 7% of the remaining primary ovarian tumors.[318] Although data are not readily available for cattle, ovarian cancer of epithelial origin is not prominently observed post slaughter or represented in the literature. In contrast, SCSTs represent greater than 95% of bovine ovarian neoplasia.[13,50] For cancer animal modelling, the bovine provides an opportunity for SCSTs oncology research. Also, because cattle appear to be highly resistant to malignant forms of epithelial cancer such as mammary cancer,[319] bovids should also be of interest to those wishing to understand the species-specific risk factors for epithelial origin ovarian cancer.

Ovarian SCSTs in humans may be manifested through a wide variety of presentations. SCSTs are generally classified as granulosa cell tumors (GCT), Sertoli stromal tumors, theca-fibroma, sex cord tumor with annular tubules (SCTAT), and mixed or unclassified SCST.[320,321] Ovarian granulosa cell tumors (GCTs) may be hormonally active sex cord stromal tumors, secreting estradiol and less commonly inhibins, and accounting for 2–8% of all human ovarian cancers.[322] GCT may show menstrual irregularities, and rarely virilization. Most human GCT arise in postmenopausal women (adult GCT); only 5% of human GCT (juvenile GCT) occurs in children and

---

Fig. 1 Developmental stages of bovine ovarian follicular dysplasia (OFD) to Sertoli-type Granulosa cell tumor (S-tGCT) Commercial beef cows (*Bos taurus* × *Bos indicus*) ovaries of individual cows. **(A)** Normal ovary from a productive cow containing a Graafian follicle (Gf), two corpus atretica (ca), and numerous primordial follicles (prf) in the outer cortex. There is also one primary follicle that is unlabeled. Corpus atreticum are scars from past atretic follicles indicating active follicular waves and primordial follicles confirm an adequate reserve pool of oocytes undergoing primordial growth activation. **(B)** Early OFD ovary containing increased numbers of secondary follicles (2ndf) with small clusters of OFD adjacent to an atretic follicle (af) in a cortex with increased fibrosis. The ratio of primordial to secondary follicles is shifted toward secondary follicles indicating a high rate of primordial follicle recruitment, a mechanism of infertility **(C)**. Late OFD with early stage S-tGCT and one small follicular cyst (sm cyst), demonstrating progression of advanced OFD to neoplasia with accompanying small cystic disease. **(D)** Advanced S-tGCT with severe stromal fibrosis in the cortex containing small foci of OFD. * indicates outer aspect of ovarian cortex, bar = 200 μm.

young women.[323] In cattle, GCT is also the most common sex cord stromal tumor that occurs unilaterally, with rapid growth, and can be observed by gross examination in slaughterhouses.[15,324–328]

Sertoli cells in the testis and ovaries both originate from bi-potential precursor linage cells during fetal development.[329,330] Other than Sertoli cells that could be found in ovarian tumors, Leydig cells can also be found in hormone secreting SCSTs that lead to virilization.[316] However, the Sertoli-Leydig cell tumor, also known as androblastomas, is extremely rare in human sex cord stromal tumors and has a different molecular pathology of GCT.[331] "Sertoli-type granulosa cell tumors" in cattle share characteristics of "sex cord tumor with annular tubules" in humans.[316] In cows, these are indolent benign small tumors, that occur bilaterally and result in infertility.[313,332] Sertoli-type GCTs in cattle have been documented in many places including United States,[13,313] Japan,[314] and Spain,[15] but require histologic examination to confirm.

Late stage bovine OFD, ovaries show progression to Sertoli-type GCT (Fig. 1). Our current unpublished study showed that the progression of OFD represents transcriptional as well as morphologic masculization in the ovary. The underlying etiology and predisposing factors for bovine OFD are yet to be determined but endocrine disrupting chemicals, granulosa cells toxicity, genetic predisposition, and advanced aging are in the differential list. It has been reported that endocrine disrupting chemicals can induce the insulin-like growth factor-1 secretion and increase the proliferation of cancer- and non-cancer- granulosa cells.[333]

13. Ovarian disease biomarkers

"Liquid biopsy" is a term used to describe detections of biomarkers circulating in blood (such as proteins, miRNA, extracellular vesicles containing proteins, mRNA, DNA fragments, non-coding RNAs, and lipids, etc.). Inhibin B and AMH are viable biomarkers for bovine and human ovarian STSC tumors.[50,334] Micro-vesicles released from tumor cells produce unique miRNA profiles for various grades of epithelial ovarian cancer.[335,336] Circulating serum miRNAs are also used to detect subtypes of polycystic ovary syndrome in humans.[337] MiR-21 in follicular fluid can be used to assess PCOS hyperandrogenism in women.[42] Serum protein biomarker for early detection of human ovarian cancer of epithelial origin have been duplicated using iTRAQ technique in rats[338] and differentiation of malignant *vs.* benign types of ovarian cancer.[339]

Ovarian disease assessment with use of specific serum-based miRNAs in humans includes 11 for PCOS,[340] 11 for POI[341] and nine for ovarian cancer.[131,342] Studies using miRNA to assess reproduction in cattle include one study that identified approximately 350 circulating miRNAs with 30 differentially expressed miRNAs associated with pregnancy status[49] and another that identified 214 circulating miRNAs with 27 miRNA associated with early pregnancy.[343] Circulating protein biomarkers for infertility include 11 specific circulating proteins for POI[344] and 11 proteins for premature ovarian failure.[345] The use of proteomics for assessment of lactation and pregnancy in dairy cattle is in the early stages, but at least 14 circulating protein biomarkers have been identified in cattle to evaluate metabolic status during pregnancy.[346]

The use of a seven-protein serum biomarker panel for early ovarian cancer in humans demonstrated 98.2% sensitivity and 98.7% specificity.[347] One of the 11 circulating proteins, APOA4, is also shown to be correlated with fertility in cattle and is an established biomarker for ovarian cancer in humans.[348] Differences in plasma and follicular fluid metabolic composition involved with carbohydrate and amino acid metabolism have been used to evaluate inactive ovaries in cattle.[349,350]

14. Ovarian therapeutic interventions

Monitoring the effects of chemotherapeutics on ovaries can be assessed in cattle.[351] Dosing chemotherapeutics can be controlled by large indwelling catheters, making continuous infusions feasible. Adhesion family G-protein-coupled receptors in the ovary regulate response to chemotherapeutic agents.[352] *GJA1* (connexin) is upregulated in cells that show a greater survival response to cisplatin, an alkylating agent.[353] Antioxidant vitamins have a positive effect on female fertility by regulating genetic expression response to ROS.[190]

Mesenchymal stem cell (MSC) therapy has therapeutic potential in treatment of POI.[354] MSC can be retrieved from the umbilical cord at birth. MSC therapy has been shown to promote the preantral follicle growth *in vitro* for humans.[355] When cultured with MSC, oocytes show increased levels of GDF9 and BMPs, and granulosa cells show increased levels of inhibin βA and TGFβ1.[355] Treatment strategies for POI have included transplantation,[356] oocyte cryopreservation[357] and *in vitro* activation via auto-transplantation.[358] Emerging modalities in stem cell therapy show promise for treatment of human ovarian diseases.[236] In cattle, MSCs are being utilized to ameliorate ovarian tissue damage caused by repeated transvaginal ultrasound-guided follicle aspiration (ovum pick-up) procedure.[45] This has translational implication as ovum pick-up is widely used in human infertility treatment.[46] The oocyte quality of the *in vitro* maturation of oocyte has been improved by adding exogenous cannabinoids (e.g., HU-210) resulting in enhancing expression of interferon tau (IFNτ) and gap junction alpha-1 protein (GJA1).[48]

15. Concerns of using bovine model for human ovarian diseases

Although humans and bovine share many similarities in reproductive characteristics, the differences between these two species are also fully acknowledged. For example, endometrial shedding in women correlated with symmetric luteal and follicular phases are different than cows.[11] Also, the length of the follicular phase is short in humans (2–3 days) in comparison to cows (12–14 days).[31] Cows on average have a slightly greater number of follicular waves during the inter-ovulatory interval.[32] Although ovulation in cattle is spontaneous, which is similar to humans, it occurs after the end of estrus in which the timing is different.[359] At the gene expression level, the expression of inhibin in human ovaries has a unique pattern which is restricted to the granulosa layer of antral and preovulatory follicles, granulosa luteal cells of CL, but not in theca and interstitial cells,[360,361] whereas in cattle, one of the inhibin subunits, INHA, is highly expressed in theca cells.[362]

16. Summary

In this article, we summarized genes/pathways and factors common to both humans and cattle ovarian physiological processes and diseases. Bovine models for ovary-related research can benefit human health and livestock animal industries simultaneously, thus representing a win-win for multiple stakeholders and a practical point when seeking funding from federal, state, user-group, and corporate agencies.

Disclosure statement

The authors have nothing to disclose.

References

1. Zimin AV, Delcher AL, Florea L, et al. A whole-genome assembly of the domestic cow, Bos taurus. *Genome Biol.* 2009;10(4):R42. https://doi.org/10.1186/gb-2009-10-4-r42.
2. Bickhart DM, McClure JC, Schnabel RD, Rosen BD, Medrano JF, Smith TPL. Symposium review: advances in sequencing technology herald a new frontier in cattle

genomics and genome-enabled selection. *J Dairy Sci.* 2020;103(6):5278–5290. https://doi.org/10.3168/jds.2019-17693.

3. Couzin-Frankel J., Mooove over humans, the cow genome is here. Science.org. doi:10.1126/article.31249.
4. McDaneld TG, Kuehn LA, Thomas MG, Pollak EJ, Keele JW. Deletion on chromosome 5 associated with decreased reproductive efficiency in female cattle. *J Anim Sci.* 2014;92:1378–1384. https://doi.org/10.2527/jas.2013-6821.
5. Aguiar LH, Hyde KA, Pedroza GH, Denicol AC. Heat stress impairs in vitro development of preantral follicles of cattle. *Anim Reprod Sci.* 2020;213:106277. https://doi.org/10.1016/j.anireprosci.2020.106277.
6. Baerwald AR, Adams GP, Pierson RA. Characterization of ovarian follicular wave dynamics in women. *Biol Reprod.* 2003;69(3):1023–1031. https://doi.org/10.1095/biolreprod.103.017772.
7. Adams GP, Pierson RA. Bovine model for study of ovarian follicular dynamics in humans. *Theriogenology.* 1995;43(1):113–120. https://doi.org/10.1016/0093-691X(94)00015-M.
8. Hummitzsch K, Hatzirodos N, Irving-Rodgers HF, et al. Morphometric analyses and gene expression related to germ cells, gonadal ridge epithelial-like cells and granulosa cells during development of the bovine fetal ovary. *PLoS One.* 2019;14(3):e0214130. https://doi.org/10.1371/journal.pone.0214130.
9. Hartanti MD, Rosario R, Hummitzsch K, et al. Could perturbed fetal development of the ovary contribute to the development of polycystic ovary syndrome in later life? *PLoS One.* 2020;15(2):e0229351. https://doi.org/10.1371/journal.pone.0229351.
10. Webber L, Davies M, Anderson R, et al. ESHRE guideline: management of women with premature ovarian insufficiency. *Hum Reprod.* 2016;31(5):926–937. https://doi.org/10.1093/humrep/dew027.
11. Adams GP, Singh J, Baerwald AR. Large animal models for the study of ovarian follicular dynamics in women. *Theriogenology.* 2012;78(8):1733–1748. https://doi.org/10.1016/j.theriogenology.2012.04.010.
12. Santos RR, Schoevers EJ, Roelen BA. Usefulness of bovine and porcine IVM/IVF models for reproductive toxicology. *Reprod Biol Endocrinol.* 2014;12:117. https://doi.org/10.1186/1477-7827-12-117.
13. McEntee K. *Reproductive Pathology of Domestic Mammals.* Academic Press Inc; 1990:401.
14. Gard JA, Roberts J, Braden T, et al. 119 assessment of ovarian follicular dysplasia utilizing ultrasound and histologic examination. *Reprod Fertil Dev.* 2017;29(1):168. https://doi.org/10.1071/RDv29n1Ab119.
15. Pérez-Martínez C, Durán-Navarrete AJ, García-Fernández RA, Espinosa-Alvarez J, Escudero Diez A, García-Iglesias MJ. Biological characterization of ovarian granulosa cell tumours of slaughtered cattle: assessment of cell proliferation and oestrogen receptors. *J Comp Pathol.* 2004;130(2-3):117–123. https://doi.org/10.1016/j.jcpa.2003.09.007.
16. Belotti EM, Sacco SC, Stassi AF, et al. Characterization of an incipient granulosa cell tumour in a Holstein cow: steroid hormone receptors and coregulators expression. *Anat Histol Embryol.* 2021. https://doi.org/10.1111/ahe.12762.
17. Kübar H, Jalakas M. Pathological changes in the reproductive organs of cows and heifers culled because of infertility. *J Vet Med A Physiol Pathol Clin Med.* 2002;49(7):365–372. https://doi.org/10.1046/j.1439-0442.2002.00459.x.
18. Teh APP, Kitahara G, Izzati UZ, et al. Immunohistochemical and morphological features of bovine granulosa cell tumours in relation to growth pattern and folliculogenesis. *J Comp Pathol.* 2021;187:40–51. https://doi.org/10.1016/j.jcpa.2021.06.008.
19. Hammond ER, Green MP, Shelling AN, Berg MC, Peek JC, Cree LM. Oocyte mitochondrial deletions and heteroplasmy in a bovine model of ageing and ovarian stimulation. *Mol Hum Reprod.* 2016;22(4):261–271. https://doi.org/10.1093/molehr/gaw003.

20. Li L, Wang Z. Ovarian aging and osteoporosis. *Adv Exp Med Biol.* 2018;1086: 199–215. https://doi.org/10.1007/978-981-13-1117-8_13.
21. Zhang J, Chen Q, Du D, et al. Can ovarian aging be delayed by pharmacological strategies? *Aging (Albany NY).* 2019;11(2):817–832. https://doi.org/10.18632/aging.101784.
22. Parkes WS, Amargant F, Zhou LT, Villanueva CE, Duncan FE, Pritchard MT. Hyaluronan and collagen are prominent extracellular matrix components in bovine and porcine ovaries. *Genes (Basel).* 2021;12(8). https://doi.org/10.3390/genes12081186.
23. Yapura J, Mapletoft RJ, Pierson R, et al. A bovine model for examining the effects of an aromatase inhibitor on ovarian function in women. *Fertil Steril.* 2011;96(2):434–438.e3. https://doi.org/10.1016/j.fertnstert.2011.05.038.
24. Campbell BK, Souza C, Gong J, et al. Domestic ruminants as models for the elucidation of the mechanisms controlling ovarian follicle development in humans. *Reprod Suppl.* 2003;61:429–443.
25. Baerwald AR, Adams GP, Pierson RA. A new model for ovarian follicular development during the human menstrual cycle. *Fertil Steril.* 2003;80(1):116–122. https://doi.org/10.1016/s0015-0282(03)00544-2.
26. Ginther OJ, Beg MA, Bergfelt DR, Donadeu FX, Kot K. Follicle selection in monovular species. *Biol Reprod.* 2001;65(3):638–647. https://doi.org/10.1095/biolreprod65.3.638.
27. Malhi PS, Adams GP, Singh J. Bovine model for the study of reproductive aging in women: follicular, luteal, and endocrine characteristics. *Biol Reprod.* 2005;73(1):45–53. https://doi.org/10.1095/biolreprod.104.038745.
28. Mihm M, Evans AC. Mechanisms for dominant follicle selection in monovulatory species: a comparison of morphological, endocrine and intraovarian events in cows, mares and women. *Reprod Domest Anim.* 2008;43(suppl 2):48–56. https://doi.org/10.1111/j.1439-0531.2008.01142.x.
29. Malhi PS, Adams GP, Pierson RA, Singh J. Bovine model of reproductive aging: response to ovarian synchronization and superstimulation. *Theriogenology.* 2006;66(5):1257–1266. https://doi.org/10.1016/j.theriogenology.2006.02.051.
30. Malhi PS, Adams GP, Mapletoft RJ, Singh J. Oocyte developmental competence in a bovine model of reproductive aging. *Reproduction.* 2007;134(2):233–239. https://doi.org/10.1530/rep-07-0021.
31. Abedal-Majed MA, Cupp AS. Livestock animals to study infertility in women. *Anim Front.* 2019;9(3):28–33. https://doi.org/10.1093/af/vfz017.
32. Ginther OJ. Intraovarianism. Local mechanisms that affect follicle and luteal dynamics in heifers and women. *Biol Reprod.* 2020;102(2):265–275. https://doi.org/10.1093/biolre/ioz199.
33. Jitjumnong J, Moonmanee T, Sudwan P, et al. Associations among thermal biology, preovulatory follicle diameter, follicular and luteal vascularities, and sex steroid hormone concentrations during preovulatory and postovulatory periods in tropical beef cows. *Anim Reprod Sci.* 2020;213:106281. https://doi.org/10.1016/j.anireprosci.2020.106281.
34. Umehara T, Richards JS, Shimada M. The stromal fibrosis in aging ovary. *Aging (Albany NY).* 2018;10(1):9–10. https://doi.org/10.18632/aging.101370.
35. Nobre HC, Gard Schunelle J, Edmondson M, et al. *Ultrasonographic Examination of Cows Affected with Ovarian Follicular Dysplasia.* Auburn University; 2019. http://hdl.handle.net/10415/6701.
36. Hilton JL, Baerwald AR, Sarty GE, Adams GP, Pierson RA. Magnetic resonance image attributes of the bovine corpus luteum during development and regression. *Anat Rec A Discov Mol Cell Evol Biol.* 2003;272(1):454–459. https://doi.org/10.1002/ar.a.10050.

37. Warren L, Murawski M, Wilk K, Zieba DA, Bartlewski PM. Suitability of antral follicle counts and computer-assisted analysis of ultrasonographic and magnetic resonance images for estimating follicular reserve in porcine, ovine and bovine ovaries ex situ. *Exp Biol Med (Maywood).* 2015;240(5):576–584. https://doi.org/10.1177/1535370214560971.
38. Martinez IYH, Santos APC, Bottino MP, et al. Molecular and endocrine factors involved in future dominant follicle dynamics during the induction of luteolysis in Bos indicus cows. *Theriogenology.* 2018;111:78–83. https://doi.org/10.1016/j.theriogenology.2018.01.019.
39. Ferrazza RA, Garcia HDM, Schmidt E, et al. Quantitative proteomic profiling of bovine follicular fluid during follicle development. *Biol Reprod.* 2017;97(6):835–849. https://doi.org/10.1093/biolre/iox148.
40. da Silveira JC, Andrade GM, Del Collado M, et al. Supplementation with small-extracellular vesicles from ovarian follicular fluid during in vitro production modulates bovine embryo development. *PLoS One.* 2017;12(6):e0179451. https://doi.org/10.1371/journal.pone.0179451.
41. Magata F, Horiuchi M, Echizenya R, et al. Lipopolysaccharide in ovarian follicular fluid influences the steroid production in large follicles of dairy cows. *Anim Reprod Sci.* 2014;144(1-2):6–13. https://doi.org/10.1016/j.anireprosci.2013.11.005.
42. Naji M, Aleyasin A, Nekoonam S, Arefian E, Mahdian R, Amidi F. Differential expression of miR-93 and miR-21 in granulosa cells and follicular fluid of polycystic ovary syndrome associating with different phenotypes. *Sci Rep.* 2017;7(1):14671. https://doi.org/10.1038/s41598-017-13250-1.
43. Herzog K, Strüve K, Kastelic JP, et al. Escherichia coli lipopolysaccharide administration transiently suppresses luteal structure and function in diestrous cows. *Reproduction.* 2012;144(4):467–476. https://doi.org/10.1530/rep-12-0138.
44. Zhang HY, Zhu FF, Zhu YJ, Hu YJ, Chen X. Effects of IL-18 on the proliferation and steroidogenesis of bovine theca cells: possible roles in the pathogenesis of polycystic ovary syndrome. *J Cell Mol Med.* 2021;25(2):1128–1139. https://doi.org/10.1111/jcmm.16179.
45. Malard PF, Peixer MAS, Grazia JG, et al. Intraovarian injection of mesenchymal stem cells improves oocyte yield and in vitro embryo production in a bovine model of fertility loss. *Sci Rep.* 2020;10(1):8018. https://doi.org/10.1038/s41598-020-64810-x.
46. Liang T, Ruetz K, Haakman O, Vilos A, Vilos G, Abu-Rafea B. Impact of class III obesity on outcomes and complications of transvaginal ultrasound-guided oocyte pickup. *F S Rep.* 2020;1(3):270–276. https://doi.org/10.1016/j.xfre.2020.08.009.
47. Puglisi R, Cambuli C, Capoferri R, et al. Differential gene expression in cumulus oocyte complexes collected by ovum pick up from repeat breeder and normally fertile Holstein Friesian heifers. *Anim Reprod Sci.* 2013;141(1-2):26–33. https://doi.org/10.1016/j.anireprosci.2013.07.003.
48. López-Cardona AP, Sánchez-Calabuig MJ, Beltran-Breña P, et al. Exocannabinoids effect on in vitro bovine oocyte maturation via activation of AKT and ERK1/2. *Reproduction.* 2016;152(6):603–612. https://doi.org/10.1530/rep-16-0199.
49. Gebremedhn S, Salilew-Wondim D, Hoelker M, et al. Exploring maternal serum microRNAs during early pregnancy in cattle. *Theriogenology.* 2018;121:196–203. https://doi.org/10.1016/j.theriogenology.2018.08.020.
50. El-Sheikh Ali H, Kitahara G, Nibe K, et al. Plasma anti-Müllerian hormone as a biomarker for bovine granulosa-theca cell tumors: comparison with immunoreactive inhibin and ovarian steroid concentrations. *Theriogenology.* 2013;80(8):940–949. https://doi.org/10.1016/j.theriogenology.2013.07.022.
51. Mohammed ZA, Robinson RS, Harris R, et al. Detrimental effects of uterine disease and lipopolysaccharide on luteal angiogenesis. *J Endocrinol.* 2020;245(1):79–92. https://doi.org/10.1530/joe-19-0443.

52. Herzog K, Debertolis L, Kastelic JP, Schmicke M, Ulbrich SE, Bollwein H. Effects of intravenous infusion of E. coli lipopolysaccharide in early pregnant cows. *Reproduction.* 2019;157(1):65–76. https://doi.org/10.1530/rep-18-0174.
53. Aardema H, van Tol HTA, Wubbolts RW, Brouwers JFHM, Gadella BM, Roelen BAJ. Stearoyl-CoA desaturase activity in bovine cumulus cells protects the oocyte against saturated fatty acid stress. *Biol Reprod.* 2017;96(5):982–992. https://doi.org/10.1095/biolreprod.116.146159.
54. Nichols JA, Perego MC, Schütz LF, Hemple AM, Spicer LJ. Hormonal regulation of vascular endothelial growth factor A (VEGFA) gene expression in granulosa and theca cells of cattle1. *J Anim Sci.* 2019;97(7):3034–3045. https://doi.org/10.1093/jas/skz164.
55. Velásquez A, Mellisho E, Castro FO, Rodríguez-Álvarez L. Effect of BMP15 and/or AMH during in vitro maturation of oocytes from involuntarily culled dairy cows. *Mol Reprod Dev.* 2019;86(2):209–223. https://doi.org/10.1002/mrd.23096.
56. Roberts J, Steel E, Gard J, et al. Morphological characteristics of Ovarian Follicular Dysplasia (OFD) observed by ultrasound in four Florida beef herds. In: *Society for Theriogenology Annual Conference, Ashville*; 2016.
57. Campen KA, Kucharczyk KM, Bogin B, Ehrlich JM, Combelles CMH. Spindle abnormalities and chromosome misalignment in bovine oocytes after exposure to low doses of bisphenol A or bisphenol S. *Hum Reprod.* 2018;33(5):895–904. https://doi.org/10.1093/humrep/dey050.
58. Marians RC, Ng L, Blair HC, Unger P, Graves PN, Davies TF. Defining thyrotropin-dependent and -independent steps of thyroid hormone synthesis by using thyrotropin receptor-null mice. *Proceedings of the National Academy of Sciences of the United States of America.* 2002;99(24):15776–15781. https://doi.org/10.1073/pnas.242322099.
59. Kliem H, Berisha B, Meyer HH, Schams D. Regulatory changes of apoptotic factors in the bovine corpus luteum after induced luteolysis. *Mol Reprod Dev.* 2009;76(3):220–230. https://doi.org/10.1002/mrd.20946.
60. Itami N, Kawahara-Miki R, Kawana H, Endo M, Kuwayama T, Iwata H. Age-associated changes in bovine oocytes and granulosa cell complexes collected from early antral follicles. *J Assist Reprod Genet.* 2014;31(8):1079–1088. https://doi.org/10.1007/s10815-014-0251-y.
61. Nguyen LT, Reverter A, Cánovas A, et al. Global differential gene expression in the pituitary gland and the ovaries of pre- and postpubertal Brahman heifers. *J Anim Sci.* 2017;95(2):599–615. https://doi.org/10.2527/jas.2016.0921.
62. Kenngott RA, Scholz W, Sinowatz F. Ultrastructural aspects of the prenatal bovine ovary differentiation with a special focus on the interstitial cells. *Anat Histol Embryol.* 2016;45(5):357–366. https://doi.org/10.1111/ahe.12203.
63. Weller MM, Fortes MR, Porto-Neto LR, et al. Candidate gene expression in *Bos indicus* ovarian tissues: Prepubertal and postpubertal heifers in diestrus. *Front Vet Sci.* 2016;3:94. https://doi.org/10.3389/fvets.2016.00094.
64. Yang Z, Du X, Wang C, et al. Therapeutic effects of human umbilical cord mesenchymal stem cell-derived microvesicles on premature ovarian insufficiency in mice. *Stem Cell Res Ther.* 2019;10(1):250. https://doi.org/10.1186/s13287-019-1327-5.
65. Dompe C, Kulus M, Stefańska K, et al. Human granulosa cells-stemness properties, molecular cross-talk and follicular angiogenesis. *Cells.* 2021;10(6). https://doi.org/10.3390/cells10061396.
66. Cho E, Kim YY, Noh K, Ku SY. A new possibility in fertility preservation: The artificial ovary. *J Tissue Eng Regen Med.* 2019;13(8):1294–1315. https://doi.org/10.1002/term.2870.

67. Sándor S, Kubinyi E. Genetic pathways of aging and their relevance in the dog as a natural model of human aging. *Front Genet.* 2019;10:948. https://doi.org/10.3389/fgene.2019.00948.
68. Girolami F, Badino P, Spalenza V, et al. Identification of candidate biomarkers of the exposure to PCBs in contaminated cattle: a gene expression- and proteomic-based approach. *Sci Total Environ.* 2018;640-641:22–30. https://doi.org/10.1016/j.scitotenv.2018.05.284.
69. Tao HY, Zhang J, Shi J, et al. Occurrence and emission of phthalates, bisphenol A, and oestrogenic compounds in concentrated animal feeding operations in Southern China. *Ecotoxicol Environ Saf.* 2021;207:111521. https://doi.org/10.1016/j.ecoenv.2020.111521.
70. Petro EM, Covaci A, Leroy JL, Dirtu AC, De Coen W, Bols PE. Occurrence of endocrine disrupting compounds in tissues and body fluids of Belgian dairy cows and its implications for the use of the cow as a model to study endocrine disruption. *Sci Total Environ.* 2010;408(22):5423–5428. https://doi.org/10.1016/j.scitotenv.2010.07.051.
71. Liu M, Hummitzsch K, Hartanti MD, et al. Analysis of expression of candidate genes for polycystic ovary syndrome in adult and fetal human and fetal bovine ovaries. *Biol Reprod.* 2020;103(4):840–853. https://doi.org/10.1093/biolre/ioaa119.
72. Vidal VP, Chaboissier MC, de Rooij DG, Schedl A. Sox9 induces testis development in XX transgenic mice. *Nat Genet.* 2001;28(3):216–217. https://doi.org/10.1038/90046.
73. Georges A, Auguste A, Bessière L, Vanet A, Todeschini AL, Veitia RA. FOXL2: a central transcription factor of the ovary. *J Mol Endocrinol.* 2014;52(1):R17–R33. https://doi.org/10.1530/jme-13-0159.
74. Thanatsis N, Kaponis A, Koika V, Georgopoulos NA, Decavalas GO. Reduced Foxo3a, FoxL2, and p27 mRNA expression in human ovarian tissue in premature ovarian insufficiency. *Hormones (Athens).* 2019;18(4):409–415. https://doi.org/10.1007/s42000-019-00134-4.
75. Pepene CE. Soluble platelet/endothelial cell adhesion molecule (sPECAM)-1 is increased in polycystic ovary syndrome and related to endothelial dysfunction. *Gynecol Endocrinol.* 2012;28(5):370–374. https://doi.org/10.3109/09513590.2011.632792.
76. Ying Y, Liu XM, Marble A, Lawson KA, Zhao GQ. Requirement of Bmp8b for the generation of primordial germ cells in the mouse. *Mol Endocrinol.* 2000;14(7):1053–1063. https://doi.org/10.1210/mend.14.7.0479.
77. Soto DA, Ross PJ. Similarities between bovine and human germline development revealed by single-cell RNA sequencing. *Reproduction.* 2021;161(3):239–253. https://doi.org/10.1530/rep-20-0313.
78. Yang D, Yang X, Dai F, et al. The role of bone morphogenetic protein 4 in ovarian function and diseases. *Reprod Sci.* 2021. https://doi.org/10.1007/s43032-021-00600-8.
79. Wrobel KH, Süss F. Identification and temporospatial distribution of bovine primordial germ cells prior to gonadal sexual differentiation. *Anat Embryol.* 1998;197(6):451–467. https://doi.org/10.1007/s004290050156.
80. Mamsen LS, Brøchner CB, Byskov AG, Møllgard K. The migration and loss of human primordial germ stem cells from the hind gut epithelium towards the gonadal ridge. *Int J Dev Biol.* 2012;56(10-12):771–778. https://doi.org/10.1387/ijdb.120202lm.
81. Nilsson EE, Skinner MK. Progesterone regulation of primordial follicle assembly in bovine fetal ovaries. *Mol Cell Endocrinol.* 2009;313(1-2):9–16. https://doi.org/10.1016/j.mce.2009.09.004.
82. Xu J, Zheng T, Hong W, Ye H, Hu C, Zheng Y. Mechanism for the decision of ovarian surface epithelial stem cells to undergo neo-oogenesis or ovarian tumorigenesis. *Cell Physiol Biochem.* 2018;50(1):214–232. https://doi.org/10.1159/000494001.

83. Banet N, Kurman RJ. Two types of ovarian cortical inclusion cysts: proposed origin and possible role in ovarian serous carcinogenesis. *Int J Gynecol Pathol.* 2015;34(1):3–8. https://doi.org/10.1097/pgp.0000000000000120.
84. Sills ES, Wood SH. Autologous activated platelet-rich plasma injection into adult human ovary tissue: molecular mechanism, analysis, and discussion of reproductive response. *Biosci Rep.* 2019;39(6). https://doi.org/10.1042/BSR20190805.
85. Franchi FF, Hernandes MP, Coalho Ferreira AL, et al. Fractal analysis and histomolecular phenotyping provides insights into extracellular matrix remodeling in the developing bovine fetal ovary. *Biochem Biophys Res Commun.* 2020;523(4):823–828. https://doi.org/10.1016/j.bbrc.2020.01.025.
86. Jones RE, Lopez KH. Chapter 5—Sexual differentiation. In: Jones RE, Lopez KH, eds. *Human Reproductive Biology*. 4th ed. Academic Press; 2014:87–102.
87. Fortune JE, Yang MY, Muruvi W. The earliest stages of follicular development: follicle formation and activation. *Soc Reprod Fertil Suppl.* 2010;67:203–216. https://doi.org/10.7313/upo9781907284991.018.
88. Gondos B, Westergaard L, Byskov AG. Initiation of oogenesis in the human fetal ovary: ultrastructural and squash preparation study. *Am J Obstet Gynecol.* 1986;155(1):189–195. https://doi.org/10.1016/0002-9378(86)90109-2.
89. Konishi I, Fujii S, Okamura H, Parmley T, Mori T. Development of interstitial cells and ovigerous cords in the human fetal ovary: an ultrastructural study. *J Anat.* 1986;148:121–135.
90. Lundgaard Riis M, Nielsen JE, Hagen CP, et al. Accelerated loss of oogonia and impaired folliculogenesis in females with Turner syndrome start during early fetal development. *Hum Reprod.* 2021;36(11):2992–3002. https://doi.org/10.1093/humrep/deab210.
91. Conti M, Andersen CB, Richard FJ, Shitsukawa K, Tsafriri A. Role of cyclic nucleotide phosphodiesterases in resumption of meiosis. *Mol Cell Endocrinol.* 1998;145(1-2):9–14. https://doi.org/10.1016/s0303-7207(98)00187-7.
92. Bukovsky A, Gupta SK, Virant-Klun I, et al. Study origin of germ cells and formation of new primary follicles in adult human and rat ovaries. *Methods Mol Biol.* 2008;450:233–265. https://doi.org/10.1007/978-1-60327-214-8_16.
93. Mugavero MJ, May M, Harris R, et al. Does short-term virologic failure translate to clinical events in antiretroviral-naïve patients initiating antiretroviral therapy in clinical practice? *Aids.* 2008;22(18):2481–2492. https://doi.org/10.1097/QAD.0b013e328318f130.
94. Wagner M, Yoshihara M, Douagi I, et al. Single-cell analysis of human ovarian cortex identifies distinct cell populations but no oogonial stem cells. *Nat Commun.* 2020;11(1):1147. https://doi.org/10.1038/s41467-020-14936-3.
95. Duffin K, Bayne RA, Childs AJ, Collins C, Anderson RA. The forkhead transcription factor FOXL2 is expressed in somatic cells of the human ovary prior to follicle formation. *Mol Hum Reprod.* 2009;15:771–777. https://doi.org/10.1093/molehr/gap065.
96. Hua J, Xu B, Yang Y, et al. Follicle Online: an integrated database of follicle assembly, development and ovulation. *Database (Oxford).* 2015;2015:bav036. https://doi.org/10.1093/database/bav036.
97. Tanaka Y, Nakada K, Moriyoshi M, Sawamukai Y. Appearance and number of follicles and change in the concentration of serum FSH in female bovine fetuses. *Reproduction.* 2001;121(5):777–782.
98. Santos SS, Ferreira MA, Pinto JA, et al. Characterization of folliculogenesis and the occurrence of apoptosis in the development of the bovine fetal ovary. *Theriogenology.* 2013;79(2):344–350. https://doi.org/10.1016/j.theriogenology.2012.09.026.
99. Yadav PK, Tiwari M, Gupta A, et al. Germ cell depletion from mammalian ovary: possible involvement of apoptosis and autophagy. *J Biomed Sci.* 2018;25(1):36. https://doi.org/10.1186/s12929-018-0438-0.

100. Baker TG. A quantitative and cytological study of germ cells in human ovaries. *Proc R Soc Lond B Biol Sci.* 1963;158:417–433. https://doi.org/10.1098/rspb.1963.0055.
101. Park SU, Walsh L, Berkowitz KM. Mechanisms of ovarian aging. *Reproduction.* 2021;162(2):R19–r33. https://doi.org/10.1530/rep-21-0022.
102. Kallen A, Polotsky AJ, Johnson J. Untapped reserves: controlling primordial follicle growth activation. *Trends Mol Med.* 2018;24(3):319–331. https://doi.org/10.1016/j.molmed.2018.01.008.
103. Glister C, Satchell L, Knight PG. Changes in expression of bone morphogenetic proteins (BMPs), their receptors and inhibin co-receptor betaglycan during bovine antral follicle development: inhibin can antagonize the suppressive effect of BMPs on thecal androgen production. *Reproduction.* 2010;140(5):699–712. https://doi.org/10.1530/rep-10-0216.
104. Spicer LJ, Schutz LF, Aad PY. Effects of bone morphogenetic protein 4, gremlin, and connective tissue growth factor on estradiol and progesterone production by bovine granulosa cells. *J Anim Sci.* 2021;99(11). https://doi.org/10.1093/jas/skab318.
105. Siu MK, Cheng CY. The blood-follicle barrier (BFB) in disease and in ovarian function. *Adv Exp Med Biol.* 2012;763:186–192. https://doi.org/10.1007/978-1-4614-4711-5_9.
106. Wang R, Song B, Wu J, Zhang Y, Chen A, Shao L. Potential adverse effects of nanoparticles on the reproductive system. *Int J Nanomed.* 2018;13:8487–8506. https://doi.org/10.2147/ijn.S170723.
107. Das D, Arur S. Conserved insulin signaling in the regulation of oocyte growth, development, and maturation. *Mol Reprod Dev.* 2017;84(6):444–459. https://doi.org/10.1002/mrd.22806.
108. Persani L, Rossetti R, Cacciatore C, Fabre S. Genetic defects of ovarian TGF-β-like factors and premature ovarian failure. *J Endocrinol Invest.* 2011;34(3):244–251. https://doi.org/10.1007/bf03347073.
109. Passos MJ, Vasconcelos GL, Silva AW, et al. Accelerated growth of bovine preantral follicles in vitro after stimulation with both FSH and BMP-15 is accompanied by ultrastructural changes and increased atresia. *Theriogenology.* 2013;79(9):1269–1277. https://doi.org/10.1016/j.theriogenology.2013.02.023.
110. Poole DH, Ocón-Grove OM, Johnson AL. Anti-Müllerian hormone (AMH) receptor type II expression and AMH activity in bovine granulosa cells. *Theriogenology.* 2016;86(5):1353–1360. https://doi.org/10.1016/j.theriogenology.2016.04.078.
111. Ilha GF, Rovani MT, Gasperin BG, et al. Regulation of anti-Müllerian hormone and its receptor expression around follicle deviation in cattle. *Reprod Domest Anim.* 2016;51(2):188–194. https://doi.org/10.1111/rda.12662.
112. Alward KJ, Bohlen JF. Overview of Anti-Müllerian hormone (AMH) and association with fertility in female cattle. *Reprod Domest Anim.* 2020;55(1):3–10. https://doi.org/10.1111/rda.13583.
113. Tang X, Ma L, Guo S, Liang M, Jiang Z. High doses of FSH induce autophagy in bovine ovarian granulosa cells via the AKT/mTOR pathway. *Reprod Domest Anim.* 2021;56(2):324–332. https://doi.org/10.1111/rda.13869.
114. Nilsson EE, Skinner MK. Bone morphogenetic protein-4 acts as an ovarian follicle survival factor and promotes primordial follicle development. *Biol Reprod.* 2003;69(4):1265–1272. https://doi.org/10.1095/biolreprod.103.018671.
115. da Cunha EV, Melo LRF, Sousa GB, et al. Effect of bone morphogenetic proteins 2 and 4 on survival and development of bovine secondary follicles cultured in vitro. *Theriogenology.* 2018;110:44–51. https://doi.org/10.1016/j.theriogenology.2017.12.032.
116. Lan Y, Zhang S, Gong F, Lu C, Lin G, Hu L. The mitochondrial DNA copy number of cumulus granulosa cells may be related to the maturity of oocyte cytoplasm. *Hum Reprod.* 2020;35(5):1120–1129. https://doi.org/10.1093/humrep/deaa085.

117. Sirard MA. Toward building the cow folliculome. *Anim Reprod Sci*. 2014;149(1-2): 90–97. https://doi.org/10.1016/j.anireprosci.2014.06.025.
118. Yamamoto T, Iwata H, Goto H, et al. Effect of maternal age on the developmental competence and progression of nuclear maturation in bovine oocytes. *Mol Reprod Dev*. 2010;77(7):595–604. https://doi.org/10.1002/mrd.21188.
119. Sugimoto M, Sasaki S, Gotoh Y, et al. Genetic variants related to gap junctions and hormone secretion influence conception rates in cows. *Proc Natl Acad Sci USA*. 2013;110(48):19495–19500. https://doi.org/10.1073/pnas.1309307110.
120. Zhou Y, Li H, Lin C, et al. Perfluorooctanoic acid (PFOA) inhibits the gap junction intercellular communication and induces apoptosis in human ovarian granulosa cells. *Reprod Toxicol*. 2020;98:125–133. https://doi.org/10.1016/j.reprotox.2020.09.005.
121. Regassa A, Rings F, Hoelker M, et al. Transcriptome dynamics and molecular cross-talk between bovine oocyte and its companion cumulus cells. *BMC Genomics*. 2011;12:57. https://doi.org/10.1186/1471-2164-12-57.
122. Dangudubiyyam SV, Ginther OJ. Relationship between more follicles in right than left ovary in recently born calves and right ovary propensity for ovulation in cattle. *Reprod Biol*. 2019;19(4):363–367. https://doi.org/10.1016/j.repbio.2019.09.005.
123. Smith P, Wilhelm D, Rodgers RJ. Development of mammalian ovary. *J Endocrinol*. 2014;221(3):R145–R161. https://doi.org/10.1530/joe-14-0062.
124. Guo Y, Sun J, Lai D. Role of microRNAs in premature ovarian insufficiency. *Reprod Biol Endocrinol*. 2017;15(1):38. https://doi.org/10.1186/s12958-017-0256-3.
125. Schultz KAP, Stewart DR, Kamihara J, et al. DICER1 tumor predisposition. In: Adam MP, Ardinger HH, Pagon RA, et al., eds. *GeneReviews(®)*. Seattle: University of Washington; 1993. Copyright © 1993-2021, University of Washington, Seattle. GeneReviews is a registered trademark of the University of Washington, Seattle. All rights reserved.
126. Yuan S, Ortogero N, Wu Q, Zheng H, Yan W. Murine follicular development requires oocyte DICER, but not DROSHA. *Biol Reprod*. 2014;91(2):39. https://doi.org/10.1095/biolreprod.114.119370.
127. Qin L, Chen J, Tang L, et al. Significant role of dicer and miR-223 in adipose tissue of polycystic ovary syndrome patients. *Biomed Res Int*. 2019;2019:9193236. https://doi.org/10.1155/2019/9193236.
128. Han X, Xue R, Yuan HJ, et al. MicroRNA-21 plays a pivotal role in the oocyte-secreted factor-induced suppression of cumulus cell apoptosis. *Biol Reprod*. 2017;96(6):1167–1180. https://doi.org/10.1093/biolre/iox044.
129. Bartolucci AF, Uliasz T, Peluso JJ. MicroRNA-21 as a regulator of human cumulus cell viability and its potential influence on the developmental potential of the oocyte. *Biol Reprod*. 2020;103(1):94–103. https://doi.org/10.1093/biolre/ioaa058.
130. Liu HY, Zhang YY, Zhu BL, et al. miR-21 regulates the proliferation and apoptosis of ovarian cancer cells through PTEN/PI3K/AKT. *Eur Rev Med Pharmacol Sci*. 2019;23(10):4149–4155. https://doi.org/10.26355/eurrev_201905_17917.
131. Taylor DD, Gercel-Taylor C. MicroRNA signatures of tumor-derived exosomes as diagnostic biomarkers of ovarian cancer. *Gynecol Oncol*. 2008;110(1):13–21. https://doi.org/10.1016/j.ygyno.2008.04.033.
132. Li X, Xie J, Wang Q, Cai H, Xie C, Fu X. miR-21 and pellino-1 expression profiling in autoimmune premature ovarian insufficiency. *J Immunol Res*. 2020;2020:3582648. https://doi.org/10.1155/2020/3582648.
133. Tahir MS, Nguyen LT, Schulz BL, et al. Proteomics recapitulates ovarian proteins relevant to puberty and fertility in brahman heifers (*Bos indicus* L.). *Genes (Basel)*. 2019;10(11). https://doi.org/10.3390/genes10110923.
134. Fleming R, Kelsey TW, Anderson RA, Wallace WH, Nelson SM. Interpreting human follicular recruitment and antimüllerian hormone concentrations throughout life. *Fertil Steril*. 2012;98(5):1097–1102. https://doi.org/10.1016/j.fertnstert.2012.07.1114.

135. Jaffe LA, Egbert JR. Regulation of mammalian oocyte meiosis by intercellular communication within the ovarian follicle. *Annu Rev Physiol.* 2017;79:237–260. https://doi.org/10.1146/annurev-physiol-022516-034102.
136. Hunter AG, Moor RM. Stage-dependent effects of inhibiting ribonucleic acids and protein synthesis on meiotic maturation of bovine oocytes in vitro. *J Dairy Sci.* 1987;70(8):1646–1651. https://doi.org/10.3168/jds.S0022-0302(87)80192-3.
137. Llarena N, Hine C. Reproductive longevity and aging: geroscience approaches to maintain long-term ovarian fitness. *J Gerontol A Biol Sci Med Sci.* 2021;76(9):1551–1560. https://doi.org/10.1093/gerona/glaa204.
138. De Felici M, Klinger FG. PI3K/PTEN/AKT signaling pathways in germ cell development and their involvement in germ cell tumors and ovarian dysfunctions. *Int J Mol Sci.* 2021;22(18). https://doi.org/10.3390/ijms22189838.
139. Zhou Y, Hong F, Wu N, et al. Suppression of ovarian follicle development by nano TiO(2) is associated with TGF-β-mediated signaling pathways. *J Biomed Mater Res A.* 2019;107(2):414–422. https://doi.org/10.1002/jbm.a.36558.
140. Roness H, Gavish Z, Cohen Y, Meirow D. Ovarian follicle burnout: a universal phenomenon? *Cell Cycle.* 2013;12(20):3245–3246. https://doi.org/10.4161/cc.26358.
141. Cha KY, Chian RC. Maturation in vitro of immature human oocytes for clinical use. *Hum Reprod Update.* 1998;4(2):103–120. https://doi.org/10.1093/humupd/4.2.103.
142. Hassaneen A, Naniwa Y, Suetomi Y, et al. Immunohistochemical characterization of the arcuate kisspeptin/neurokinin B/dynorphin (KNDy) and preoptic kisspeptin neuronal populations in the hypothalamus during the estrous cycle in heifers. *J Reprod Dev.* 2016;62(5):471–477. https://doi.org/10.1262/jrd.2016-075.
143. Macedo GG, Mingoti RD, Batista EOS, et al. Profile of LH release in response to intramuscular treatment with kisspeptin in Bos indicus and Bos taurus prepubertal heifers. *Theriogenology.* 2019;125:64–70. https://doi.org/10.1016/j.theriogenology.2018.10.011.
144. Koizumi M, Kadokawa H. Positive correlations of age and parity with plasma anti-Müllerian hormone concentrations in Japanese Black cows. *J Reprod Dev.* 2017;63(2):205–209. https://doi.org/10.1262/jrd.2016-088.
145. Adams GP. Comparative patterns of follicle development and selection in ruminants. *J Reprod Fertil Suppl.* 1999;54:17–32.
146. Xu Z, Garverick HA, Smith GW, Smith MF, Hamilton SA, Youngquist RS. Expression of follicle-stimulating hormone and luteinizing hormone receptor messenger ribonucleic acids in bovine follicles during the first follicular wave. *Biol Reprod.* 1995;53(4):951–957. https://doi.org/10.1095/biolreprod53.4.951.
147. Yamashita H, Murayama C, Takasugi R, Miyamoto A, Shimizu T. BMP-4 suppresses progesterone production by inhibiting histone H3 acetylation of StAR in bovine granulosa cells in vitro. *Mol Cell Biochem.* 2011;348(1-2):183–190. https://doi.org/10.1007/s11010-010-0653-9.
148. Astapova O, Minor BMN, Hammes SR. Physiological and pathological androgen actions in the ovary. *Endocrinology.* 2019;160(5):1166–1174. https://doi.org/10.1210/en.2019-00101.
149. Rajakoski E. The ovarian follicular system in sexually mature heifers with special reference to seasonal, cyclical, end left-right variations. *Acta Endocrinol Suppl (Copenh).* 1960;34(suppl 52):1–68.
150. Pierson RA, Ginther OJ. Ultrasonography of the bovine ovary. *Theriogenology.* 1984;21(3):495–504. https://doi.org/10.1016/0093-691x(84)90411-4.
151. Queenan JT, O'Brien GD, Bains LM, Simpson J, Collins WP, Campbell S. Ultrasound scanning of ovaries to detect ovulation in women. *Fertil Steril.* 1980;34(2):99–105.
152. Fleischer AC, Daniell JF, Rodier J, Lindsay AM, James Jr AE. Sonographic monitoring of ovarian follicular development. *J Clin Ultrasound.* 1981;9(6):275–280. https://doi.org/10.1002/jcu.1870090605.

153. Pierson RA, Ginther OJ. Follicular populations during the estrous cycle in heifers. I. Influence of day. *Anim Reprod Sci.* 1987;14(3):165–176. https://doi.org/10.1016/0378-4320(87)90080-7.
154. Pierson RA, Ginther OJ. Ovarian follicular populations during early pregnancy in heifers. *Theriogenology.* 1986;26(5):649–659. https://doi.org/10.1016/0093-691x(86)90173-1.
155. Pierson RA, Ginther OJ. Follicular populations during the estrous cycle in heifers. II. Influence of right and left sides and intraovarian effect of the corpus luteum. *Anim Reprod Sci.* 1987;14(3):177–186. https://doi.org/10.1016/0378-4320(87)90081-9.
156. Pierson RA, Ginther OJ. Follicular populations during the estrous cycle in heifers. III. Time of selection of the ovulatory follicle. *Anim Reprod Sci.* 1988;16(2):81–95. https://doi.org/10.1016/0378-4320(88)90029-2.
157. Sirois J, Fortune JE. Ovarian follicular dynamics during the estrous cycle in heifers monitored by real-time ultrasonography. *Biol Reprod.* 1988;39(2):308–317. https://doi.org/10.1095/biolreprod39.2.308.
158. Savio JD, Keenan L, Boland MP, Roche JF. Pattern of growth of dominant follicles during the oestrous cycle of heifers. *J Reprod Fertil.* 1988;83(2):663–671. https://doi.org/10.1530/jrf.0.0830663.
159. Andreotti RF, Thompson GH, Janowitz W, Shapiro AG, Zusmer NR. Endovaginal and transabdominal sonography of ovarian follicles. *J Ultrasound Med.* 1989;8(10):555–560. https://doi.org/10.7863/jum.1989.8.10.555.
160. Pache TD, Wladimiroff JW, de Jong FH, Hop WC, Fauser BC. Growth patterns of nondominant ovarian follicles during the normal menstrual cycle. *Fertil Steril.* 1990;54(4):638–642. https://doi.org/10.1016/s0015-0282(16)53821-7.
161. de Mello Bianchi PH, Serafini P, Monteiro da Rocha A, et al. Review: follicular waves in the human ovary: a new physiological paradigm for novel ovarian stimulation protocols. *Reprod Sci.* 2010;17(12):1067–1076. https://doi.org/10.1177/1933719110366483.
162. Knopf L, Kastelic JP, Schallenberger E, Ginther OJ. Ovarian follicular dynamics in heifers: test of two-wave hypothesis by ultrasonically monitoring individual follicles. *Domest Anim Endocrinol.* 1989;6(2):111–119. https://doi.org/10.1016/0739-7240(89)90040-4.
163. Aboelenain M, Kawahara M, Balboula AZ, et al. Status of autophagy, lysosome activity and apoptosis during corpus luteum regression in cattle. *J Reprod Dev.* 2015;61(3):229–236. https://doi.org/10.1262/jrd.2014-135.
164. Kinnear HM, Tomaszewski CE, Chang FL, et al. The ovarian stroma as a new frontier. *Reproduction.* 2020;160(3):R25–r39. https://doi.org/10.1530/rep-19-0501.
165. Pérez-Martínez C, García-Fernández RA, Escudero A, Ferreras MC, García-Iglesias MJ. Expression of cytokeratins and vimentin in normal and neoplastic tissue from the bovine female reproductive tract. *J Comp Pathol.* 2001;124(1):70–78. https://doi.org/10.1053/jcpa.2000.0430.
166. Theocharis AD, Skandalis SS, Gialeli C, Karamanos NK. Extracellular matrix structure. *Adv Drug Deliv Rev.* 2016;97:4–27. https://doi.org/10.1016/j.addr.2015.11.001.
167. Zhou F, Shi LB, Zhang SY. Ovarian fibrosis: a phenomenon of concern. *Chin Med J (Engl).* 2017;130(3):365–371. https://doi.org/10.4103/0366-6999.198931.
168. Han P, Relav L, Price CA. Regulation of the early growth response-1 binding protein NAB2 in bovine granulosa cells and effect on connective tissue growth factor expression. *Mol Cell Endocrinol.* 2020;518:111041. https://doi.org/10.1016/j.mce.2020.111041.
169. Hatzirodos N, Irving-Rodgers HF, Hummitzsch K, Harland ML, Morris SE, Rodgers RJ. Transcriptome profiling of granulosa cells of bovine ovarian follicles during growth from small to large antral sizes. *BMC Genom.* 2014;15:24. https://doi.org/10.1186/1471-2164-15-24.

170. Trevisol E, Mogollón García HD, Ackermann CL, et al. Partial luteolysis during early diestrus in cattle downregulates VEGFA expression and reduces large luteal cell and corpus luteum sizes and plasma progesterone concentration. *Theriogenology*. 2020; 158:188–195. https://doi.org/10.1016/j.theriogenology.2020.09.015.
171. Stassi AF, Gasser F, Velázquez MML, et al. Contribution of the VEGF system to the follicular persistence associated with bovine cystic ovaries. *Theriogenology*. 2019;138:52–65. https://doi.org/10.1016/j.theriogenology.2019.07.002.
172. Hansen KR, Hodnett GM, Knowlton N, Craig LB. Correlation of ovarian reserve tests with histologically determined primordial follicle number. *Fertil Steril*. 2011;95 (1):170–175. https://doi.org/10.1016/j.fertnstert.2010.04.006.
173. Tutt DAR, Silvestri G, Serrano-Albal M, et al. Analysis of bovine blastocysts indicates ovarian stimulation does not induce chromosome errors, nor discordance between inner-cell mass and trophectoderm lineages. *Theriogenology*. 2021;161:108–119. https://doi.org/10.1016/j.theriogenology.2020.11.021.
174. Stevenson JC. A woman's journey through the reproductive, transitional and postmenopausal periods of life: impact on cardiovascular and musculo-skeletal risk and the role of estrogen replacement. *Maturitas*. 2011;70(2):197–205. https://doi.org/10.1016/j.maturitas.2011.05.017.
175. Jamil Z, Fatima SS, Ahmed K, Malik R. Anti-Mullerian hormone: above and beyond conventional ovarian reserve markers. *Dis Markers*. 2016;2016:5246217. https://doi.org/10.1155/2016/5246217.
176. Santoro N, Johnson J. Diagnosing the onset of menopause. *JAMA*. 2019;322 (8):775–776. https://doi.org/10.1001/jama.2019.6250.
177. Llerena Cari E, Hagen-Lillevik S, Giornazi A, et al. Integrated stress response control of granulosa cell translation and proliferation during normal ovarian follicle development. *Mol Hum Reprod*. 2021;27(8). https://doi.org/10.1093/molehr/gaab050.
178. Erickson BH, Reynolds RA, Murphree RL. Ovarian characteristics and reproductive performance of the aged cow. *Biol Reprod*. 1976;15(4):555–560. https://doi.org/10.1095/biolreprod15.4.555.
179. Alvarez RH, Bayeux BM, Joaquim DA, Watanabe YF, Humblot P. Antral follicle count, oocyte production and embryonic developmental competence of senescent Nellore (Bos indicus) cows. *Theriogenology*. 2021;174:27–35. https://doi.org/10.1016/j.theriogenology.2021.08.016.
180. Hunt PA, Hassold TJ. Human female meiosis: what makes a good egg go bad? *Trends Genet*. 2008;24(2):86–93. https://doi.org/10.1016/j.tig.2007.11.010.
181. Dias FC, Khan MI, Adams GP, Sirard MA, Singh J. Granulosa cell function and oocyte competence: super-follicles, super-moms and super-stimulation in cattle. *Anim Reprod Sci*. 2014;149(1-2):80–89. https://doi.org/10.1016/j.anireprosci.2014.07.016.
182. Bosch E, Labarta E, Kolibianakis E, Rosen M, Meldrum D. Regimen of ovarian stimulation affects oocyte and therefore embryo quality. *Fertil Steril*. 2016;105 (3):560–570. https://doi.org/10.1016/j.fertnstert.2016.01.022.
183. Lodde V, Luciano AM, Musmeci G, et al. A nuclear and cytoplasmic characterization of bovine oocytes reveals that cysteamine partially rescues the embryo development in a model of low ovarian reserve. *Animals (Basel)*. 2021;11(7). https://doi.org/10.3390/ani11071936.
184. Dupont J, Scaramuzzi RJ, Reverchon M. The effect of nutrition and metabolic status on the development of follicles, oocytes and embryos in ruminants. *Animal*. 2014;8 (7):1031–1044. https://doi.org/10.1017/s1751731114000937.
185. Morrison AE, Fleming S, Levy MJ. A review of the pathophysiology of functional hypothalamic amenorrhoea in women subject to psychological stress, disordered eating, excessive exercise or a combination of these factors. *Clin Endocrinol (Oxf)*. 2021;95 (2):229–238. https://doi.org/10.1111/cen.14399.

186. Dupont J, Scaramuzzi RJ. Insulin signalling and glucose transport in the ovary and ovarian function during the ovarian cycle. *Biochem J.* 2016;473(11):1483–1501. https://doi.org/10.1042/bcj20160124.
187. Silvestris E, Lovero D, Palmirotta R. Nutrition and female fertility: an interdependent correlation. *Front Endocrinol (Lausanne).* 2019;10:346. https://doi.org/10.3389/fendo.2019.00346.
188. Yenuganti VR, Koczan D, Vanselow J. Genome wide effects of oleic acid on cultured bovine granulosa cells: evidence for the activation of pathways favoring folliculo-luteal transition. *BMC Genomics.* 2021;22(1):486. https://doi.org/10.1186/s12864-021-07817-6.
189. Correa-Calderón A, Hernández-Rivera JA, Avendaño-Reyes L, Diaz-Molina R, Macias-Cruz U. Progesterone supplementation in Holstein heifers subjected to cooling and timed AI during summer: physiological and reproductive variables and thyroid hormone concentrations. *Trop Anim Health Prod.* 2021;53(2):249. https://doi.org/10.1007/s11250-021-02688-1.
190. Luderer U. Ovarian toxicity from reactive oxygen species. *Vitam Horm.* 2014; 94:99–127. https://doi.org/10.1016/b978-0-12-800095-3.00004-3.
191. Nazari A, Dirandeh E, Ansari-Pirsaraei Z, Deldar H. Antioxidant levels, copper and zinc concentrations were associated with postpartum luteal activity, pregnancy loss and pregnancy status in Holstein dairy cows. *Theriogenology.* 2019;133:97–103. https://doi.org/10.1016/j.theriogenology.2019.04.034.
192. Dodé C, Hardelin JP. Kallmann syndrome. *Eur J Hum Genet.* 2009;17(2):139–146. https://doi.org/10.1038/ejhg.2008.206.
193. Cree LM, Hammond ER, Shelling AN, Berg MC, Peek JC, Green MP. Maternal age and ovarian stimulation independently affect oocyte mtDNA copy number and cumulus cell gene expression in bovine clones. *Hum Reprod.* 2015;30(6):1410–1420. https://doi.org/10.1093/humrep/dev066.
194. Mishra SR. Behavioural, physiological, neuro-endocrine and molecular responses of cattle against heat stress: an updated review. *Trop Anim Health Prod.* 2021;53(3):400. https://doi.org/10.1007/s11250-021-02790-4.
195. Roth Z. Symposium review: reduction in oocyte developmental competence by stress is associated with alterations in mitochondrial function. *J Dairy Sci.* 2018;101(4):3642–3654. https://doi.org/10.3168/jds.2017-13389.
196. Ticianelli JS, Emanuelli IP, Satrapa RA, et al. Gene expression profile in heat-shocked Holstein and Nelore oocytes and cumulus cells. *Reprod Fertil Dev.* 2017;29(9):1787–1802. https://doi.org/10.1071/rd16154.
197. Aljaff P, Rasheed BO, Salh DM. Assessment of heavy metals in livers of cattle and chicken by spectroscopic method. *IOSR J Appl Phys.* 2014;6(1):23–26.
198. Xu G, Liu S, Huang M, Jiang X, Yang M. Cadmium induces apoptosis of human granulosa cell line KGN via mitochondrial dysfunction-mediated pathways. *Ecotoxicol Environ Saf.* 2021;220:112341. https://doi.org/10.1016/j.ecoenv.2021.112341.
199. Kim K, Pollack AZ, Nobles CJ, et al. Associations between blood cadmium and endocrine features related to PCOS-phenotypes in healthy women of reproductive age: a prospective cohort study. *Environ Health.* 2021;20(1):64. https://doi.org/10.1186/s12940-021-00749-4.
200. Soares MB, Izaguirry AP, Vargas LM, Mendez AS, Spiazzi CC, Santos FW. Catechins are not major components responsible for the beneficial effect of Camellia sinensis on the ovarian δ-ALA-D activity inhibited by cadmium. *Food Chem Toxicol.* 2013; 55:463–469. https://doi.org/10.1016/j.fct.2013.01.032.
201. Silva LA, de Mello MRB, Oliveira Pião D, et al. Effects of experimental exposure to zearalenone on reproductive system morphometry, plasma oestrogen levels, and oocyte quality of beef heifer. *Reprod Domest Anim.* 2021;56(5):775–782. https://doi.org/10.1111/rda.13917.

202. Takagi M, Mukai S, Kuriyagawa T, et al. Detection of zearalenone and its metabolites in naturally contaminated follicular fluids by using LC/MS/MS and in vitro effects of zearalenone on oocyte maturation in cattle. *Reprod Toxicol.* 2008;26(2):164–169. https://doi.org/10.1016/j.reprotox.2008.08.006.
203. Wang S, Panter KE, Gaffield W, Evans RC, Bunch TD. Effects of steroidal glycoalkaloids from potatoes (Solanum tuberosum) on in vitro bovine embryo development. *Anim Reprod Sci.* 2005;85(3-4):243–250. https://doi.org/10.1016/j.anireprosci.2004.06.002.
204. Bai C, Liu H, Liu Y, et al. Diploid oocyte formation and tetraploid embryo development induced by cytochalasin B in bovine. *Cell Reprogram.* 2011;13(1):37–45. https://doi.org/10.1089/cell.2010.0038.
205. Kinkade CW, Rivera-Núñez Z, Gorcyzca L, Aleksunes LM, Barrett ES. Impact of fusarium-derived mycoestrogens on female reproduction: a systematic review. *Toxins (Basel).* 2021;13(6). https://doi.org/10.3390/toxins13060373.
206. Sleiman HK, de Oliveira JM, Langoni de Freitas GB. Isoflavones alter male and female fertility in different development windows. *Biomed Pharmacother.* 2021;140: 111448. https://doi.org/10.1016/j.biopha.2021.111448.
207. Adams NR. Detection of the effects of phytoestrogens on sheep and cattle. *J Anim Sci.* 1995;73(5):1509–1515. https://doi.org/10.2527/1995.7351509x.
208. Konieczna A, Rachoń D, Owczarek K, et al. Serum bisphenol A concentrations correlate with serum testosterone levels in women with polycystic ovary syndrome. *Reprod Toxicol.* 2018;82:32–37. https://doi.org/10.1016/j.reprotox.2018.09.006.
209. Guo Z, Qiu H, Wang L, et al. Association of serum organochlorine pesticides concentrations with reproductive hormone levels and polycystic ovary syndrome in a Chinese population. *Chemosphere.* 2017;171:595–600. https://doi.org/10.1016/j.chemosphere.2016.12.127.
210. Młynarczuk J, Wróbel MH, Kotwica J. Adverse influence of coumestrol on secretory function of bovine luteal cells in the first trimester of pregnancy. *Environ Toxicol.* 2013;28(7):411–418. https://doi.org/10.1002/tox.20735.
211. Lovekamp TN, Davis BJ. Mono-(2-ethylhexyl) phthalate suppresses aromatase transcript levels and estradiol production in cultured rat granulosa cells. *Toxicol Appl Pharmacol.* 2001;172(3):217–224. https://doi.org/10.1006/taap.2001.9156.
212. Liao C, Liu F, Alomirah H, et al. Bisphenol S in urine from the United States and seven Asian countries: occurrence and human exposures. *Environ Sci Technol.* 2012;46 (12):6860–6866. https://doi.org/10.1021/es301334j.
213. Paoli D, Pallotti F, Dima AP, et al. Phthalates and bisphenol A: presence in blood serum and follicular fluid of Italian women undergoing assisted reproduction techniques. *Toxics.* 2020;8(4). https://doi.org/10.3390/toxics8040091.
214. Sabry R, Apps C, Reiter-Saunders JA, et al. BPA and BPS affect connexin 37 in bovine cumulus cells. *Genes (Basel).* 2021;12(2). https://doi.org/10.3390/genes12020321.
215. Lupton SJ, Huwe JK, Smith DJ, Dearfield KL, Johnston JJ. Distribution and excretion of perfluorooctane sulfonate (PFOS) in beef cattle (*Bos taurus*). *J Agric Food Chem.* 2014;62(5):1167–1173. https://doi.org/10.1021/jf404355b.
216. Calafat AM, Wong LY, Kuklenyik Z, Reidy JA, Needham LL. Polyfluoroalkyl chemicals in the U.S. population: data from the National Health and Nutrition Examination Survey (NHANES) 2003-2004 and comparisons with NHANES 1999-2000. *Environ Health Perspect.* 2007;115(11):1596–1602. https://doi.org/10.1289/ehp.10598.
217. Barrett ES, Chen C, Thurston SW, et al. Perfluoroalkyl substances and ovarian hormone concentrations in naturally cycling women. *Fertil Steril.* 2015;103(5). https://doi.org/10.1016/j.fertnstert.2015.02.001. 1261-70.e3.
218. Lupton SJ, Dearfield KL, Johnston JJ, Wagner S, Huwe JK. Perfluorooctane sulfonate plasma half-life determination and long-term tissue distribution in beef cattle (*Bos*

taurus). *J Agric Food Chem*. 2015;63(51):10988–10994. https://doi.org/10.1021/acs.jafc.5b04565.

219. Paulose T, Tannenbaum LV, Borgeest C, Flaws JA. Methoxychlor-induced ovarian follicle toxicity in mice: dose and exposure duration-dependent effects. *Birth Defects Res B Dev Reprod Toxicol*. 2012;95(3):219–224. https://doi.org/10.1002/bdrb.21007.
220. Alarcón R, Ingaramo PI, Rivera OE, et al. Neonatal exposure to a glyphosate-based herbicide alters the histofunctional differentiation of the ovaries and uterus in lambs. *Mol Cell Endocrinol*. 2019;482:45–56. https://doi.org/10.1016/j.mce.2018.12.007.
221. Cho HW, Lee S, Min KJ, et al. Advances in the treatment and prevention of chemotherapy-induced ovarian toxicity. *Int J Mol Sci*. 2020;21(20). https://doi.org/10.3390/ijms21207792.
222. Ruth KS, Perry JR, Henley WE, Melzer D, Weedon MN, Murray A. Events in early life are associated with female reproductive ageing: a UK biobank study. *Sci Rep*. 2016;6:24710. https://doi.org/10.1038/srep24710.
223. Spears N, Lopes F, Stefansdottir A, et al. Ovarian damage from chemotherapy and current approaches to its protection. *Hum Reprod*. 2019;25(6):673–693. https://doi.org/10.1093/humupd/dmz027.
224. Himelstein-Braw R, Peters H, Faber M. Morphological study of the ovaries of leukaemic children. *Br J Cancer*. 1978;38(1):82–87. https://doi.org/10.1038/bjc.1978.166.
225. Roness H, Kashi O, Meirow D. Prevention of chemotherapy-induced ovarian damage. *Fertil Steril*. 2016;105(1):20–29. https://doi.org/10.1016/j.fertnstert.2015.11.043.
226. Meirow D, Dor J, Kaufman B, et al. Cortical fibrosis and blood-vessels damage in human ovaries exposed to chemotherapy. Potential mechanisms of ovarian injury. *Hum Reprod*. 2007;22(6):1626–1633. https://doi.org/10.1093/humrep/dem027.
227. Kafi M, Azari M, Chashnigir O, et al. Inherent inferior quality of follicular fluid in repeat breeder heifers as evidenced by low rates of in vitro production of bovine embryos. *Theriogenology*. 2017;102:29–34. https://doi.org/10.1016/j.theriogenology.2017.07.011.
228. Romito A, Bove S, Romito I, et al. Ovarian reserve after chemotherapy in breast cancer: a systematic review and meta-analysis. *J Pers Med*. 2021;11(8). https://doi.org/10.3390/jpm11080704.
229. Chon SJ, Umair Z, Yoon MS. Premature ovarian insufficiency: past, present, and future. *Front Cell Dev Biol*. 2021;9. https://doi.org/10.3389/fcell.2021.672890, 672890.
230. Rudnicka E, Kruszewska J, Klicka K, et al. Premature ovarian insufficiency—aetiopathology, epidemiology, and diagnostic evaluation. *Prz Menopauzalny*. 2018;17(3):105–108. https://doi.org/10.5114/pm.2018.78550.
231. Lakhal B, Braham R, Berguigua R, et al. Cytogenetic analyses of premature ovarian failure using karyotyping and interphase fluorescence in situ hybridization (FISH) in a group of 1000 patients. *Clin Genet*. 2010;78(2):181–185. https://doi.org/10.1111/j.1399-0004.2009.01359.x.
232. França MM, Mendonca BB. Genetics of ovarian insufficiency and defects of folliculogenesis. *Best Pract Res Clin Endocrinol Metab*. 2021;14:101594. https://doi.org/10.1016/j.beem.2021.101594.
233. Mayer A, Fouquet B, Pugeat M, Misrahi M. BMP15 "knockout-like" effect in familial premature ovarian insufficiency with persistent ovarian reserve. *Clin Genet*. 2017;92(2):208–212. https://doi.org/10.1111/cge.12970.
234. Eskenazi S, Bachelot A, Hugon-Rodin J, et al. Next generation sequencing should be proposed to every woman with "idiopathic" primary ovarian insufficiency. *J Endocr Soc*. 2021;5(7):bvab032. https://doi.org/10.1210/jendso/bvab032.
235. Santoro N. Mechanisms of premature ovarian failure. *Ann Endocrinol (Paris)*. 2003;64(2):87–92.

236. Shareghi-Oskoue O, Aghebati-Maleki L, Yousefi M. Transplantation of human umbilical cord mesenchymal stem cells to treat premature ovarian failure. *Stem Cell Res Ther.* 2021;12(1):454. https://doi.org/10.1186/s13287-021-02529-w.
237. Nelson LM. Clinical practice. Primary ovarian insufficiency. *N Engl J Med.* 2009;360 (6):606–614. https://doi.org/10.1056/NEJMcp0808697.
238. Piedade KC, Spencer H, Persani L, Nelson LM. Optimizing fertility in primary ovarian insufficiency: case report and literature review. *Front Genet.* 2021;12:676262. https://doi.org/10.3389/fgene.2021.676262.
239. Ferrarini E, De Marco G, Orsolini F, et al. Characterization of a novel mutation V136L in bone morphogenetic protein 15 identified in a woman affected by POI. *J Ovarian Res.* 2021;14(1):85. https://doi.org/10.1186/s13048-021-00836-7.
240. Fortune JE. The early stages of follicular development: activation of primordial follicles and growth of preantral follicles. *Anim Reprod Sci.* 2003;78(3-4):135–163. https://doi.org/10.1016/s0378-4320(03)00088-5.
241. Manasa P, Sidhanth C, Krishnapriya S, Vasudevan S, Ganesan TS. Oncogenes in high grade serous adenocarcinoma of the ovary. *Genes Cancer.* 2020;11(3-4):122–136. https://doi.org/10.18632/genesandcancer.206.
242. Ortega MS, Wohlgemuth S, Tribulo P, Siqueira LG, Cole JB, Hansen PJ. A single nucleotide polymorphism in COQ9 affects mitochondrial and ovarian function and fertility in Holstein cows. *Biol Reprod.* 2017;96(3):652–663. https://doi.org/10.1093/biolre/iox004.
243. Kasippillai T, MacArthur DG, Kirby A, et al. Mutations in eIF4ENIF1 are associated with primary ovarian insufficiency. *J Clin Endocrinol Metab.* 2013;98(9):E1534–E1539. https://doi.org/10.1210/jc.2013-1102.
244. Tripurani SK, Wee G, Lee KB, Smith GW, Wang L, Jianboyao. MicroRNA-212 post-transcriptionally regulates oocyte-specific basic-helix-loop-helix transcription factor, factor in the germline alpha (FIGLA), during bovine early embryogenesis. *PLoS One.* 2013;8(9):e76114. https://doi.org/10.1371/journal.pone.0076114.
245. Yenuganti VR, Vanselow J. Oleic acid induces down-regulation of the granulosa cell identity marker FOXL2, and up-regulation of the Sertoli cell marker SOX9 in bovine granulosa cells. *Reprod Biol Endocrinol.* 2017;15(1):57. https://doi.org/10.1186/s12958-017-0276-z.
246. Sharma A, Baddela VS, Roettgen V, et al. Effects of dietary fatty acids on bovine oocyte competence and granulosa cells. *Front Endocrinol (Lausanne).* 2020;11:87. https://doi.org/10.3389/fendo.2020.00087.
247. Tang X, Wang P, He Y, et al. Primitive neuroectodermal tumor in female genital tract: a clinicopathologic study. *Zhonghua Bing Li Xue Za Zhi.* 2012;41:729–732. https://doi.org/10.3760/cma.j.issn.0529-5807.2012.11.004.
248. Michalovic L, Currin L, Gutierrez K, et al. Granulosa cells of prepubertal cattle respond to gonadotropin signaling and upregulate genes that promote follicular growth and prevent cell apoptosis. *Mol Reprod Dev.* 2018;85:909–920. https://doi.org/10.1002/mrd.23066.
249. Paulini F, Melo EO. Effects of growth and differentiation factor 9 and bone morphogenetic protein 15 overexpression on the steroidogenic metabolism in bovine granulosa cells in vitro. *Reprod Domest Anim.* 2021;56(6):837–847. https://doi.org/10.1111/rda.13923.
250. Fu L, Zhang M, Mastrantoni K, Perfetto M, Wei S, Yao J. Bovine Lhx8, a germ cell-specific nuclear factor, interacts with figla. *PLoS One.* 2016;11(10):e0164671. https://doi.org/10.1371/journal.pone.0164671.
251. Tripurani SK, Lee KB, Wang L, et al. A novel functional role for the oocyte-specific transcription factor newborn ovary homeobox (NOBOX) during early embryonic development in cattle. *Endocrinology.* 2011;152(3):1013–1023. https://doi.org/10.1210/en.2010-1134.

252. Walther N, Jansen M, Akbary W, Ivell R. Differentiation-specific action of orphan nuclear receptor NR5A1 (SF-1): transcriptional regulation in luteinizing bovine theca cells. *Reprod Biol Endocrinol.* 2006;4:64. https://doi.org/10.1186/1477-7827-4-64.
253. Terzaghi L, Luciano AM, Dall'Acqua PC, Modina SC, Peluso JJ, Lodde V. PGRMC1 localization and putative function in the nucleolus of bovine granulosa cells and oocytes. *Reproduction.* 2018;155(3):273–282. https://doi.org/10.1530/rep-17-0534.
254. Luciano AM, Corbani D, Lodde V, et al. Expression of progesterone receptor membrane component-1 in bovine reproductive system during estrous cycle. *Eur J Histochem.* 2011;55(3):e27. https://doi.org/10.4081/ejh.2011.e27.
255. Kowalik MK, Kotwica J. Progesterone receptor membrane component 1 (PGRMC1) gene expression in corpus luteum during the estrous cycle in cows. *Reprod Biol.* 2008;8 (3):291–297. https://doi.org/10.1016/s1642-431x(12)60019-9.
256. Weller M, Fortes MRS, Marcondes MI, et al. Effect of maternal nutrition and days of gestation on pituitary gland and gonadal gene expression in cattle. *J Dairy Sci.* 2016;99 (4):3056–3071. https://doi.org/10.3168/jds.2015-9673.
257. Yan C, Wang P, DeMayo J, et al. Synergistic roles of bone morphogenetic protein 15 and growth differentiation factor 9 in ovarian function. *Mol Endocrinol.* 2001;15 (6):854–866. https://doi.org/10.1210/mend.15.6.0662.
258. Sun M, Sheng Y, Ma Z, Chen Z, Tang R. Correlation analysis between polycystic ovary syndrome susceptibility genes and metabolic phenotypes. *Zhonghua Fu Chan Ke Za Zhi.* 2014;49(6):441–445.
259. Fonseca PAS, Dos Santos FC, Lam S, et al. Genetic mechanisms underlying spermatic and testicular traits within and among cattle breeds: systematic review and prioritization of GWAS results. *J Anim Sci.* 2018;96(12):4978–4999. https://doi.org/10.1093/jas/sky382.
260. Knochenhauer ES, Key TJ, Kahsar-Miller M, Waggoner W, Boots LR, Azziz R. Prevalence of the polycystic ovary syndrome in unselected black and white women of the southeastern United States: a prospective study. *J Clin Endocrinol Metab.* 1998;83 (9):3078–3082. https://doi.org/10.1210/jcem.83.9.5090.
261. Goldzieher JW. Polycystic ovarian disease. *Fertil Steril.* 1981;35(4):371–394. https://doi.org/10.1016/s0015-0282(16)45429-4.
262. Farquhar CM, Birdsall M, Manning P, Mitchell JM, France JT. The prevalence of polycystic ovaries on ultrasound scanning in a population of randomly selected women. *Aust N Z J Obstet Gynaecol.* 1994;34(1):67–72. https://doi.org/10.1111/j.1479-828X.1994.tb01041.x.
263. Clayton RN, Ogden V, Hodgkinson J, et al. How common are polycystic ovaries in normal women and what is their significance for the fertility of the population? *Clin Endocrinol (Oxf).* 1992;37(2):127–134. https://doi.org/10.1111/j.1365-2265.1992.tb02296.x.
264. Polson DW, Adams J, Wadsworth J, Franks S. Polycystic ovaries—a common finding in normal women. *Lancet.* 1988;1(8590):870–872. https://doi.org/10.1016/s0140-6736(88)91612-1.
265. Lujan ME, Chizen DR, Pierson RA. Diagnostic criteria for polycystic ovary syndrome: pitfalls and controversies. *J Obstet Gynaecol Can.* 2008;30(8):671–679. https://doi.org/10.1016/s1701-2163(16)32915-2.
266. Osuka S, Nakanishi N, Murase T, et al. Animal models of polycystic ovary syndrome: a review of hormone-induced rodent models focused on hypothalamus-pituitary-ovary axis and neuropeptides. *Reprod Med Biol.* 2019;18(2):151–160. https://doi.org/10.1002/rmb2.12262.
267. Ryu Y, Kim SW, Kim YY, Ku SY. Animal models for human polycystic ovary syndrome (PCOS) focused on the use of indirect hormonal perturbations: a review of the literature. *Int J Mol Sci.* 2019;20(11). https://doi.org/10.3390/ijms20112720.

268. Franks S. Do animal models of polycystic ovary syndrome help to understand its pathogenesis and management? Yes, but their limitations should be recognized. *Endocrinology*. 2009;150(9):3983–3985. https://doi.org/10.1210/en.2009-0652.
269. Rosenfield RL, Ehrmann DA. The pathogenesis of polycystic ovary syndrome (PCOS): the hypothesis of PCOS as functional ovarian hyperandrogenism revisited. *Endocr Rev*. 2016;37(5):467–520. https://doi.org/10.1210/er.2015-1104.
270. Shi D, Dyck MK, Uwiera RR, Russell JC, Proctor SD, Vine DF. A unique rodent model of cardiometabolic risk associated with the metabolic syndrome and polycystic ovary syndrome. *Endocrinology*. 2009;150(9):4425–4436. https://doi.org/10.1210/en.2008-1612.
271. Couse JF, Bunch DO, Lindzey J, Schomberg DW, Korach KS. Prevention of the polycystic ovarian phenotype and characterization of ovulatory capacity in the estrogen receptor-alpha knockout mouse. *Endocrinology*. 1999;140:5855–5865. https://doi.org/10.1210/endo.140.12.7222.
272. Fisher CR, Graves KH, Parlow AF, Simpson ER. Characterization of mice deficient in aromatase (ArKO) because of targeted disruption of the cyp19 gene. *Proceedings of the National Academy of Sciences of the United States of America*. 1998;95(12):6965–6970. https://doi.org/10.1073/pnas.95.12.6965.
273. Risma KA, Hirshfield AN, Nilson JH. Elevated luteinizing hormone in prepubertal transgenic mice causes hyperandrogenemia, precocious puberty, and substantial ovarian pathology. *Endocrinology*. 1997;138(8):3540–3547. https://doi.org/10.1210/endo.138.8.5313.
274. Blank SK, McCartney CR, Marshall JC. The origins and sequelae of abnormal neuroendocrine function in polycystic ovary syndrome. *Hum Reprod*. 2006;12(4):351–361. https://doi.org/10.1093/humupd/dml017.
275. Goodarzi MO, Dumesic DA, Chazenbalk G, Azziz R. Polycystic ovary syndrome: etiology, pathogenesis and diagnosis. *Nat Rev Endocrinol*. 2011;7(4):219–231. https://doi.org/10.1038/nrendo.2010.217.
276. Jones MR, Goodarzi MO. Genetic determinants of polycystic ovary syndrome: progress and future directions. *Fertil Steril*. 2016;106(1):25–32. https://doi.org/10.1016/j.fertnstert.2016.04.040.
277. Khalaf M, Morera J, Bourret A, et al. BMP system expression in GCs from polycystic ovary syndrome women and the in vitro effects of BMP4, BMP6, and BMP7 on GC steroidogenesis. *Eur J Endocrinol*. 2013;168(3):437–444. https://doi.org/10.1530/eje-12-0891.
278. Park JH, Choi TS. Polycystic ovary syndrome (PCOS)-like phenotypes in the d-galactose-induced aging mouse model. *Biochem Biophys Res Commun*. 2012;427(4):701–704. https://doi.org/10.1016/j.bbrc.2012.09.099.
279. Bernuci MP, Szawka RE, Helena CV, Leite CM, Lara HE, Anselmo-Franci JA. Locus coeruleus mediates cold stress-induced polycystic ovary in rats. *Endocrinology*. 2008;149(6):2907–2916. https://doi.org/10.1210/en.2007-1254.
280. Kafali H, Iriadam M, Ozardali I, Demir N. Letrozole-induced polycystic ovaries in the rat: a new model for cystic ovarian disease. *Arch Med Res*. 2004;35(2):103–108. https://doi.org/10.1016/j.arcmed.2003.10.005.
281. Diamanti-Kandarakis E, Dunaif A. Insulin resistance and the polycystic ovary syndrome revisited: an update on mechanisms and implications. *Endocr Rev*. 2012;33(6):981–1030. https://doi.org/10.1210/er.2011-1034.
282. Mohammadi M. Oxidative stress and polycystic ovary syndrome: a brief review. *Int J Prev Med*. 2019;10:86. https://doi.org/10.4103/ijpvm.IJPVM_576_17.
283. Murri M, Luque-Ramírez M, Insenser M, Ojeda-Ojeda M, Escobar-Morreale HF. Circulating markers of oxidative stress and polycystic ovary syndrome (PCOS): a systematic review and meta-analysis. *Hum Reprod*. 2013;19(3):268–288. https://doi.org/10.1093/humupd/dms059.

284. González F, Rote NS, Minium J, Kirwan JP. Reactive oxygen species-induced oxidative stress in the development of insulin resistance and hyperandrogenism in polycystic ovary syndrome. *J Clin Endocrinol Metab.* 2006;91(1):336–340. https://doi.org/10.1210/jc.2005-1696.
285. Ho SC, Liu JH, Wu RY. Establishment of the mimetic aging effect in mice caused by D-galactose. *Biogerontology.* 2003;4(1):15–18. https://doi.org/10.1023/a:1022417102206.
286. Du C, Chen X. Transcriptome profiling of oocytes at the germinal vesicle stage from women from Mongolia with polycystic ovary syndrome. *Int J Gen Med.* 2021; 14:4469–4478. https://doi.org/10.2147/ijgm.S321853.
287. Liu Q, Kong L, Zhang J, et al. Involvement of GJA1 and gap junctional intercellular communication between cumulus cells and oocytes from women with PCOS. *Biomed Res Int.* 2020;2020:5403904. https://doi.org/10.1155/2020/5403904.
288. Nelson ST, Martin AD, Østerås O. Risk factors associated with cystic ovarian disease in Norwegian dairy cattle. *Acta Vet Scand.* 2010;52(1):60. https://doi.org/10.1186/1751-0147-52-60.
289. Casida LE, McShan WH, Meyer RK. Effects of an unfractionated pituitary extract upon cystic ovaries and nymphomania in cows. *J Anim Sci.* 1944;3(3):273–282. https://doi.org/10.2527/jas1944.33273x.
290. Garm O. Investigations on cystic ovarian degeneration in the cow, with special regard to etiology and pathogenesis. *Cornell Vet.* 1949;39(1):39–52.
291. Zulu VC, Sawamukai Y, Nakada K, Kida K, Moriyoshi M. Relationship among insulin-like growth factor-I, blood metabolites and postpartum ovarian function in dairy cows. *J Vet Med Sci.* 2002;64:879–885. https://doi.org/10.1292/jvms.64.879.
292. Ortega HH, Marelli BE, Rey F, et al. Molecular aspects of bovine cystic ovarian disease pathogenesis. *Reproduction.* 2015;149(6):R251–R264. https://doi.org/10.1530/REP-14-0618.
293. Vanholder T, Opsomer G, de Kruif A. Aetiology and pathogenesis of cystic ovarian follicles in dairy cattle: a review. *Reprod Nutr Dev.* 2006;46(2):105–119. https://doi.org/10.1051/rnd:2006003.
294. López-Gatius F, Santolaria P, Yániz J, Fenech M, López-Béjar M. Risk factors for postpartum ovarian cysts and their spontaneous recovery or persistence in lactating dairy cows. *Theriogenology.* 2002;58(8):1623–1632. https://doi.org/10.1016/s0093-691x(02)01046-4.
295. Jan MH, Das GK, Singh J, et al. Oxidative stress during cystic ovarian disease in water buffalo. *Indian J Anim Sci.* 2015;85(4):356–360.
296. Day F, Karaderi T, Jones MR, et al. Large-scale genome-wide meta-analysis of polycystic ovary syndrome suggests shared genetic architecture for different diagnosis criteria. *PLoS Genet.* 2018;14. https://doi.org/10.1371/journal.pgen.1007813, e1007813.
297. Peng Y, Zhang W, Yang P, et al. ERBB4 confers risk for polycystic ovary syndrome in Han Chinese. *Sci Rep.* 2017;7:42000. https://doi.org/10.1038/srep42000.
298. Sağsöz H, Ketani MA, Saruhan BG. Expression of the erbB/HER receptor family in the bovine uterus during the sexual cycle and the relation of this family to serum sex steroids. *Biotech Histochem.* 2012;87(2):105–116. https://doi.org/10.3109/10520295.2011.556666.
299. Pau CT, Mosbruger T, Saxena R, Welt CK. Phenotype and tissue expression as a function of genetic risk in polycystic ovary syndrome. *PLoS One.* 2017;12(1):e0168870. https://doi.org/10.1371/journal.pone.0168870.
300. Cherry JA, Hou X, Rueda BR, Davis JS, Townson DH. Microvascular endothelial cells of the bovine corpus luteum: a comparative examination of the estrous cycle and pregnancy. *J Reprod Dev.* 2008;54(3):183–191. https://doi.org/10.1262/jrd.19182.

301. Goodarzi MO, Jones MR, Li X, et al. Replication of association of DENND1A and THADA variants with polycystic ovary syndrome in European cohorts. *J Med Genet.* 2012;49(2):90–95. https://doi.org/10.1136/jmedgenet-2011-100427.
302. Li T, Zhao H, Zhao X, et al. Identification of YAP1 as a novel susceptibility gene for polycystic ovary syndrome. *J Med Genet.* 2012;49(4):254–257. https://doi.org/10.1136/jmedgenet-2011-100727.
303. Plewes MR, Hou X, Zhang P, et al. Yes-associated protein 1 is required for proliferation and function of bovine granulosa cells in vitro. *Biol Reprod.* 2019;101(5):1001–1017. https://doi.org/10.1093/biolre/ioz139.
304. Summers A, Pohlmeier W, Sargent K, et al. A bovine model for polycystic ovary syndrome. *Soc Stud Reprod Ann Meet.* 2013;P53.
305. Owens LA, Abbara A, Lerner A, et al. The direct and indirect effects of kisspeptin-54 on granulosa lutein cell function. *Hum Reprod.* 2018;33(2):292–302. https://doi.org/10.1093/humrep/dex357.
306. Opsomer G, Wensing T, Laevens H, Coryn M, de Kruif A. Insulin resistance: the link between metabolic disorders and cystic ovarian disease in high yielding dairy cows? *Anim Reprod Sci.* 1999;56(3-4):211–222. https://doi.org/10.1016/s0378-4320(99)00048-2.
307. Amweg AN, Salvetti NR, Stangaferro ML, et al. Ovarian localization of 11β-hydroxysteroid dehydrogenase (11βHSD): effects of ACTH stimulation and its relationship with bovine cystic ovarian disease. *Domest Anim Endocrinol.* 2013;45(3):126–140. https://doi.org/10.1016/j.domaniend.2013.07.001.
308. Rey F, Rodríguez FM, Salvetti NR, et al. Insulin-like growth factor-II and insulin-like growth factor-binding proteins in bovine cystic ovarian disease. *J Comp Pathol.* 2010;142(2-3):193–204. https://doi.org/10.1016/j.jcpa.2009.11.002.
309. Roberge S, Brown JL, Reeves JJ. Elevated inhibin concentration in the follicular fluid of dairy cows with chronic cystic ovarian disease. *Theriogenology.* 1993;40(4):809–818. https://doi.org/10.1016/0093-691x(93)90216-r.
310. Nestler JE, Jakubowicz DJ, de Vargas AF, Brik C, Quintero N, Medina F. Insulin stimulates testosterone biosynthesis by human thecal cells from women with polycystic ovary syndrome by activating its own receptor and using inositolglycan mediators as the signal transduction system. *J Clin Endocrinol Metab.* 1998;83(6):2001–2005. https://doi.org/10.1210/jcem.83.6.4886.
311. Ovalle F, Azziz R. Insulin resistance, polycystic ovary syndrome, and type 2 diabetes mellitus. *Fertil Steril.* 2002;77(6):1095–1105. https://doi.org/10.1016/s0015-0282(02)03111-4.
312. Filippou P, Homburg R. Is foetal hyperexposure to androgens a cause of PCOS? *Hum Reprod.* 2017;23(4):421–432. https://doi.org/10.1093/humupd/dmx013.
313. Norris HJ, Taylor HB, Garner FM. Comparative pathology of ovarian neoplasms. II. Gonadal stromal tumors of bovine species. *Pathol Vet.* 1969;6(1):45–58. https://doi.org/10.1177/030098586900600104.
314. Teh A, Izzati UZ, Mori K, et al. Histological and immunohistochemical evaluation of granulosa cells during different stages of folliculogenesis in bovine ovaries. *Reprod Domest Anim.* 2018;53(3):569–581. https://doi.org/10.1111/rda.13132.
315. Roberts JF. *Ovarian Follicular Dysplasia in Florida Beef Cattle. DH Barron Reproductive and Perinatal Biology Seminar. D.H. Barron Reproductive and Perinatal Biology Seminar.* University of Florida; 2020.
316. Lim D, Oliva E. Ovarian sex cord-stromal tumours: an update in recent molecular advances. *Pathology.* 2018;50(2):178–189. https://doi.org/10.1016/j.pathol.2017.10.008.
317. Sandru F, Petca A, Dumitrascu MC, Petca RC, Carsote M. Peutz-Jeghers syndrome: Skin manifestations and endocrine anomalies (review). *Exp Ther Med.* 2021;22(6):1387. https://doi.org/10.3892/etm.2021.10823.

318. Haroon S, Zia A, Idrees R, Memon A, Fatima S, Kayani N. Clinicopathological spectrum of ovarian sex cord-stromal tumors; 20 years' retrospective study in a developing country. *J Ovarian Res*. 2013;6(1):87. https://doi.org/10.1186/1757-2215-6-87.
319. Seluanov A, Gladyshev VN, Vijg J, Gorbunova V. Mechanisms of cancer resistance in long-lived mammals. *Nat Rev Cancer*. 2018;18(7):433–441. https://doi.org/10.1038/s41568-018-0004-9.
320. Onder S, Hurdogan O, Bayram A, Yilmaz I, Sozen H, Yavuz E. The role of FOXL2, SOX9, and β-catenin expression and DICER1 mutation in differentiating sex cord tumor with annular tubules from other sex cord tumors of the ovary. *Virchows Arch*. 2021;479(2):317–324. https://doi.org/10.1007/s00428-021-03052-2.
321. Fuller PJ, Leung D, Chu S. Genetics and genomics of ovarian sex cord-stromal tumors. *Clin Genet*. 2017;91(2):285–291. https://doi.org/10.1111/cge.12917.
322. Pectasides D, Pectasides E, Psyrri A. Granulosa cell tumor of the ovary. *Cancer Treat Rev*. 2008;34(1):1–12. https://doi.org/10.1016/j.ctrv.2007.08.007.
323. Calaminus G, Wessalowski R, Harms D, Göbel U. Juvenile granulosa cell tumors of the ovary in children and adolescents: results from 33 patients registered in a prospective cooperative study. *Gynecol Oncol*. 1997;65(3):447–452. https://doi.org/10.1006/gyno.1997.4695.
324. Lagerlof N, Boyd H. Ovarian hypoplasia and other abnormal conditions in the sexual organs of cattle of the Swedish Highland breed; results of postmortem examination of over 6,000 cows. *Cornell Vet*. 1953;43(1):64–79.
325. Izquierdo N, Zhelev V, Angelov AK. Tumors of the ovaries in zebus. *Vet Med Nauki*. 1983;20(2):71–78. Tumori na iaĭchnitsite pri zebu.
326. Zemjanis R, Larson LL, Bhalla RP. Clinical incidence of genital abnormalities in the cow. *J Am Vet Med Assoc*. 1961;139:1015–1018.
327. Li J, Bao R, Peng S, Zhang C. The molecular mechanism of ovarian granulosa cell tumors. *J Ovarian Res*. 2018;11(1):13. https://doi.org/10.1186/s13048-018-0384-1.
328. Färkkilä A, Anttonen M, Pociuviene J, et al. Vascular endothelial growth factor (VEGF) and its receptor VEGFR-2 are highly expressed in ovarian granulosa cell tumors. *Eur J Endocrinol*. 2011;164(1):115–122. https://doi.org/10.1530/eje-10-0849.
329. Kocer A, Reichmann J, Best D, Adams IR. Germ cell sex determination in mammals. *Mol Hum Reprod*. 2009;15(4):205–213. https://doi.org/10.1093/molehr/gap008.
330. Huang C-CJ, Yao HHC. Inactivation of Dicer1 in steroidogenic factor 1-positive cells reveals tissue-specific requirement for Dicer1 in adrenal, testis, and ovary. *BMC Dev Biol*. 2010;10. doi:66 1186/1471-213x-10-66.
331. Schultz KA, Pacheco MC, Yang J, et al. Ovarian sex cord-stromal tumors, pleuropulmonary blastoma and DICER1 mutations: a report from the International Pleuropulmonary Blastoma Registry. *Gynecol Oncol*. 2011;122(2):246–250. https://doi.org/10.1016/j.ygyno.2011.03.024.
332. Soto N, Iñiguez G, López P, et al. Anti-Mullerian hormone and inhibin B levels as markers of premature ovarian aging and transition to menopause in type 1 diabetes mellitus. *Hum Reprod*. 2009;24:2838–2844. https://doi.org/10.1093/humrep/dep276.
333. Gogola J, Hoffmann M, Ptak A. Persistent endocrine-disrupting chemicals found in human follicular fluid stimulate IGF1 secretion by adult ovarian granulosa cell tumor spheroids and thereby increase proliferation of non-cancer ovarian granulosa cells. *Toxicol In Vitro*. 2020;65:104769. https://doi.org/10.1016/j.tiv.2020.104769.
334. Haltia UM, Hallamaa M, Tapper J, et al. Roles of human epididymis protein 4, carbohydrate antigen 125, inhibin B and anti-Müllerian hormone in the differential diagnosis and follow-up of ovarian granulosa cell tumors. *Gynecol Oncol*. 2017;144(1):83–89. https://doi.org/10.1016/j.ygyno.2016.11.018.
335. Giusti I, Di Francesco M, D'Ascenzo S, et al. Ovarian cancer-derived extracellular vesicles affect normal human fibroblast behavior. *Cancer Biol Ther*. 2018;19(8):722–734. https://doi.org/10.1080/15384047.2018.1451286.

336. Giusti I, D'Ascenzo S, Dolo V. Microvesicles as potential ovarian cancer biomarkers. *Biomed Res Int.* 2013;2013. https://doi.org/10.1155/2013/703048, 703048.
337. Sørensen AE, Udesen PB, Maciag G, et al. Hyperandrogenism and metabolic syndrome are Aassociated with changes in serum-derived microRNAs in women with polycystic ovary syndrome. *Front Med (Lausanne).* 2019;6:242. https://doi.org/10.3389/fmed.2019.00242.
338. Huang Y, Zhang X, Jiang W, et al. Discovery of serum biomarkers implicated in the onset and progression of serous ovarian cancer in a rat model using iTRAQ technique. *Eur J Obstet Gynecol Reprod Biol.* 2012;165(1):96–103. https://doi.org/10.1016/j.ejogrb.2012.06.031.
339. Enroth S, Berggrund M, Lycke M, et al. High throughput proteomics identifies a high-accuracy 11 plasma protein biomarker signature for ovarian cancer. *Commun Biol.* 2019;2:221. https://doi.org/10.1038/s42003-019-0464-9.
340. Sathyapalan T, David R, Gooderham NJ, Atkin SL. Increased expression of circulating miRNA-93 in women with polycystic ovary syndrome may represent a novel, non-invasive biomarker for diagnosis. *Sci Rep.* 2015;5:16890. https://doi.org/10.1038/srep16890.
341. Dang Y, Zhao S, Qin Y, Han T, Li W, Chen ZJ. MicroRNA-22-3p is down-regulated in the plasma of Han Chinese patients with premature ovarian failure. *Fertil Steril.* 2015;103(3). https://doi.org/10.1016/j.fertnstert.2014.12.106. 802-7.e1.
342. Zhu T, Gao W, Chen X, et al. A pilot study of circulating microRNA-125b as a diagnostic and prognostic biomarker for epithelial ovarian cancer. *Int J Gynecol Cancer.* 2017;27(1):3–10. https://doi.org/10.1097/igc.0000000000000846.
343. Pohler KG, Green JA, Moley LA, et al. Circulating microRNA as candidates for early embryonic viability in cattle. *Mol Reprod Dev.* 2017;84(8):731–743. https://doi.org/10.1002/mrd.22856.
344. Liu J, Huang X, Cao X, Feng X, Wang X. Serum biomarker analysis in patients with premature ovarian insufficiency. *Cytokine.* 2020;126:154876. https://doi.org/10.1016/j.cyto.2019.154876.
345. Lee DH, Pei CZ, Song JY, et al. Identification of serum biomarkers for premature ovarian failure. *Biochim Biophys Acta Proteins Proteomics.* 2019;1867(3):219–226. https://doi.org/10.1016/j.bbapap.2018.12.007.
346. Ceciliani F, Lecchi C, Urh C, Sauerwein H. Proteomics and metabolomics characterizing the pathophysiology of adaptive reactions to the metabolic challenges during the transition from late pregnancy to early lactation in dairy cows. *J Proteomics.* Apr 30 2018;178:92–106. https://doi.org/10.1016/j.jprot.2017.10.010.
347. Muinao T, Deka Boruah HP, Pal M. Multi-biomarker panel signature as the key to diagnosis of ovarian cancer. *Heliyon.* 2019;5:e02826. https://doi.org/10.1016/j.heliyon.2019.e02826.
348. Zhao C, Shu S, Bai Y, Wang D, Xia C, Xu C. Plasma protein comparison between dairy cows with inactive ovaries and estrus. *Sci Rep.* 2019;9(1):13709. https://doi.org/10.1038/s41598-019-49785-8.
349. Bai Y, Zhang F, Zhang H, Xu C, Wu L, Xia C. Follicular fluid metabolite changes in dairy cows with inactive ovary identified using untargeted metabolomics. *Biomed Res Int.* 2020;2020:9837543. https://doi.org/10.1155/2020/9837543.
350. Zhao C, Hu P, Bai YL, Xia C. Plasma metabolic differences in cows affected by inactive ovaries or normal ovarian function post partum. *Pol J Vet Sci.* 2020;23(1):59–67. https://doi.org/10.24425/pjvs.2020.132749.
351. Bedoschi G, Navarro PA, Oktay K. Chemotherapy-induced damage to ovary: mechanisms and clinical impact. *Future Oncol.* 2016;12(20):2333–2344. https://doi.org/10.2217/fon-2016-0176.

352. Henriques TB, Dos Santos DZ, Dos Santos GI, et al. Inhibition of CXCR2 plays a pivotal role in re-sensitizing ovarian cancer to cisplatin treatment. *Aging (Albany NY)*. 2021;13(10):13405–13420. https://doi.org/10.18632/aging.203074.
353. Li J, Wood 3rd WH, Becker KG, Weeraratna AT, Morin PJ. Gene expression response to cisplatin treatment in drug-sensitive and drug-resistant ovarian cancer cells. *Oncogene*. 2007;26(20):2860–2872. https://doi.org/10.1038/sj.onc.1210086.
354. Huang B, Qian C, Ding C, Meng Q, Zou Q, Li H. Fetal liver mesenchymal stem cells restore ovarian function in premature ovarian insufficiency by targeting MT1. *Stem Cell Res Ther*. 2019;10(1):362. https://doi.org/10.1186/s13287-019-1490-8.
355. Xia X, Wang T, Yin T, et al. Mesenchymal stem cells facilitate in vitro development of human preantral follicle. *Reprod Sci*. 2015;22:1367–1376. https://doi.org/10.1177/1933719115578922.
356. Mhatre P, Mhatre J. Orthotopic ovarian transplant- -review and three surgical techniques. *Pediatr Transplant*. 2006;10(7):782–787. https://doi.org/10.1111/j.1399-3046.2006.00547.x.
357. Dolmans MM, Donnez J. Fertility preservation in women for medical and social reasons: Oocytes vs ovarian tissue. *Best Pract Res Clin Obstet Gynaecol*. 2021; 70:63–80. https://doi.org/10.1016/j.bpobgyn.2020.06.011.
358. Zhai J, Zhang J, Zhang L, et al. Autotransplantation of the ovarian cortex after in-vitro activation for infertility treatment: a shortened procedure. *Hum Reprod*. 2021;36 (8):2134–2147. https://doi.org/10.1093/humrep/deab143.
359. Royston JP. Basal body temperature, ovulation and the risk of conception, with special reference to the lifetimes of sperm and egg. *Biometrics*. 1982;38(2):397–406.
360. Yamoto M, Minami S, Nakano R, Kobayashi M. Immunohistochemical localization of inhibin/activin subunits in human ovarian follicles during the menstrual cycle. *J Clin Endocrinol Metab*. 1992;74(5):989–993. https://doi.org/10.1210/jcem.74.5.1569176.
361. Yamoto M, Minami S, Nakano R. Immunohistochemical localization of inhibin subunits in human corpora lutea during menstrual cycle and pregnancy. *J Clin Endocrinol Metab*. 1991;73(3):470–477. https://doi.org/10.1210/jcem-73-3-470.
362. Laird M, Glister C, Cheewasopit W, Satchell LS, Bicknell AB, Knight PG. 'Free' inhibin alpha subunit is expressed by bovine ovarian theca cells and its knockdown suppresses androgen production. *Sci Rep*. 2019;9(1):19793. https://doi.org/10.1038/s41598-019-55829-w.

CHAPTER FIVE

Mutations in rhodopsin, endothelin B receptor, and CC chemokine receptor 5 in large animals: Modeling human diseases

Ren-Lei Ji and Ya-Xiong Tao*
Department of Anatomy, Physiology and Pharmacology, College of Veterinary Medicine, Auburn University, Auburn, AL, United States
*Corresponding author: e-mail address: taoyaxi@auburn.edu

Contents

Abstract

G protein-coupled receptors (GPCRs) are the largest family of cell membrane receptors involved in modulating almost all physiological processes by transducing extracellular signals into the cytoplasm. Dysfunctions of GPCR-regulated signaling result in diverse human diseases, making GPCRs the most popular drug targets for human medicine. Large animals share higher similarities (in physiology and metabolism) with humans than rodents. Similar to findings in human genetics, diverse diseases caused by mutations in GPCR genes have also been discovered in large animals. Rhodopsin, endothelin B receptor, and C—C chemokine receptor type 5 have been shown to be involved in human retinitis pigmentosa, Hirschsprung disease, and HIV infection/AIDS, respectively, and several mutations of these GPCRs have also been identified from large animals. The large animals with naturally occurring mutations of these GPCRs provide an opportunity to gain a better understanding of the pathogenesis of human diseases, and can be used for preclinical trials of therapies for human diseases. In this review, we aim to summarize the naturally occurring mutations of these three GPCRs in large animals and humans.

Progress in Molecular Biology and Translational Science, Volume 189
ISSN 1877-1173
https://doi.org/10.1016/bs.pmbts.2022.02.003

Abbreviations

CCR5 C-C chemokine receptor type 5
cGMP cyclic guanosine monophosphate
ET endothelin
ET_BR endothelin B receptor
GPCR G protein-coupled receptor
HIV human immunodeficiency virus
RHO Rhodopsin
SIV simian immunodeficiency virus
TMD transmembrane domain
WT wild-type

1. Introduction

G protein-coupled receptors (GPCRs) are the largest family of cell surface proteins, with over 800 members identified in humans.[1–3] GPCRs have seven transmembrane domains (TMDs) connected by the extracellular N termini, alternating extracellular and intracellular loops, and the intracellular C termini. Based on structural features, GPCRs are subdivided into six groups: Family A (rhodopsin-like), Family B (secretin-like), Family C (metabotropic glutamate), Family D (pheromone), Family E (cAMP), and Family F (Frizzled).[1] GPCRs regulate an extremely wide range of physiological processes by binding to various external signals or ligands, including light, pheromones, odors, hormones, and neurotransmitters. Since the discovery of the first naturally occurring mutation in GPCR resulting in human disease, that of rhodopsin causing retinitis pigmentosa,[4] diverse human diseases caused by GPCR mutations have been reported (for reviews, see Refs. [5–12] for examples). GPCRs have been exploited as the most popular drug targets, accounting for ~35% of all approved drugs and over 50% of total pharmacotherapies.[13–15]

Animal models are powerful tools for exploring the modality of transmission and pathogenesis of human diseases, determining the efficacy of novel vaccines and drugs, as well as used for preclinical trials of therapies for human diseases. Compared to small animals such as rodents, large animals, such as non-human primates, dogs, pigs, and horses, show higher similarities to humans, including organ size, anatomy, physiology, and metabolism.[16–20] Diverse diseases caused by GPCR mutations have also been observed in large animals, showing similar phenotypes with those of humans.[21–26]

The large animals with naturally occurring GPCR mutations provide natural models to gain a better understanding of disease pathophysiology and tools for developing new therapeutics.

Rhodopsin (the dim-light photoreceptor, RHO), endothelin B receptor (ET_BR), C—C chemokine receptor type 5 (CCR5), are three GPCRs that have been shown to be involved in human retinitis pigmentosa, Hirschsprung disease, human immunodeficiency virus (HIV) infection, respectively.[4,27,28] Several variants of these three GPCRs have also been discovered from large animals. In this chapter, we summarize the naturally occurring mutations of these three GPCRs associated with diseases in large animals and humans.

2. Rhodopsin mutations and retinitis pigmentosa

Rhodopsin, a light-sensitive receptor, is one of the best studied GPCRs. Human *RHO* gene has five exons and four introns located on chromosome 3q22.1.[29] Rhodopsin is primarily expressed in the rod cells of the retina.[30,31] Rhodopsin consists of a covalent complex between opsin and 11-*cis*-retinal (a covalently-bound chromophore produced from vitamin A in the retina). Photons are absorbed by rhodopsin, catalyze the isomerization of 11-*cis*-retinal to all-*trans*-retinal, resulting in the conversion of inactive rhodopsin to metarhodopsin II (active rhodopsin). Metarhodopsin II interacts with and activates a heterotrimeric G protein, transducin (the rod-specific heterotrimeric G protein) in rod cells, initiating a cyclic guanosine monophosphate (cGMP) second messenger cascade[32,33] (Fig. 1).

Retinitis pigmentosa is the most prevalent inherited retinal degenerative disease affecting 1/3000–1/7000 individuals, manifested as night blindness (nyctalopia) to total blindness.[34,35] Retinitis pigmentosa is characterized by the gradual loss of rod photoreceptor cells followed by cone degeneration. The first mutation in the GPCR associated with human disease was discovered in rhodopsin, P23H, which results in autosomal dominant retinitis pigmentosa.[4,36] Since then, over 150 *RHO* mutations have been identified and are associated with retinitis pigmentosa.[37] It is estimated that roughly 15% of all retinal degenerative diseases are caused by *RHO* mutations, and the largest subtype of retinitis pigmentosa is autosomal dominant retinitis pigmentosa.[35] Recently, more than 200 *RHO* mutations are identified from 141,456 humans and have been registered in gnomAD v2.1.1 database (https://gnomad.broadinstitute.org/) (Fig. 2) based on Ref. [39].

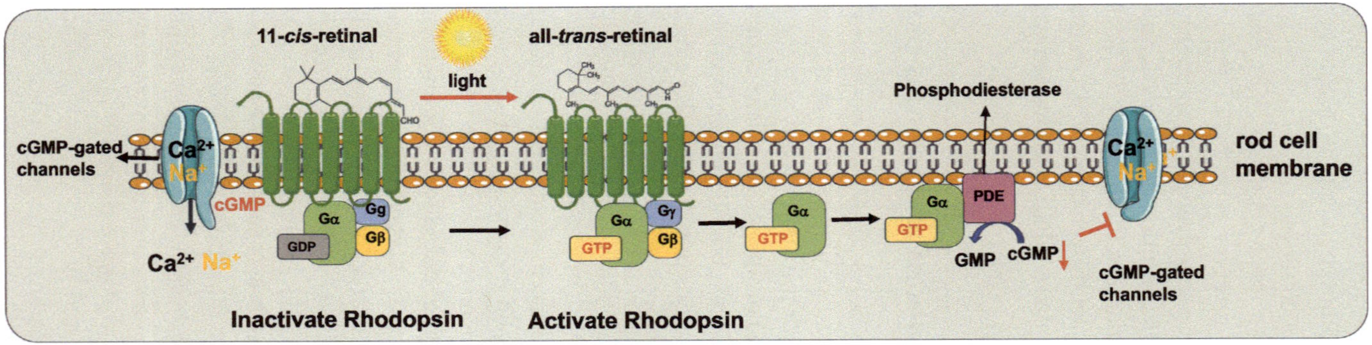

Fig. 1 The rhodopsin-regulated signal transduction cascade in the rod cells. Photons are absorbed by rhodopsin, catalyzing the isomerization from 11-*cis*-retinal to all-*trans*-retinal, resulting in the conversion of inactive rhodopsin to active rhodopsin. Active rhodopsin further interacts with and activates a heterotrimeric G-protein in rod cells, transducin. Cyclic GMP is hydrolyzed by the cGMP phosphodiesterase (PDE). The reduced cGMP will contribute to close the cGMP-gated channels, resulting in decreased influx of Na^+ and Ca^{2+}.

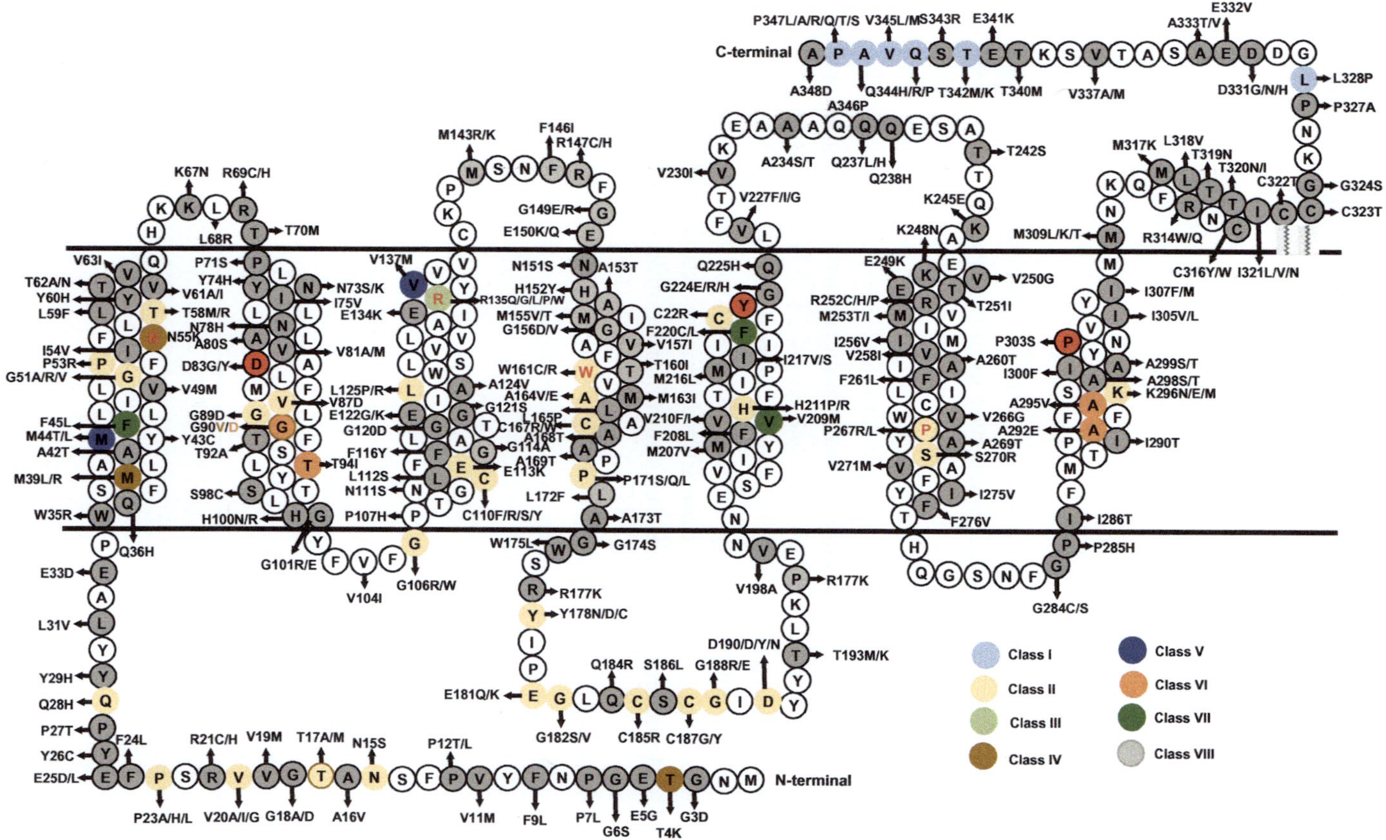

Fig. 2 Naturally occurring *RHO* mutations identified. Frameshift mutations were not included in the figure. The most conserved residues in transmembrane domains (TMDs) were labeled with red font. The circles filled with different colors indicated different Class mutations according to Athanasiou et al.[38]

Rhodopsin mutations may cause receptor misfolding, mistrafficking, or aberrant signaling.[40] Based on the biochemical and cellular properties, *RHO* mutations have been sorted into eight classes, including folding normally but are not transported to the outer segment (Class I), misfolding (Class II), disrupted vesicular traffic and endocytosis (Class III), altered stability and posttranslational modification (Class IV), altered transducin activation (Class V), constitutive activation (Class VI), dimerization deficiency (Class VII), and variants with unknown defect or not studied in detail (Class VIII).[38,41] Among these classes of mutations, Class II is the major subtype.[38,42]

Retinitis pigmentosa is also mirrored in animals, where naturally occurring retinal degeneration has been observed in mice, rats, cats, and dogs.[16] However, the dog is the only animal known to express a naturally occurring *RHO* mutation, providing a natural model to explore potential mechanisms of neuronal injury, repair, degeneration, and compensation, as well as supporting the gene-specific clinical trial of light reduction in humans with autosomal dominant retinitis pigmentosa caused by *RHO* mutations.[26,43] Amino acid sequences of dog rhodopsin share 93% identity with that of human. Naturally occurring T4R *RHO* was identified from English Mastiff dogs with a naturally occurring dominant retinal degeneration[26] (Table 1). Mutation at the same position in human opsin (T4K) also causes retinitis pigmentosa.[51] English Mastiff dogs with *RHO* mutation with a slowly progressive retinal degeneration share similar features as human autosomal dominant retinitis pigmentosa caused by Class B *RHO* mutations

Table 1 Diseases associated with rhodopsin, endothelin B receptor, and *CCR5* mutations.

Receptor	Diseases	Species	Mutations	References
Rhodopsin	Retinitis pigmentosa	Dog	T4R	26,43
Endothelin B receptor	Hirschsprung disease	Horse	I118K	21,25,44–46
C-C chemokine receptor type 5 (CCR5)	HIV infection	Monkey	CCR5Δ24, CCR5Δ2, T9I, D13N, Y14N, S38T, N57S, Q93R, Q93K, V147M, R163G, and L352F	23,47–50

(refers to a milder phenotype, with retained rod function at least in certain retinal regions).[52] T4R dogs show very prolonged recovery from light exposure, similar to phenotypes of humans with seven pathogenic *RHO* mutations, including T17M, P23H, T58R, V87D, G106R, T193M, and D190G.[52–54]

Normal rhodopsin expression, retinal structure, receptor activation, and signaling are observed in young dogs with T4R variant, suggesting that the pathogenesis is not associated with abnormal development of photoreceptor.[26] T4R mutant is not defective in mature rhodopsin protein localization to the photoreceptor outer segment and displays normal signaling during phototransduction activation.[26] The residue T4 in rhodopsin is very conserved in mammals, and might be a glycosylation site.[55] T4R mutation might alter the conserved glycosylation site (NGT to NGR) in rhodopsin.[56] Glycosylation is required for ER associated degradation of mutant rhodopsin.[57] T4R in dog causes a light dependent degeneration,[56] similar to human mutations, including T4K, T4N, and T17M.[58] Subsequent studies suggest that T4R mutant is structurally unstable in the absence of chromophore and has been proposed to be Class IV.[56]

The therapeutic strategies aim to correct misfolding, increase stability, reduce rhodopsin aggregation, promote the degradation, decrease basal activity of rhodopsin variants, and enhance cell survival. Pharmacological chaperone, molecular chaperone, cell death inhibitors, and neurotrophic factors, could be potentially used.[38,41] Bright-light exposure might promote degradation of Class II and Class IV mutants, and dark rearing further exacerbate this process,[59] indicating that avoiding exposure to bright light, especially after periods of darkness, might be a useful method for rhodopsin retinitis pigmentosa patients.[43,56]

Dietary supplements of retinoids and vitamin A might be beneficial for the misfolding or instability of Class II and IV mutants against an age-related decline in the supply of 11-*cis*-retinal.[56,60] Pharmacological chaperone, as potential therapeutic agent, is able to promote or stabilize folding for Class II and Class IV mutations, including retinoids, 11-*cis*-7-ring retinal, β-ionone, 1-(3,5-dimethyl-1H-pyrazol-4-yl)ethenone, YC-001 (non-retinoid molecule), and 5,8-epoxy-13-*cis*-retinoic acid.[41,61–64] For example, retinoids promote the stability of T4R mutant in dog.[56] Indeed, pharmacological chaperones have been extensively studied in many GPCRs as well as other proteins as therapeutics for treating protein conformational diseases, diseases caused by protein misfolding.[65–67] Inverse agonists could decrease the constitutive activities of Class VI *RHO* mutations, including retinylamine analog and YC-001.[63,68] Additionally, cell death inhibitors and neurotrophic factors prolong cell survival and show protective effects on retina degeneration.[38]

3. *EDNRB* nutation and Hirschsprung disease

ET_BR is a Family A GPCR that transduces signals of the three endothelins (ET-1, ET-2, and ET-3).[69,70] ET_BR is encoded by the *EDNRB* gene located in 13q22.3.[71] Among the three endothelins, ET-3, a 21-amino acid peptide, is the major ligand for ET_BR. Activated ET_BR, primarily coupled to Gq, leads to phosphatidyl inositol and calcium generation. *EDNRB* is expressed widely, predominantly in endothelial cells, as well as in the vascular smooth muscle, aorta, coronary arteries, mesenteric arteries, human mammary arteries and veins of different animal species.[72–74] *EDNRB* is also expressed in glia and neurons of the brainstem, ventricular myocardium, the atrium, the atrioventricular conducting tissue, airway smooth muscle, renal tubules and collecting duct epithelial cells, osteoblasts, and hepatocytes.[73,75,76] Due its wide expression, the ET-ET_BR system plays vital roles in vascular smooth muscle, kidney, nervous, and vascular systems (Fig. 3).[77,78] ET_BR is also involved in regulating several cancers and considered as a new target in cancer immune therapy (Fig. 3).[79]

Hirschsprung disease is a congenital disease with aganglionic megacolon phenotype, lack of enteric ganglia, and absence of innervation to the lower gastrointestinal tract.[77,80] The prevalence of Hirschsprung disease is estimated to be at 1/5000 births.[81–83] Males are more likely to be affected than females.[82,83] Isolated Hirschsprung disease has been associated with several genes, including *EDNRB* and *EDN3*. ET_BR and ET-3 play a key role in the migration and development of the enteric nervous system.[84,85]

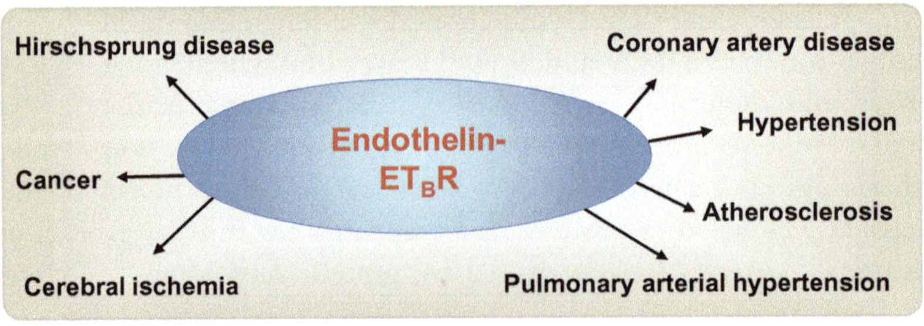

Fig. 3 Diseases related to endothelin-ET_BR system.

Genetic studies have also demonstrated the critical role of the ET pathway in Hirschsprung disease. Piebald and piebald lethal mouse models with naturally occurring mutation of *Ednrb* also show similar phenotype with human Hirschsprung disease and defect in neural crest cell differentiation and migration.[85–87]

Mutations in *EDNRB* were shown to be associated with human Hirschsprung disease, and have been demonstrated in approximately 5% of this disease.[27,88–91] W276C ET_BR variant was first identified from Mennonite kindred with a high incidence of Hirschsprung disease.[27] Since then, 46 additional *EDNRB* mutations have been reported, including M1V, L17P, K15X, P48L, K56T, G57S, C90R, N104I, C109R, V111Q, G115R, M132I, R133X, N137Y, P156S, P156R, I157V, M173T, C174Y, A183G, V185M, G186R, S196N, R201X, W226C, W226X, T244N, D246F, R253X, V260F, W275X, W276C, Y293fs, Y293L, S305N, A310T, R319W, C335S, L361S, M374I, N378fs, N378S, N378I, P383L, S390R, and R434C[8,89–108] (Fig. 4). Other *EDNRB* mutations recently registered in gnomAD v2.1.1 database (https://gnomad.broadinstitute.org/) are listed in Fig. 4.

Lethal white foal syndrome in horse provides a new animal model for human Hirschsprung disease.[109] Mutation in horse *ENDRB* has been shown to be associated with horse Hirschsprung disease.[21,25,44–46] The diseased horse *ENDRB* contained a dinucleotide TC to AG change, resulting in a missense mutation in ET_BR protein: I118K[21,25,44–46] (Table 1). I118 in ET_BR, a highly conserved residue in all species, is located at TMD1, indicating that I118K mutant may affect receptor function. All Overo Lethal white foal syndrome foals are homozygous for the I118K mutation, and no homozygous adults are found, indicating that homozygosity was restricted to lethal white foals. Color patterns of most heterozygotes (I118K) are frame overo.[21,25,44,46] All of the homozygous foals' parents are heterozygous, suggesting inheriting a copy from each heterozygous parent, similar to rodents.[86] Intestinal aganglionosis is the cause of death in the homozygous I118K horse. However, in human, both heterozygous and homozygous *EDNRB* mutation carriers display aganglionic megacolon, while heterozygous horses do not. Overo Lethal white foal syndrome phenotype also occurs in mice, rats, and humans with Hirschsprung disease.[27,85,110] Naturally occurring mutations and disruption of *EDNRB* or its ligand, *EDN3*, result in a phenotype similar to Overo Lethal white foal syndrome, suggesting that the ET signaling pathway is crucial for development and migration of neural crest cells that ultimately form melanocytes and enteric neurons.

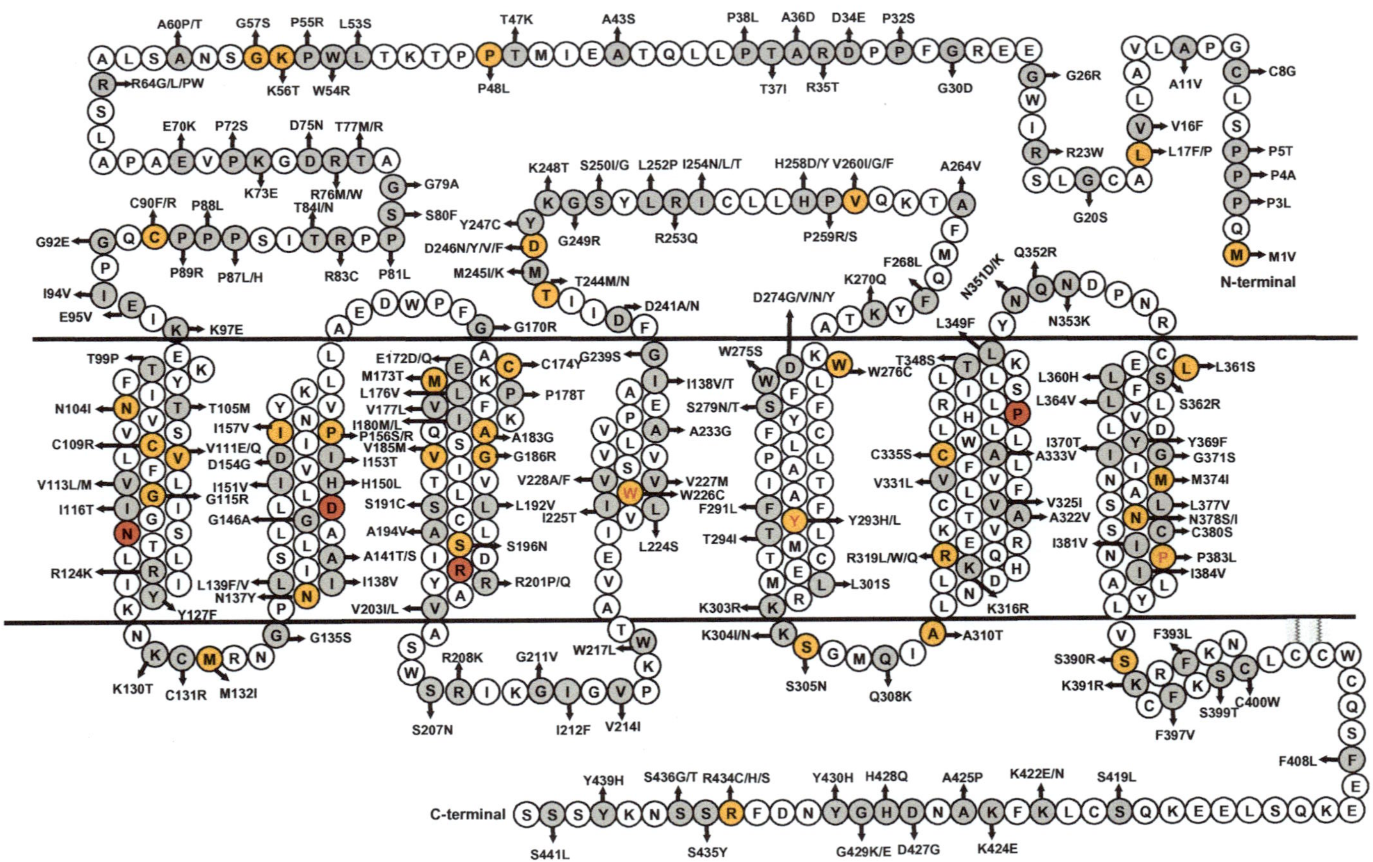

Fig. 4 Naturally occurring *EDNRB* mutations identified. Yellow color indicated mutation identified previously. The circles filled with gray color were the missense and nonsense mutations in gnomAD database v2.1.1 (https://gnomad.broadinstitute.org/). Frameshift mutations were not included in the figure. The most conserved residues in transmembrane domains (TMDs) were labeled with red font.

Although numerous *EDNRB* mutations have been identified in human, horse, rats, and mice, there are few functional studies and therapeutic strategies targeting these mutations. Based on the life cycle of GPCRs, loss of-function mutations can be sorted into five classes, including defective expression (Class I), disrupted trafficking to the cell surface (Class II), disrupted ligand binding (Class III), defective signaling (Class IV), and mutations with no known defects (Class V).[8,65,111] Among human *EDNRB* mutations characterized, C109R, A183G, W276C, and P383L (Class II) are retained intracellularly.[93,112,113] W276C and S390R (Class IV) bind ligand with normal affinity but have decreased signaling[27,93,114]; G57S, N137Y, and R319W show biased signaling[107,113]; L17P and N104I (Class V) have similar functions as the wild-type (WT) ET_BR.[93,107]

Misfolding and intracellular retention of mutant receptors (Class II) have been reported as the primary defect in human diseases caused by GPCR mutations, and small molecule hydrophobic ligands can serve as pharmacological chaperones, increasing the cell surface expression of Class II mutants, and are potential therapeutics for diseases caused by protein misfolding.[8,65,66,115] Misfolding in C109R and P383L might be corrected by pharmacological chaperones. For receptor variants with defective signaling or biased signaling (Class IV), allosteric ligands or biased agonists might be effective ways to correct these mutations. In addition, three ET_BR selective agonists, Sarafotoxin 6c,[116] IRL-1620 ([Suc-[Glu^9,Ala^{11},^{15}]-endothelin-1(8–12)]),[117] and BQ3020 ([Ala^{11},^{15}]Ac-ET-l6–21),[118] show high affinity for ET_BR, especially considering that IRL-1620 provides long-term neuroprotection and decreases neurological damage in a rat model.[119] Hence, we hypothesize that these selective agonists might be used to correct some Class III and IV *EDNRB* mutations.

Interestingly, in an inbred Cameroon sheep flock, there is a deletion of about 110 kb encompassing the entire *EDNRB* gene in some hypopigmented animals.[120] However, these animals have morphologically normal ganglion cells in appropriate number and distribution, different from the animal models described above.[120]

4. *CCR5* mutations and susceptibility to HIV infection

CCR5 belongs to the beta chemokine receptor family of GPCRs. CCR5 is encoded by *CCR5* gene located on the short arm at position 21 on chromosome 3.[121,122] CCR5 binds several highly active endogenous chemokine ligands, including chemokine (C—C motif) ligand 3 (CCL3, also known as MIP-1α, macrophage inflammatory protein-1α), CCL4 (MIP-1β), and CCL5 (also known as RANTES, regulated on activation, normal T cell expressed and secreted).[121,123,124] CCR5 is primarily expressed on the surface of leukocytes, including monocytes and antigen-presenting cells (macrophages, dendritic, natural killer, memory, and effector T cells).[125–127] CCR5 is a critical receptor for the activation and mobilization of leukocytes. Additionally, CCR5 activation promotes the release of cytokines and chemokines,[123,127] as well as immune response by promoting the activation and proliferation of T cells and macrophages.[128,129]

CCR5 mediates the transmission of human immunodeficiency virus type 1 (HIV-1) infection, acting as the major coreceptor.[121,130–132] HIV, an RNA virus, is encased in an envelope made of a lipid membrane (containing the envelope (Env) glycoprotein) derived from host cells (Fig. 5). During HIV entry into human cells, gp120 glycoprotein of HIV Env binds to the CD4 receptor (the primary receptor for HIV binding on the host cell surface). Binding of gp120-CD4 receptor bimolecular complex to CCR5 results in changed Env conformation that induces the membrane-spanning gp41 subunit of Env to penetrate the host cell membrane and further stimulates the fusion of the cellular and viral membrane (Fig. 5).

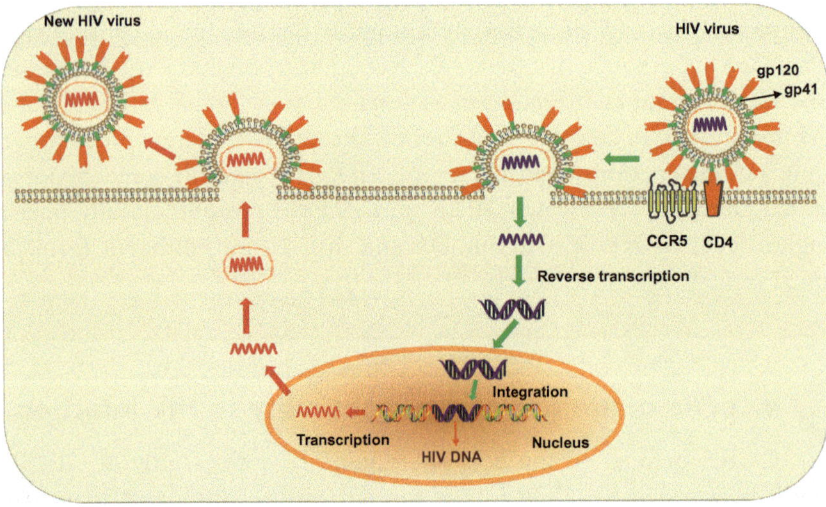

Fig. 5 HIV replication cycle.

Loss-of-function *CCR5* mutations are associated with high but incomplete resistance to HIV infection upon exposure, indicating that CCR5 has essential role in AIDS pathogenesis. The first mutation identified was a 32-base pair deletion (CCR5Δ32), resulting in a frameshift mutation and premature termination of the receptor with only four TMDs that is not expressed on the cell surface (Class II mutation, intracellularly retained) and cannot function as a chemokine receptor.[28,133] The mutant receptor also has dominant negative activity on the WT CCR5 trafficking.[134] In Caucasians, CCR5Δ32 mutations exists at 10% of the population, with about 18% heterozygous and 1% homozygous.[28,135] More than 200 *CCR5* variants have been registered in gnomAD v2.1.1 database (https://gnomad.broadinstitute.org/) (Fig. 6).

African non-human primates are the wild reservoirs of simian immunodeficiency virus (SIV). Phylogenetic analyses indicates that SIVs are the ancestors of HIVs (HIV-1 and HIV-2).[136,137] The similarities between SIV and HIV in biological feature and genomic structure, and infection of the macaques with SIVs, or with chimeric viruses having both SIV and HIV sequences (SHIVs), make non-human primates the most relevant animal models for studying HIV-1 infection and AIDS. Additionally, many of non-human primate genes related with immune response to SIV are similar to the human genes involved in controlling the immune response to HIV-1. Non-human primates infected with SIV and SHIV provide attractive models to explore effects of coreceptor usage on $CD4^+$ T cell depletion, viral replication, and disease. High identity in CCR5 and similar variants identified between human and non-human primates also provide an opportunity to study receptor structure and function. Non-human primates are also used as animal models for anti-HIV-1 drug and vaccine development.

Mutations in *CCR5* have also been identified from several species of monkeys. Similar negative selection pressure undergoing in humans, likely to avoid SIV infection, results in favoring the mutant alleles. Like human CCR5Δ32, CCR5Δ24 allele, with 24 bp deletion, corresponding to inframe deletion of eight amino acids in TMD4, is common in monkeys[23,47,48] (Table 1). The CCR5Δ24 variant is not expressed on the cell surface (Class II) and is considered as an inactive viral coreceptor.[23,47,48]

Peripheral blood mononuclear cells from sooty mangabey monkeys heterozygous for the deletion allele have lower CCR5 expression.[23] Approximate 3–4% frequency is observed in the institute-housed and wild-living monkeys for CCR5Δ24 allele.[23,47,48] However, out of 15 red-capped mangabeys, eleven were homozygous (frequency 86.6%) and four were

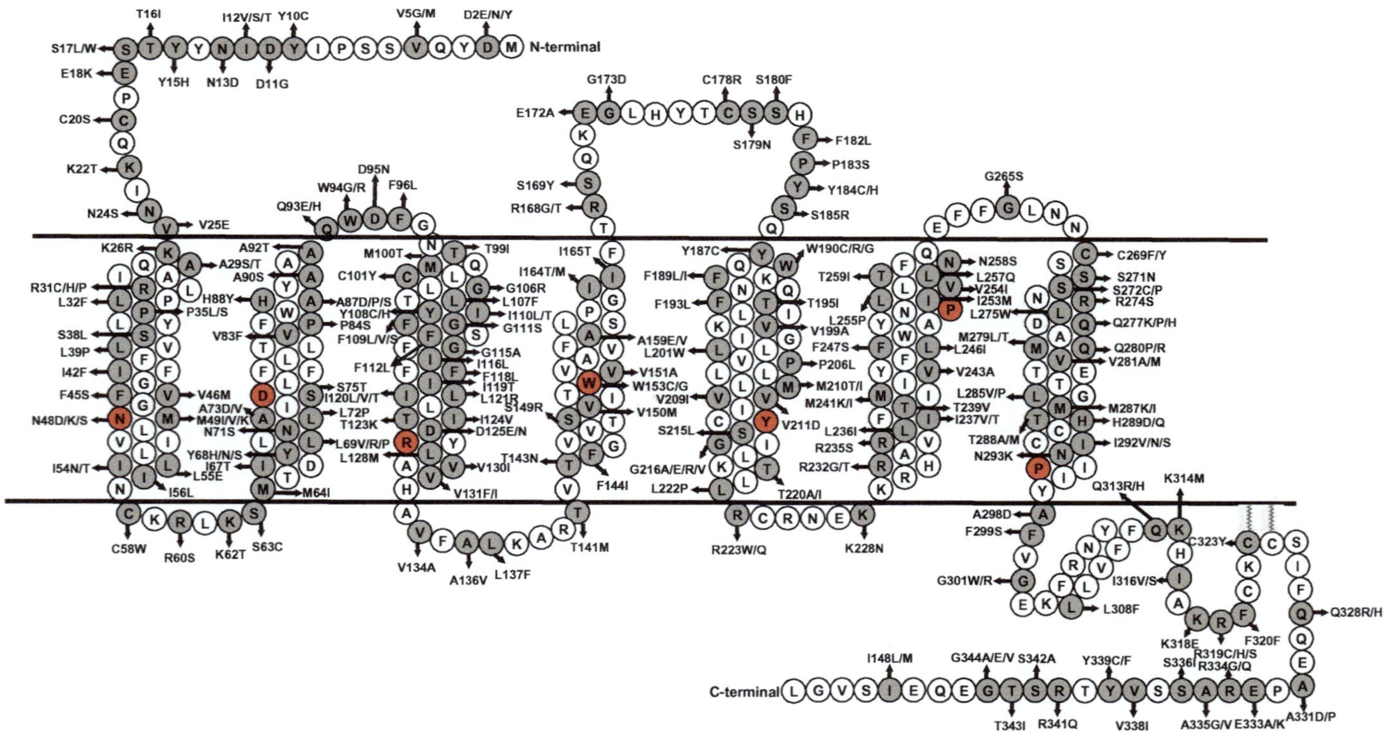

Fig. 6 Naturally occurring *CCR5* mutations identified. The circles filled with gray color were the missense and nonsense mutations in gnomAD database v2.1.1 (https://gnomad.broadinstitute.org/). Frameshift mutations were not included in the figure. The most conserved residues in transmembrane domains (TMDs) were labeled with red font.

heterozygous for CCR5Δ24 allele, and no WT homozygote was found in 6 zoo-housed (American zoo colony) and 8 household (Gabon) red-capped mangabeys.[47] Red-capped mangabey CCR5Δ24 is resistant to entry of SIVmac, SIVcpz, SIVmne, R5-tropic, HIV-2, or HIV-1.[47] One of 11 homozygotes is naturally infected with R2b-tropic SIV, in which CCR2b, not CCR5, is the major coreceptor. The higher frequency of CCR5Δ24 in red-capped mangabeys compared to that in sooty mangabey monkeys indicates that SIVrcm might acquire CCR2b-tropism in response to defective *CCR5* alleles in the host.[47]

Similar to monkey CCR5Δ24, CCR5Δ2 was identified from sooty mangabey monkeys, which contains a two base pair deletion at nucleotides 466 and 467 of the coding sequence, corresponding to TMD4 of the sooty mangabey monkey CCR5[48] (Table 1). This deletion results in a frameshift and truncated protein with 110 missense amino acids. Similar to human CCR5Δ32 and sooty mangabey monkey/red-capped mangabey CCR5Δ24, CCR5Δ2 is absent on the cell surface (Class II) and confers resistance to SIV entry in vitro. The frequency of CCR5Δ2 allele is 26% in 202 sooty mangabey monkeys and 9% in 33 wild-living sooty mangabey monkeys.[48] In addition, 65% of homozygous WT *CCR5* sooty mangabey monkeys are naturally infected with SIVsmm, and 62% of *CCR5* heterozygous animals are infected with SIVsmm, and 50% of *CCR5*-null allele animals are naturally infected with SIVsmm. Plasma viral loads are only slightly decreased in the heterozygous and homozygous mutant animals compared to WT animals.[48] These results clearly indicated that SIVsmm does not use CCR5 as the entry coreceptor in naturally infected and *CCR5*-null sooty mangabey monkeys.[48] Rather, SIVsmm envelope glycoproteins can penetrate cells using other GPCRs, such as G protein-coupled receptor 15, CXC chemokine receptor 6, and G protein-coupled receptor 1.[48]

Other *CCR5* mutations, including T9I, D13N, Y14N, S38T, N57S, Q93R, Q93K, V147M, R163G, and L352F, were identified from African green monkey[49,50] (Table 1). African green monkey D13N mutant (normal cell surface expression and signal transduction) is inactive as a coreceptor and strongly inhibits SIVagm infection.[49,50] Human CCR5 varies from WT African green monkey CCR5 in that position 13 contains N rather than D. D13N in macaque CCR5 significantly reduces CD4-independent binding of SIVmac gp120 and infection of SIVmac, whereas WT CCR5 in human (contains N13) is a highly efficient coreceptor.[138,139]

The Y14N mutation was frequently identified, with 7 being homozygous and 4 being heterozygous out of a total of 12 monkeys, and among these, the only WT homozygote and two of heterozygotes are naturally infected with SIVagm, but none of the Y14N homozygotes.[49] Subsequent studies suggest that Y14N, with normal cell surface expression, reduces signaling and inhibits SIVagm and HIV-1 infection in vitro.[49,50] The Y14N substitution removes a critical tyrosine sulfation site, resulting in an NYT consensus (the potential *N*-linked glycosylation) at this position. Previously, studies have found that tyrosine sulfation is important for HIV-1 infection, occurring at the Y3, Y10, Y14, and Y15 positions (common in both human and African green monkey CCR5s).[140] Additionally, *N*-glycosylated CCR5 is inactive for both signaling and coreceptor function. However, ungylcosylated CCR5 (Y14N) mutant (40% of the total) is fully active in chemokine binding and decreases signal transduction.[50,141]

Q93R and Q93K (decreased cell surface expression) in African green monkey CCR5 show similar signal transduction properties as the WT receptor but strongly inhibit infections by the SIVagm.[50] S38T has decreased cell surface expression and has no effect on SIVagm infection.[50] African green monkey L352F CCR5 has normal cell surface expression and is an active coreceptor.[49] All other mutations have normal cell surface expression as the WT receptor.[49,50] Thus, CCR5 variants might strongly inhibit coreceptor functions but have a negligible effect on chemokine binding or signaling.[50]

CCR5 is considered as a potentially safe target for inhibiting HIV infection.[142] The R5-tropic HIV strains that cause new infections cannot penetrate $CD4^+$ T cells and macrophages when CCR5 protein is missing on the cell surface.[28,133,143] The loss-of-function *CCR5* variants show protection against HIV infection. Thus, CCR5 antagonists provide a potential therapeutic strategy for HIV infection. Numerous CCR5 antagonists have been developed, including non-peptide small molecule compounds, monoclonal antibodies, chemokine derivatives, and peptide compounds.[124,144,145] Maraviroc (a small molecule CCR5 antagonist developed by Pfizer) is the only approved drug for use by the FDA.[146–148]

5. Conclusions

GPCR regulates almost all physiological processes, and proves to be important therapeutic targets. The diverse human diseases caused by GPCR mutations have been documented. Several GPCR mutations have also been identified from large animals, resulting in diseases with similar phenotypes of humans. Naturally occurring mutations in dog *RHO*, horse *EDNRB*, and monkey *CCR5*, have been discovered and show similar characteristics with those of humans. Naturally occurring GPCR mutations in these animals provide an opportunity to explore receptor structure and function of human, as well as provide the natural model for understanding of pathophysiology of human diseases and testing the efficacy for novel vaccines and drugs.

Acknowledgment

Ren-Lei Ji received fellowship from China Scholarship Council, People's Republic of China.

Conflict of interest

The authors declare that there is no conflict of interest that would prejudice the impartiality of this study.

References

1. Fredriksson R, Lagerstrom MC, Lundin LG, Schioth HB. The G-protein-coupled receptors in the human genome form five main families. Phylogenetic analysis, paralogon groups, and fingerprints. *Mol Pharmacol.* 2003;63(6):1256–1272.
2. Lefkowitz RJ. A brief history of G-protein coupled receptors (Nobel lecture). *Angew Chem Int Ed Engl.* 2013;52(25):6366–6378.
3. Tao YX. Molecular chaperones and G protein-coupled receptor maturation and pharmacology. *Mol Cell Endocrinol.* 2020;511:110862.
4. Dryja TP, McGee TL, Reichel E, et al. A point mutation of the rhodopsin gene in one form of retinitis pigmentosa. *Nature.* 1990;343(6256):364–366.
5. Spiegel AM, Weinstein LS, Shenker A. Abnormalities in G protein-coupled signal transduction pathways in human disease. *J Clin Invest.* 1993;92(3):1119–1125.
6. Spiegel AM, Weinstein LS. Inherited diseases involving G proteins and G protein-coupled receptors. *Annu Rev Med.* 2004;55:27–39.
7. Schoneberg T, Schulz A, Biebermann H, Hermsdorf T, Rompler H, Sangkuhl K. Mutant G-protein-coupled receptors as a cause of human diseases. *Pharmacol Ther.* 2004;104(3):173–206.
8. Tao YX. Inactivating mutations of G protein-coupled receptors and diseases: Structure-function insights and therapeutic implications. *Pharmacol Ther.* 2006; 111(3):949–973.
9. Tao YX. Constitutive activation of G protein-coupled receptors and diseases: Insights into mechanism of activation and therapeutics. *Pharmacol Ther.* 2008;120:129–148.
10. Vassart G, Costagliola S. G protein-coupled receptors: Mutations and endocrine diseases. *Nat Rev Endocrinol.* 2011;7(6):362–372.
11. Schoneberg T, Liebscher I. Mutations in G protein-coupled receptors: Mechanisms, pathophysiology and potential therapeutic approaches. *Pharmacol Rev.* 2021;73 (1):89–119.
12. Liu T, Ji RL, Tao YX. Naturally occurring mutations in G protein-coupled receptors associated with obesity and type 2 diabetes mellitus. *Pharmacol Ther.* 2022;108044.

13. Zhang R, Xie X. Tools for GPCR drug discovery. *Acta Pharmacol Sin*. 2012; 33(3):372–384.
14. Goswami DB, Ogawa LM, Ward JM, Miller GM, Vallender EJ. Large-scale polymorphism discovery in macaque G-protein coupled receptors. *BMC Genomics*. 2013; 14(1):703.
15. Sriram K, Insel PA. G protein-coupled receptors as targets for approved drugs: How many targets and how many drugs? *Mol Pharmacol*. 2018;93(4):251–258.
16. Fletcher EL, Jobling AI, Vessey KA, Luu C, Guymer RH, Baird PN. Animal models of retinal disease. *Prog Mol Biol Transl Sci*. 2011;100:211–286.
17. Prabhakar S. Translational research challenges: Finding the right animal models. *J Invest Med*. 2012;60(8):1141–1146.
18. Hatziioannou T, Evans DT. Animal models for HIV/AIDS research. *Nat Rev Microbiol*. 2012;10(12):852–867.
19. Rogers CS. Engineering large animal species to model human diseases. *Curr Protoc Hum Genet*. 2016;90. 15 19 11–15 19 14.
20. Zhao J, Lai L, Ji W, Zhou Q. Genome editing in large animals: Current status and future prospects. *Natl Sci Rev*. 2019;6(3):402–420.
21. Santschi EM, Purdy AK, Valberg SJ, Vrotsos PD, Kaese H, Mickelson JR. Endothelin receptor B polymorphism associated with lethal white foal syndrome in horses. *Mamm Genome*. 1998;9(4):306–309.
22. Kijas JM, Wales R, Tornsten A, Chardon P, Moller M, Andersson L. Melanocortin receptor 1 (*MC1R*) mutations and coat color in pigs. *Genetics*. 1998;150(3):1177–1185.
23. Palacios E, Digilio L, McClure HM, et al. Parallel evolution of CCR5-null phenotypes in humans and in a natural host of simian immunodeficiency virus. *Curr Biol*. 1998; 8(16):943–946.
24. Kim KS, Larsen N, Short T, Plastow G, Rothschild MF. A missense variant of the porcine melanocortin-4 receptor (MC4R) gene is associated with fatness, growth, and feed intake traits. *Mamm Genome*. 2000;11(2):131–135.
25. Vrotsos PD, Santschi EM, Mickelson JR. The impact of the mutation causing overo lethal white syndrome on white patterning in horses. *Paper Presented at: AAEP Proceedings*. 2001;47:385–391.
26. Kijas JW, Cideciyan AV, Aleman TS, et al. Naturally occurring *rhodopsin* mutation in the dog causes retinal dysfunction and degeneration mimicking human dominant retinitis pigmentosa. *Proc Natl Acad Sci U S A*. 2002;99(9):6328–6333.
27. Puffenberger EG, Hosoda K, Washington SS, et al. A missense mutation of the endothelin-B receptor gene in multigenic Hirschsprung's disease. *Cell*. 1994;79 (7):1257–1266.
28. Samson M, Libert F, Doranz BJ, et al. Resistance to HIV-1 infection in caucasian individuals bearing mutant alleles of the CCR-5 chemokine receptor gene. *Nature*. 1996;382(6593):722–725.
29. Nathans J, Hogness DS. Isolation and nucleotide sequence of the gene encoding human rhodopsin. *Proc Natl Acad Sci U S A*. 1984;81(15):4851–4855.
30. Wald G. Molecular basis of visual excitation. *Science*. 1968;162(3850):230–239.
31. Baylor D. How photons start vision. *Proc Natl Acad Sci U S A*. 1996;93(2):560–565.
32. Fesenko EE, Kolesnikov SS, Lyubarsky AL. Induction by cyclic GMP of cationic conductance in plasma membrane of retinal rod outer segment. *Nature*. 1985;313 (6000):310–313.
33. Cobbs WH, Barkdoll 3rd AE, Pugh Jr EN. Cyclic GMP increases photocurrent and light sensitivity of retinal cones. *Nature*. 1985;317(6032):64–66.
34. Berson EL, Retinitis pigmentosa. The Friedenwald lecture. *Invest Ophthalmol Vis Sci*. 1993;34(5):1659–1676.

35. Hartong DT, Berson EL, Dryja TP. Retinitis pigmentosa. *Lancet.* 2006;368(9549): 1795–1809.
36. Dryja TP, McGee TL, Hahn LB, et al. Mutations within the rhodopsin gene in patients with autosomal dominant retinitis pigmentosa. *N Engl J Med.* 1990;323(19): 1302–1307.
37. Stenson PD, Mort M, Ball EV, Shaw K, Phillips A, Cooper DN. The human gene mutation database: Building a comprehensive mutation repository for clinical and molecular genetics, diagnostic testing and personalized genomic medicine. *Hum Genet.* 2014;133(1):1–9.
38. Athanasiou D, Aguila M, Bellingham J, et al. The molecular and cellular basis of rhodopsin retinitis pigmentosa reveals potential strategies for therapy. *Prog Retin Eye Res.* 2018;62:1–23.
39. Lek M, Karczewski KJ, Minikel EV, et al. Analysis of protein-coding genetic variation in 60,706 humans. *Nature.* 2016;536(7616):285–297.
40. Malanson KM, Lem J. Rhodopsin-mediated retinitis pigmentosa. *Prog Mol Biol Transl Sci.* 2009;88:1–31.
41. Mendes HF, van der Spuy J, Chapple JP, Cheetham ME. Mechanisms of cell death in rhodopsin retinitis pigmentosa: Implications for therapy. *Trends Mol Med.* 2005;11(4):177–185.
42. Rakoczy EP, Kiel C, McKeone R, Stricher F, Serrano L. Analysis of disease-linked rhodopsin mutations based on structure, function, and protein stability calculations. *J Mol Biol.* 2011;405(2):584–606.
43. Cideciyan AV, Jacobson SG, Aleman TS, et al. *In vivo* dynamics of retinal injury and repair in the *rhodopsin* mutant dog model of human retinitis pigmentosa. *Proc Natl Acad Sci U S A.* 2005;102(14):5233–5238.
44. Metallinos DL, Bowling AT, Rine J. A missense mutation in the endothelin-B receptor gene is associated with lethal white foal syndrome: An equine version of Hirschsprung disease. *Mamm Genome.* 1998;9(6):426–431.
45. Yang GC, Croaker D, Zhang AL, Manglick P, Cartmill T, Cass D. A dinucleotide mutation in the endothelin-B receptor gene is associated with lethal white foal syndrome (LWFS); a horse variant of Hirschsprung disease. *Hum Mol Genet.* 1998; 7(6):1047–1052.
46. Santschi EM, Vrotsos PD, Purdy AK, Mickelson JR. Incidence of the endothelin receptor B mutation that causes lethal white foal syndrome in white-patterned horses. *Am J Vet Res.* 2001;62(1):97–103.
47. Chen Z, Kwon D, Jin Z, et al. Natural infection of a homozygous Δ24 CCR5 red-capped mangabey with an R2b-tropic simian immunodeficiency virus. *J Exp Med.* 1998;188(11):2057–2065.
48. Riddick NE, Hermann EA, Loftin LM, et al. A novel CCR5 mutation common in sooty mangabeys reveals SIVsmm infection of CCR5-null natural hosts and efficient alternative coreceptor use *in vivo*. *PLoS Pathog.* 2010;6(8):e1001064.
49. Kuhmann SE, Platt EJ, Kozak SL, Kabat D. Polymorphisms in the CCR5 genes of African green monkeys and mice implicate specific amino acids in infections by simian and human immunodeficiency viruses. *J Virol.* 1997;71(11):8642–8656.
50. Kuhmann SE, Madani N, Diop OM, et al. Frequent substitution polymorphisms in African green monkey CCR5 cluster at critical sites for infections by simian immunodeficiency virus SIVagm, implying ancient virus-host coevolution. *J Virol.* 2001;75 (18):8449–8460.
51. Bunge S, Wedemann H, David D, et al. Molecular analysis and genetic mapping of the rhodopsin gene in families with autosomal dominant retinitis pigmentosa. *Genomics.* 1993;17(1):230–233.

52. Cideciyan AV, Hood DC, Huang Y, et al. Disease sequence from mutant rhodopsin allele to rod and cone photoreceptor degeneration in man. *Proc Natl Acad Sci U S A*. 1998;95(12):7103–7108.
53. Jacobson SG, Kemp CM, Sung CH, Nathans J. Retinal function and rhodopsin levels in autosomal dominant retinitis pigmentosa with rhodopsin mutations. *Am J Ophthalmol*. 1991;112(3):256–271.
54. Kemp CM, Jacobson SG, Roman AJ, Sung CH, Nathans J. Abnormal rod dark adaptation in autosomal dominant retinitis pigmentosa with proline-23-histidine rhodopsin mutation. *Am J Ophthalmol*. 1992;113(2):165–174.
55. Ferretti L, Karnik SS, Khorana HG, Nassal M, Oprian DD. Total synthesis of a gene for bovine rhodopsin. *Proc Natl Acad Sci U S A*. 1986;83(3):599–603.
56. Zhu L, Jang GF, Jastrzebska B, et al. A naturally occurring mutation of the opsin gene (T4R) in dogs affects glycosylation and stability of the G protein-coupled receptor. *J Biol Chem*. 2004;279(51):53828–53839.
57. Saliba RS, Munro PM, Luthert PJ, Cheetham ME. The cellular fate of mutant rhodopsin: Quality control, degradation and aggresome formation. *J Cell Sci*. 2002;115 (Pt 14):2907–2918.
58. Tam BM, Moritz OL. The role of rhodopsin glycosylation in protein folding, trafficking, and light-sensitive retinal degeneration. *J Neurosci*. 2009;29(48):15145–15154.
59. Vaughan DK, Coulibaly SF, Darrow RM, Organisciak DT. A morphometric study of light-induced damage in transgenic rat models of retinitis pigmentosa. *Invest Ophthalmol Vis Sci*. 2003;44(2):848–855.
60. Li T, Sandberg MA, Pawlyk BS, et al. Effect of vitamin A supplementation on rhodopsin mutants threonine-17 → methionine and proline-347 → serine in transgenic mice and in cell cultures. *Proc Natl Acad Sci U S A*. 1998;95(20):11933–11938.
61. Noorwez SM, Kuksa V, Imanishi Y, et al. Pharmacological chaperone-mediated in vivo folding and stabilization of the P23H-opsin mutant associated with autosomal dominant retinitis pigmentosa. *J Biol Chem*. 2003;278(16):14442–14450.
62. Noorwez SM, Ostrov DA, McDowell JH, Krebs MP, Kaushal S. A high-throughput screening method for small-molecule pharmacologic chaperones of misfolded rhodopsin. *Invest Ophthalmol Vis Sci*. 2008;49(7):3224–3230.
63. Chen Y, Chen Y, Jastrzebska B, et al. A novel small molecule chaperone of rod opsin and its potential therapy for retinal degeneration. *Nat Commun*. 2018;9(1):1976.
64. Behnen P, Felline A, Comitato A, et al. A small chaperone improves folding and routing of rhodopsin mutants linked to inherited blindness. *iScience*. 2018;4:1–19.
65. Tao YX, Conn PM. Chaperoning G protein-coupled receptors: From cell biology to therapeutics. *Endocr Rev*. 2014;35(4):602–647.
66. Tao YX, Conn PM. Pharmacoperones as novel therapeutics for diverse protein conformational diseases. *Physiol Rev*. 2018;98(2):697–725.
67. Marinko JT, Huang H, Penn WD, Capra JA, Schlebach JP, Sanders CR. Folding and misfolding of human membrane proteins in health and disease: From single molecules to cellular proteostasis. *Chem Rev*. 2019;119(9):5537–5606.
68. Yang T, Snider BB, Oprian DD. Synthesis and characterization of a novel retinylamine analog inhibitor of constitutively active rhodopsin mutants found in patients with autosomal dominant retinitis pigmentosa. *Proc Natl Acad Sci U S A*. 1997;94(25): 13559–13564.
69. Inoue A, Yanagisawa M, Kimura S, et al. The human endothelin family: Three structurally and pharmacologically distinct isopeptides predicted by three separate genes. *Proc Natl Acad Sci U S A*. 1989;86(8):2863–2867.
70. Shihoya W, Nishizawa T, Okuta A, et al. Activation mechanism of endothelin ETB receptor by endothelin-1. *Nature*. 2016;537(7620):363–368.

71. Cyr C, Huebner K, Druck T, Kris R. Cloning and chromosomal localization of a human endothelin ETA receptor. *Biochem Biophys Res Commun.* 1991; 181(1):184–190.
72. Sakurai T, Yanagisawa M, Takuwa Y, et al. Cloning of a cDNA encoding a non-isopeptide-selective subtype of the endothelin receptor. *Nature.* 1990;348(6303): 732–735.
73. Schiffrin EL. Endothelin: Potential role in hypertension and vascular hypertrophy. *Hypertension.* 1995;25(6):1135–1143.
74. Hynynen MM, Khalil RA. The vascular endothelin system in hypertension—recent patents and discoveries. *Recent Pat Cardiovasc Drug Discov.* 2006;1(1):95–108.
75. Gariepy CE, Ohuchi T, Williams SC, Richardson JA, Yanagisawa M. Salt-sensitive hypertension in endothelin-B receptor-deficient rats. *J Clin Invest.* 2000;105(7): 925–933.
76. Kedzierski RM, Yanagisawa M. Endothelin system: The double-edged sword in health and disease. *Annu Rev Pharmacol Toxicol.* 2001;41:851–876.
77. Mazzuca MQ, Khalil RA. Vascular endothelin receptor type B: Structure, function and dysregulation in vascular disease. *Biochem Pharmacol.* 2012;84(2):147–162.
78. Davenport AP, Hyndman KA, Dhaun N, et al. Endothelin. *Pharmacol Rev.* 2016;68 (2):357–418.
79. Kandalaft LE, Facciabene A, Buckanovich RJ, Coukos G. Endothelin B receptor, a new target in cancer immune therapy. *Clin Cancer Res.* 2009;15(14):4521–4528.
80. Whitehouse FR, Kernohan JW. Myenteric plexus in congenital megacolon; study of 11 cases. *Arch Intern Med (Chic).* 1948;82(1):75–111.
81. Passarge E. The genetics of Hirschsprung's disease. Evidence for heterogeneous etiology and a study of sixty-three families. *N Engl J Med.* 1967;276(3):138–143.
82. Badner JA, Sieber WK, Garver KL, Chakravarti A. A genetic study of Hirschsprung disease. *Am J Hum Genet.* 1990;46(3):568–580.
83. Amiel J, Sproat-Emison E, Garcia-Barcelo M, et al. Hirschsprung disease, associated syndromes and genetics: A review. *J Med Genet.* 2008;45(1):1–14.
84. Baynash AG, Hosoda K, Giaid A, et al. Interaction of endothelin-3 with endothelin-B receptor is essential for development of epidermal melanocytes and enteric neurons. *Cell.* 1994;79(7):1277–1285.
85. Hosoda K, Hammer RE, Richardson JA, et al. Targeted and natural (piebald-lethal) mutations of endothelin-B receptor gene produce megacolon associated with spotted coat color in mice. *Cell.* 1994;79(7):1267–1276.
86. Lane PW. Association of megacolon with two recessive spotting genes in the mouse. *J Hered.* 1966;57(1):29–31.
87. Yamada T, Ohtani S, Sakurai T, Tsuji T, Kunieda T, Yanagisawa M. Reduced expression of the endothelin receptor type B gene in piebald mice caused by insertion of a retroposon-like element in intron 1. *J Biol Chem.* 2006;281(16):10799–10807.
88. Chakravarti A. Endothelin receptor-mediated signaling in hirschsprung disease. *Hum Mol Genet.* 1996;5(3):303–307.
89. Amiel J, Attie T, Jan D, et al. Heterozygous endothelin receptor B (EDNRB) mutations in isolated Hirschsprung disease. *Hum Mol Genet.* 1996;5(3):355–357.
90. Kusafuka T, Wang Y, Puri P. Novel mutations of the endothelin-B receptor gene in isolated patients with Hirschsprung's disease. *Hum Mol Genet.* 1996;5(3):347–349.
91. Auricchio A, Casari G, Staiano A, Ballabio A. Endothelin-B receptor mutations in patients with isolated Hirschsprung disease from a non-inbred population. *Hum Mol Genet.* 1996;5(3):351–354.
92. Attie T, Till M, Pelet A, et al. Mutation of the endothelin-receptor B gene in Waardenburg-Hirschsprung disease. *Hum Mol Genet.* 1995;4(12):2407–2409.

93. Tanaka H, Moroi K, Iwai J, et al. Novel mutations of the endothelin B receptor gene in patients with Hirschsprung's disease and their characterization. *J Biol Chem.* 1998;273(18):11378–111383.
94. Svensson PJ, Tapper-Persson M, Anvret M, Molander ML, Eng C, Nordenskjold A. Mutations in the endothelin-receptor B gene in Hirschsprung disease in Sweden. *Clin Genet.* 1999;55(3):215–217.
95. Syrris P, Carter ND, Patton MA. Novel nonsense mutation of the endothelin-B receptor gene in a family with Waardenburg-Hirschsprung disease. *Am J Med Genet.* 1999;87(1):69–71.
96. Sangkhathat S, Chiengkriwate P, Kusafuka T, Patrapinyokul S, Fukuzawa M. Novel mutation of endothelin-B receptor gene in Waardenburg-Hirschsprung disease. *Pediatr Surg Int.* 2005;21(12):960–963.
97. Soufir N, Meziani R, Lacapere JJ, et al. Association between endothelin receptor B nonsynonymous variants and melanoma risk. *J Natl Cancer Inst.* 2005;97(17): 1297–1301.
98. Sangkhathat S, Kusafuka T, Chengkriwate P, Patrapinyokul S, Sangthong B, Fukuzawa M. Mutations and polymorphisms of Hirschsprung disease candidate genes in Thai patients. *J Hum Genet.* 2006;51(12):1126–1132.
99. Chen WC, Chang SS, Sy ED, Tsai MC. A *De novo* novel mutation of the *EDNRB* gene in a Taiwanese boy with Hirschsprung disease. *J Formos Med Assoc.* 2006;105(4): 349–354.
100. Lin YC, Lai HS, Hsu WM, Lee PI, Chen HL, Chang MH. Mutation analysis of endothelin-B receptor gene in patients with Hirschsprung disease in Taiwan. *J Pediatr Gastroenterol Nutr.* 2008;46(1):36–40.
101. Sanchez-Mejias A, Fernandez RM, Lopez-Alonso M, Antinolo G, Borrego S. New roles of *EDNRB* and *EDN3* in the pathogenesis of Hirschsprung disease. *Genet Med.* 2010;12(1):39–43.
102. Jabeen R, Babar ME, Ahmad J, Awan AR. Novel mutations of endothelin-B receptor gene in Pakistani patients with Waardenburg syndrome. *Mol Biol Rep.* 2012; 39(1):785–788.
103. Cui L, Wong EH, Cheng G, et al. Genetic analyses of a three generation family segregating Hirschsprung disease and iris heterochromia. *PLoS One.* 2013;8(6): e66631.
104. Granstrom AL, Markljung E, Fink K, et al. A novel stop mutation in the *EDNRB* gene in a family with Hirschsprung's disease associated with multiple sclerosis. *J Pediatr Surg.* 2014;49(4):622–625.
105. Doubaj Y, Pingault V, Elalaoui SC, et al. A novel mutation in the endothelin B receptor gene in a moroccan family with shah-waardenburg syndrome. *Mol Syndromol.* 2015; 6(1):44–49.
106. Widowati T, Melhem S, Patria SY, et al. *RET* and *EDNRB* mutation screening in patients with Hirschsprung disease: Functional studies and its implications for genetic counseling. *Eur J Hum Genet.* 2016;24(6):823–829.
107. Issa S, Bondurand N, Faubert E, et al. *EDNRB* mutations cause Waardenburg syndrome type II in the heterozygous state. *Hum Mutat.* 2017;38(5):581–593.
108. Varga L, Danis D, Drsata J, et al. Novel variants in *EDNRB* gene in Waardenburg syndrome type II and *SOX10* gene in PCWH syndrome. *Int J Pediatr Otorhinolaryngol.* 2021;140:110499.
109. McCabe L, Griffin LD, Kinzer A, Chandler M, Beckwith JB, McCabe ER. Overo lethal white foal syndrome: Equine model of aganglionic megacolon (Hirschsprung disease). *Am J Med Genet.* 1990;36(3):336–340.
110. Gariepy CE, Cass DT, Yanagisawa M. Null mutation of endothelin receptor type B gene in spotting lethal rats causes aganglionic megacolon and white coat color. *Proc Natl Acad Sci U S A.* 1996;93(2):867–872.

111. Tao YX, Segaloff DL. Functional characterization of melanocortin-4 receptor mutations associated with childhood obesity. *Endocrinology*. 2003;144(10):4544–4551.
112. Abe Y, Sakurai T, Yamada T, Nakamura T, Yanagisawa M, Goto K. Functional analysis of five endothelin-B receptor mutations found in human Hirschsprung disease patients. *Biochem Biophys Res Commun*. 2000;275(2):524–531.
113. Fuchs S, Amiel J, Claudel S, Lyonnet S, Corvol P, Pinet F. Functional characterization of three mutations of the endothelin B receptor gene in patients with Hirschsprung's disease: Evidence for selective loss of Gi coupling. *Mol Med*. 2001;7(2):115–124.
114. Imamura F, Arimoto I, Fujiyoshi Y, Doi T. W276 mutation in the endothelin receptor subtype B impairs Gq coupling but not Gi or go coupling. *Biochemistry*. 2000; 39(4):686–692.
115. Hou ZS, Ulloa-Aguirre A, Tao YX. Pharmacoperone drugs: Targeting misfolded proteins causing lysosomal storage-, ion channels-, and G protein-coupled receptors-associated conformational disorders. *Expert Rev Clin Pharmacol*. 2018;11(6):611–624.
116. Lal H, Woodward B, Williams KI. Actions of endothelins and sarafotoxin 6c in the rat isolated perfused lung. *Br J Pharmacol*. 1995;115(4):653–659.
117. Takai M, Umemura I, Yamasaki K, et al. A potent and specific agonist, Suc-[Glu^9, Ala^{11},^{15}]-endothelin-1(8–21), IRL 1620, for the ET_B receptor. *Biochem Biophys Res Commun*. 1992;184(2):953–959.
118. Ihara M, Saeki T, Fukuroda T, et al. A novel radioligand [^{125}I]BQ-3020 selective for endothelin (ET_B) receptors. *Life Sci*. 1992;51(6):PL47–52.
119. Leonard MG, Briyal S, Gulati A. Endothelin B receptor agonist, IRL-1620, reduces neurological damage following permanent middle cerebral artery occlusion in rats. *Brain Res*. 2011;1420:48–58.
120. Lühken G, Fleck K, Pauciullo A, Huisinga M, Erhardt G. Familiar hypopigmentation syndrome in sheep associated with homozygous deletion of the entire endothelin type-B receptor gene. *PLoS One*. 2012;7(12):e53020.
121. Alkhatib G, Combadiere C, Broder CC, et al. CC CKR5: A RANTES, MIP-1alpha, MIP-1beta receptor as a fusion cofactor for macrophage-tropic HIV-1. *Science*. 1996;272(5270):1955–1958.
122. Raport CJ, Gosling J, Schweickart VL, Gray PW, Charo IF. Molecular cloning and functional characterization of a novel human CC chemokine receptor (CCR5) for RANTES, MIP-1beta, and MIP-1alpha. *J Biol Chem*. 1996;271(29):17161–17166.
123. Flanagan CA. Receptor conformation and constitutive activity in CCR5 chemokine receptor function and HIV infection. *Adv Pharmacol*. 2014;70:215–263.
124. Qi B, Fang Q, Liu S, et al. Advances of CCR5 antagonists: From small molecules to macromolecules. *Eur J Med Chem*. 2020;208, 112819.
125. Lee B, Sharron M, Montaner LJ, Weissman D, Doms RW. Quantification of CD4, CCR5, and CXCR4 levels on lymphocyte subsets, dendritic cells, and differentially conditioned monocyte-derived macrophages. *Proc Natl Acad Sci U S A*. 1999;96(9):5215–5220.
126. Castellino F, Huang AY, Altan-Bonnet G, Stoll S, Scheinecker C, Germain RN. Chemokines enhance immunity by guiding naive CD8+ T cells to sites of CD4+ T cell-dendritic cell interaction. *Nature*. 2006;440(7086):890–895.
127. Bachelerie F, Ben-Baruch A, Burkhardt AM, et al. Update on the extended family of chemokine receptors and introducing a new nomenclature for atypical chemokine receptors. *Pharmacol Rev*. 2014;66(1):1–79.
128. Molon B, Gri G, Bettella M, et al. T cell costimulation by chemokine receptors. *Nat Immunol*. 2005;6(5):465–471.
129. Zamilpa R, Kanakia R, Jt C, et al. CC chemokine receptor 5 deletion impairs macrophage activation and induces adverse remodeling following myocardial infarction. *Am J Physiol Heart Circ Physiol*. 2011;300(4):H1418–H1426.

130. Feng Y, Broder CC, Kennedy PE, Berger EA. HIV-1 entry cofactor: Functional cDNA cloning of a seven-transmembrane, G protein-coupled receptor. *Science.* 1996;272(5263):872–877.
131. Choe H, Farzan M, Sun Y, et al. The beta-chemokine receptors CCR3 and CCR5 facilitate infection by primary HIV-1 isolates. *Cell.* 1996;85(7):1135–1148.
132. Deng H, Liu R, Ellmeier W, et al. Identification of a major co-receptor for primary isolates of HIV-1. *Nature.* 1996;381(6584):661–666.
133. Liu R, Paxton WA, Choe S, et al. Homozygous defect in HIV-1 coreceptor accounts for resistance of some multiply-exposed individuals to HIV-1 infection. *Cell.* 1996;86(3):367–377.
134. Benkirane M, Jin DY, Chun RF, Koup RA, Jeang KT. Mechanism of transdominant inhibition of CCR5-mediated HIV-1 infection by ccr5Δ32. *J Biol Chem.* 1997;272(49):30603–30606.
135. Ansari-Lari MA, Liu XM, Metzker ML, Rut AR, Gibbs RA. The extent of genetic variation in the CCR5 gene. *Nat Genet.* 1997;16(3):221–222.
136. Hirsch VM, Olmsted RA, Murphey-Corb M, Purcell RH, Johnson PR. An African primate lentivirus (SIVsm) closely related to HIV-2. *Nature.* 1989;339(6223):389–392.
137. Van Heuverswyn F, Li Y, Neel C, et al. Human immunodeficiency viruses: SIV infection in wild gorillas. *Nature.* 2006;444(7116):164.
138. Martin KA, Wyatt R, Farzan M, et al. CD4-independent binding of SIV gp120 to rhesus CCR5. *Science.* 1997;278(5342):1470–1473.
139. Edinger AL, Blanpain C, Kunstman KJ, Wolinsky SM, Parmentier M, Doms RW. Functional dissection of CCR5 coreceptor function through the use of CD4-independent simian immunodeficiency virus strains. *J Virol.* 1999;73(5):4062–4073.
140. Farzan M, Mirzabekov T, Kolchinsky P, et al. Tyrosine sulfation of the amino terminus of CCR5 facilitates HIV-1 entry. *Cell.* 1999;96(5):667–676.
141. Siciliano SJ, Kuhmann SE, Weng Y, et al. A critical site in the core of the CCR5 chemokine receptor required for binding and infectivity of human immunodeficiency virus type 1. *J Biol Chem.* 1999;274(4):1905–1913.
142. Wei X, Nielsen R. CCR5-32 is deleterious in the homozygous state in humans. *Nat Med.* 2019;25(6):909–910.
143. Dean M, Carrington M, Winkler C, et al. Genetic restriction of HIV-1 infection and progression to AIDS by a deletion allele of the CKR5 structural gene. Hemophilia growth and development study, multicenter AIDS cohort study, multicenter hemophilia cohort study, San Francisco City Cohort, ALIVE study. *Science.* 1996;273 (5283):1856–1862.
144. Miao M, De Clercq E, Li G. Clinical significance of chemokine receptor antagonists. *Expert Opin Drug Metab Toxicol.* 2020;16(1):11–30.
145. Mohamed H, Gurrola T, Berman R, et al. Targeting CCR5 as a component of an HIV-1 therapeutic strategy. *Front Immunol.* 2021;12:816515.
146. Dorr P, Westby M, Dobbs S, et al. Maraviroc (UK-427,857), a potent, orally bioavailable, and selective small-molecule inhibitor of chemokine receptor CCR5 with broad-spectrum anti-human immunodeficiency virus type 1 activity. *Antimicrob Agents Chemother.* 2005;49(11):4721–4732.
147. Gulick RM, Lalezari J, Goodrich J, et al. Maraviroc for previously treated patients with R5 HIV-1 infection. *N Engl J Med.* 2008;359(14):1429–1441.
148. Lieberman-Blum SS, Fung HB, Bandres JC. Maraviroc: A CCR5-receptor antagonist for the treatment of HIV-1 infection. *Clin Ther.* 2008;30(7):1228–1250.

CHAPTER SIX

Melanocortin-1 receptor mutations and pigmentation: Insights from large animals

Ren-Lei Ji and Ya-Xiong Tao*
Department of Anatomy, Physiology and Pharmacology, College of Veterinary Medicine, Auburn University, Auburn, AL, United States
*Corresponding author: e-mail address: taoyaxi@auburn.edu

Contents

Abstract

The melanocortin-1 receptor (MC1R) is a G protein-coupled receptor expressed in cutaneous and hair follicle melanocytes, and plays a central role in coat color determination in vertebrates. Numerous *MC1R* variants have been identified in diverse species. Some of these variants have been associated with specific hair and skin color phenotypes in humans as well as coat color in animals. Gain-of-function mutations of the *MC1R* gene cause dominant or partially dominant black/dark coat color, and loss-of-function mutations of the *MC1R* gene cause recessive or partially recessive red/yellow/pale coat color phenotypes. These have been well documented in a large number of mammals, including human, dog, cattle, horse, sheep, pig, and fox. Higher similarities between large mammals and humans makes them better models to understand pathogenesis

Progress in Molecular Biology and Translational Science, Volume 189
ISSN 1877-1173
https://doi.org/10.1016/bs.pmbts.2022.03.001

of human diseases caused by *MC1R* mutations. High identities in MC1Rs and similar variants identified in both humans and large mammals also provide an opportunity for receptor structure and function study. In this review, we aim to summarize the naturally occurring mutations of *MC1R* in humans and large animals.

Abbreviations

α-MSH	α-melanocyte stimulating hormone
ASIP	agouti-signaling protein
cAMP	cyclic adenosine monophosphate
ERK1/2	extracellular signal regulated protein kinase 1 and 2
GPCR	G protein-coupled receptor
LOF	loss-of-function
MC1R	melanocortin-1 receptor
MC4R	melanocortin-4 receptor
RF	reduced function
RHC	red hair color
TMD	transmembrane domain
WT	wild-type

1. Introduction

Coat color is a conspicuous trait in mammals, particularly in distinguishing farm animal breeds. For human, pigmentation plays a key role in skin physiology, where different types and levels of pigmentation can affect skin cancer predisposition.[1–3] In animals, coat color has intrigued scientists for centuries as a model phenotype for exploring gene action and interaction. Domestic animals, with tremendous coat color diversity, provide excellent models for exploring the genetic basis of coat color.[4] A large number of genes implicated in skin pigmentation have been identified.[5,6] Of those, melanocortin-1 receptor (MC1R) and its antagonist, agouti signaling protein (ASIP), are two major genetic determinants of pigment phenotype in mammals.

MC1R is located in *Extension* locus, and is primarily expressed on the membrane of melanocytes, with a key role in the regulation of pigmentation. Pigmentation depends on melanin, including eumelanin (black/brown) and pheomelanin (red/yellow). The *Extension* locus has epistatic interaction with *Agouti* locus (*ASIP*). Dominant *Extension* allele (E^D) leads to increased eumelanin synthesis and black coat color, while recessive allele (*e*) increases pheomelanin generation and causes red/yellow/pale pigmentation. Only the wild-type (WT) allele (E^+) enables the *Agouti* allele expression to regulate coat color.[7] MC1R activation, by agonist stimulation or constitutive activation, switches melanin synthesis from phaeomelanin to eumelanin, resulting in darker pigmentation. Therefore, constitutive activation of MC1R increases eumelanin production and dark skin or coat, whereas reduced or defective MC1R signaling caused by increased antagonist action or impaired MC1R action results in predominantly pheomelanin synthesis. Thus, MC1R signaling plays an essential role in control of skin or coat color.

Mutations affecting MC1R function determine different coat color phenotypes in various vertebrates, including human,[8] cattle,[7] horse,[9] fox,[10] pig,[11] sheep,[12] and dog.[13,14] Some of these variants have been well established to be associated with specific coat color in these animals. In this chapter, we summarize the naturally occurring *MC1R* mutations in humans and large animals.

2. MC1R and pigmentation

The human *MC1R* was first cloned by two independent groups[15,16] and predicted to encode a 317 amino acid protein as a single exon gene located on chromosome 16q24.3.[17] Further studies established that the *MC1R* gene comprises four exons and yields several transcripts. One alternative splice variant, named MC1R-002 with a 1149 bp open reading frame (ID ENST00000555427), encodes a protein of 382 amino acids with a 66-amino acid C-terminal extension.[18] Another splice variant, designated MC1R-350, shares the same sequence with MC1R and MC1R-002 up to Cys315, but has a 35-amino acid C-terminal extension different from MC1R-002.[19] The transcript (GenBank: NM_002386.4) encoding 317 amino acids with all the structural characteristics of a family A G protein-coupled receptor (GPCR) is now considered the canonical form of MC1R.[20] The *MC1R* has been cloned from 260 different species (https://www.ncbi.nlm.nih.gov/gene).

The MC1R is primarily expressed in melanocytes, particularly in the skin and hair follicles.[21] Additionally, MC1R is expressed in immune cells, including macrophage,[22] monocytes,[22] neutrophils,[23] dendritic cells,[24] and lymphocytes with antigen-presenting and cytotoxic functions.[25] According to the Human Protein Atlas database,[26] human *MC1R* is expressed in almost all 55 human tissues examined (Fig. 1A). Consistent with previous studies, *MC1R* has high expression in melanocytes and immune cells based on

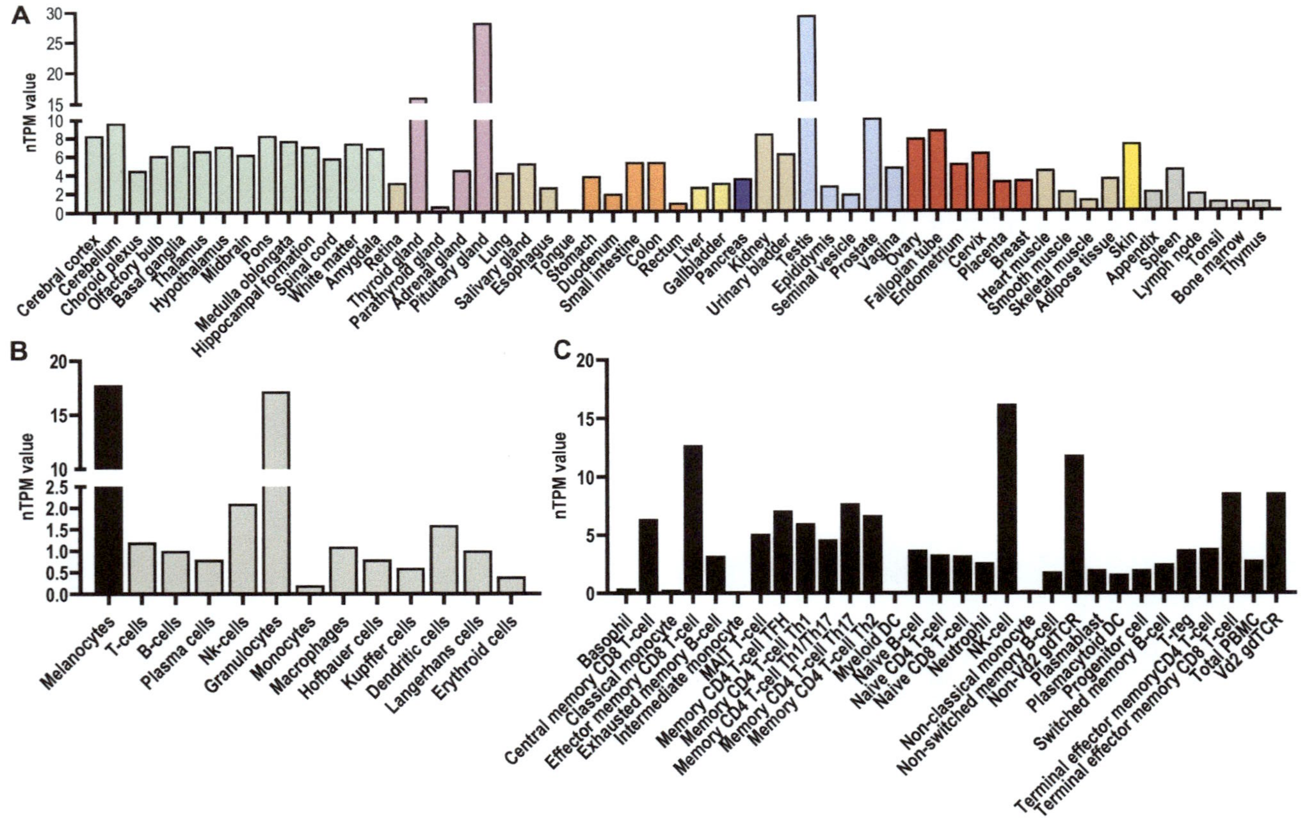

Fig. 1 *MC1R* expression in various tissues (A), single cell (B), and immune cell (C). Tissue and single cell data based on https://www.proteinatlas.org/ (Ref. 26). Immune cell data based on Monaco immune cell dataset.[27] nTPM indicates normalized protein-coding transcripts per million. Color-coding is based on tissue groups with functional features in common.

single cell and immune cell sequencing (Fig. 1B and C). Thus, MC1R has broad expression, suggesting that it has important roles in multiple physiological functions, especially in regulating pigmentation, inflammation, and immunity.[8,28,29] Additionally, MC1R is involved in activation of the DNA damage response and in antioxidant defense.[30]

MC1R is activated by its endogenous ligands, including α-MSH, β-MSH, and ACTH, ligands derived from post-translational processing of the precursor, proopiomelanocortin.[31] MC1R activation is primarily coupled to the stimulatory heterotrimeric G protein (Gαs). Activation of the MC1R stimulates adenylyl cyclase activity, increasing the intracellular level of the second messenger, cyclic adenosine monophosphate (cAMP). Most physiological effects of MC1R on melanocytes, especially induction of eumelanogenesis, are mediated by the cAMP signaling.[32,33] MC1R activation additionally induces pleiotropic signaling, including p38 stress-activated kinase, the extracellular signal regulated protein kinases 1 and 2 (ERK1/2), and AKT pathways.[20,34] ERK signaling is involved in the modulation of both proliferation and melanogenesis.[35–37] MC1R is also regulated by the inverse antagonist, Agouti or ASIP.[38–42] Binding experiments showed that α-MSH and ASIP compete for the same binding sites on the MC1R[41] (reviewed in Refs. 43, 44).

The functional role of MC1R in pigmentation has been well documented (Fig. 2). Prior to the cloning of the *MC1R*, α-MSH was shown to play an important and evolutionarily conserved role in modulating integumental pigmentation, stimulating melanogenesis in mammals, and rapid color changes in poikilotherms, fish, amphibians, and reptiles.[45–48] Genetic studies provide unequivocal evidence that the receptor (*Mc1r*) for α-MSH maps to the *extension* locus in the mouse, and loss-of-function (LOF) mutation in this locus, recessive yellow ($^{e/e}$), leads to yellow coat color.[49] Subsequent studies, including genetic analysis, extensive phenotypic association and structure-function studies, as well as signaling and regulatory mechanisms, suggest that MC1R is a major determinant of cutaneous pigmentation.[50–54]

Skin and hair colors are determined by the ratio of the yellowish-reddish sulfur-containing pheomelanin and the brown/black eumelanin.[55] Melanocytes expressing MC1R are located in the basal layer, surrounded by keratinocytes of epidermal cells. Melanocytes synthesize melanin pigments. MC1R activation in melanocytes switches melanin synthesis from phaeomelanin (red/yellow) to eumelanin (black/brown), and the eumelanin is transferred to keratinocytes, resulting in darkening of skin or hair (Fig. 2). Ultraviolet radiation stimulates the production and release of MC1R agonists from cutaneous keratinocytes and further promotes eumelanin

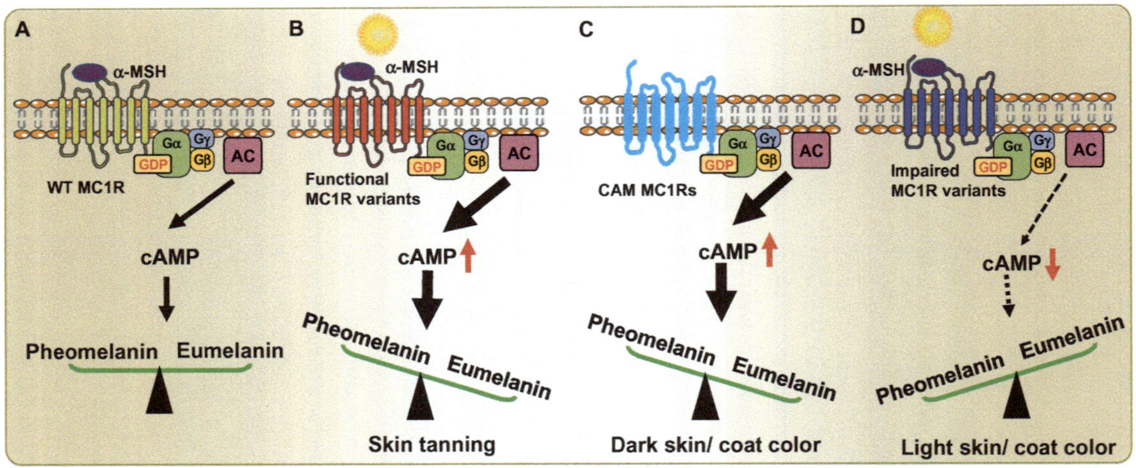

Fig. 2 Role of MC1R on pigmentation and coat color determination. (A) Functional pathway of MC1R on skin and hair pigmentation; (B) UV exposure increases eumelanin production and causes skin tanning; (C) constitutively active mutant (CAM) MC1Rs result in dark skin/coat color; (D) impaired *MC1R* variants cause light skin/coat color.

synthesis, leading to protection against ultraviolet-induced skin damage.[56–59] Numerous *MC1R* variants have been identified from human, mouse, dog, pig, horse, cattle, and fox, supporting MC1R association with pigmentation.[60] Collectively, these results demonstrate that MC1R plays vital roles in regulation of pigmentation.

3. *MC1R* variants in human

MC1R is highly polymorphic, and many of the LOF variants are associated with the RHC (red hair color) phenotype with red hair, fair skin, and poor ability to tan.[61–64] The first study on the association of specific *MC1R* polymorphisms with RHC phenotype, including fair and freckled skin, impaired or absent tanning response to UVR, and propensity to sunburn, was reported by Valverde et al.[8] Since then, numerous *MC1R* variants have been identified in individuals with various hair and skin abnormalities, including RHC, melanoma and other skin cancers.[65] The newly identified and characterized variants keep growing continuously, and approximately 200 nonsynonymous coding region variants of the *MC1R* have been described.[20,62,66] Recently, many *MC1R* variants have been registered in gnomAD v2.1.1 database (https://gnomad.broadinstitute.org/) (Fig. 3) (based on Ref. 67). We integrate all variants previously reported[20,66,69] and newly identified in database, and approximately 420 variants are shown in Fig. 3.

RHC variants have been classified into two groups by strength of association with red hair (strong (designated *R*) and weak (*r*)) RHC alleles.[70] Several *R* alleles are fairly common in Caucasians, especially R151C, R160W, and D294H, accounting for roughly 60% of all red heads. The significant heterogeneity among the Caucasian groups is observed among the three *R* alleles. The frequencies of R151C and R160W are generally lower in the French, Greek, and Italian than in the others; in addition, there is a suggestive gradient of decreasing frequency of *R* variants from Northern to Southern Europe, consistent with the general pigmentation of the populations.[62]

The lightly-pigmented populations have more *MC1R* variants than darkly-pigmented populations, in which African, Asian-Indian, and Papua New Guinean show the smallest number of identified *MC1R* variants,[71] suggesting that there is strong functional constraint on MC1R in Africa and that, in areas with less sun exposure, there is either a relaxation

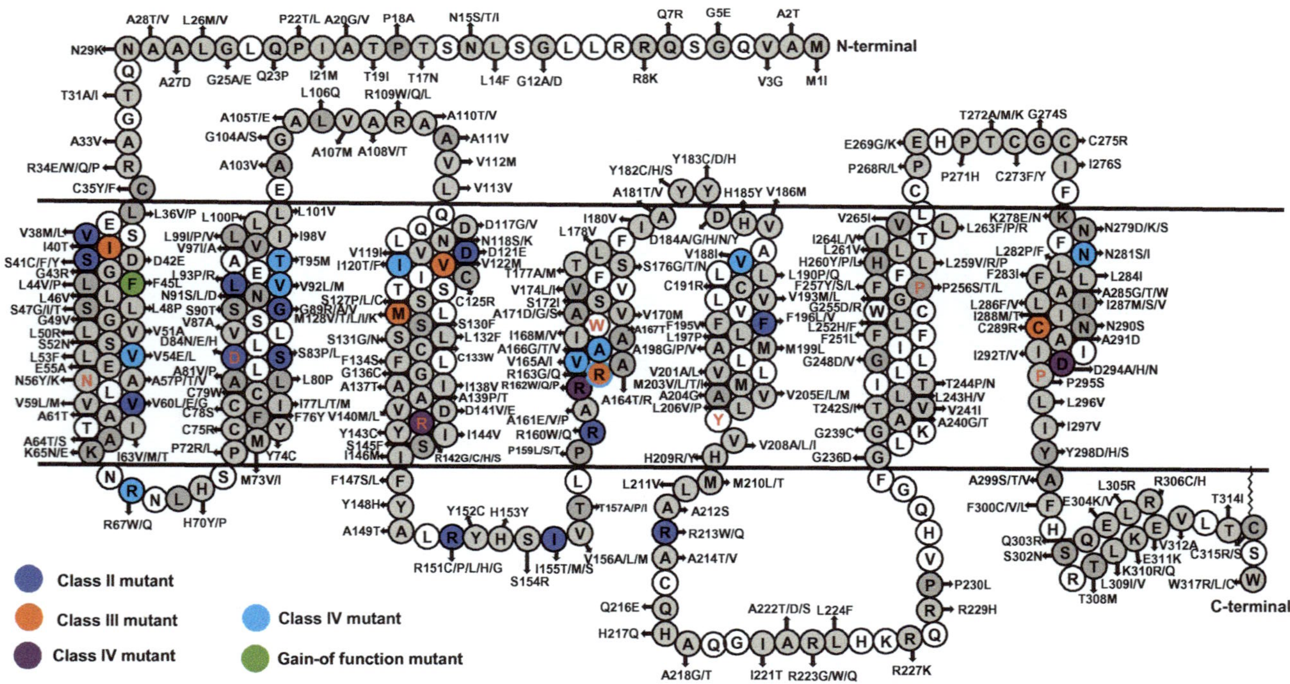

Fig. 3 Structure of human MC1R and naturally occurring variants. The circles filled with grey color were the missense and nonsense mutations from gnomAD database v2.1.1 (https://gnomad.broadinstitute.org/) (Ref. 67 and previously reported). Frameshift mutations were not included in the figure. The circles filled with different colors correspond to different classes of mutations according to Tao.[68] The most conserved residues in transmembrane domains (TMDs) are labeled with red font.

of these constraints or a selection for functionally relevant variants in non-African populations that lead to lighter skin. Importantly, this indicates that conservation of the MC1R coding sequence in Africans where UVR is strong may have been maintained by purifying selection for dark skin protecting against UVR damage.[71]

The co-segregation of one variant with RHC in the patient's family and absence of the variant in ethnically matched controls would provide solid support for the causative role of this variant in pigmentation defect. Functional characterization of the mutant receptor *in vitro* is also indispensable for establishing the role of the variant in the pathogenesis of hair and skin abnormality. Based on life cycle of GPCRs, a classification scheme for melanocortin-4 receptor (MC4R) mutants can also be applied to other GPCRs including the MC1R,[61,72–74] including defective expression (Class I), intracellular retention (Class II), defective ligand binding (Class III), defective signaling (Class IV), and mutants with no known defects (Class V).

Although numerous *MC1R* variants have been identified in subjects with various skin and hair abnormalities and are associated with RHC phenotype, only some of these variants have been studied for their pharmacology *in vitro* and can be cataloged into different classes according to this classification system (Table 1). Apparently, the largest set of the naturally occurring *MC1R* variants is Class II with 13 mutants (intracellularly retained mutants), including V38M, S41F, V60L, S83L, D84E, G89R, L93R, D121E, R151C, I155T, R160W, F196L, and R213W. This is in accordance with *MC3R* and *MC4R* mutations,[68,74,90] and moreover, misfolding and intracellular retention have been described as the primary defect in human diseases caused by GPCR mutations.[61,73] Five variants, I40T, V122M, M128I, R163Q, and C289R, are Class III mutants. Three mutants, R142H, R162P, and D294H, are Class IV mutants. Ten mutants belong to Class V (Table 1).

Several variants are not fully studied and could not be definitively classified. T157I show no signaling and P159T has decreased agonist-induced signaling.[83] C35Y was identified in a Mediterranean population, and is very likely non-functional.[75] Two mutants, C35A and C35G, obtained from lab-generated site-directed mutagenesis, are both LOF mutants.[87] Furthermore, nonsense mutations, including D42X, Y152X, Y74X, Q115X, Y148X, Y152X, R163X, Y182X, Y207X, K226X, are LOF mutants because they lack some transmembrane domains (TMDs).

Table 1 Functional analysis of human MC1R variants.

Mutation	Domain	Variant identification references	Function	Classification according to Tao[68]	Functional analyses references
V38M	TMD1	75	RF, decreased cell surface expression	II	76
I40T	TMD1	77	Normal B_{max}, decreased affinity and potency	III	77
S41F	TMD1	76	RF, decreased cell surface expression	II	76
V51A	TMD1	76	Like WT	V	
V60L	TMD1	78	RF, decreased cell surface expression	II	76
R67Q	ICL1	79	Like WT	V	80,81
S83L	TMD2	82	Defective cell surface expression	II	83
D84E	TMD2	8	RF, decreased cell surface expression	II	84
G89R	TMD2	84	Defective cell surface expression	II	85
V92M	TMD2	8	Like WT	V	84
L93R	TMD2	86	Defective cell surface expression	II	83,85
T95M	TMD2	8	Like WT	V	87
I120T	TMD3	75	Like WT	V	84
D121E	TMD3	84	RF, decreased cell surface expression	II	83
V122M	TMD3	77	RF, Similar B_{max}, decreased affinity,	III	84
M128T	TMD3	76	Decreased cell surface expression, no binding	III	77
R142H	TMD3	78	RF, normal cell surface expression and binding	IV	76
R151C	ICL2	78	RF, decreased or defected cell surface expression	II	81

Table 1 Functional analysis of human MC1R variants.—cont'd

Mutation	Domain	Variant identification references	Function	Classification according to Tao[68]	Functional analyses references
I155T	ICL2	78	Decreased cell surface expression	II	81,84
R160W	TMD4	78	RF, decreased cell surface expression	II	85
R162P	TMD4	88	Normal binding, defective signaling	IV	81
R163Q	TMD4	78	Decreased or normal cell surface expression, normal signaling or RF	V or III	88
V165I	TMD4	84	Like WT	V	80,83,85
A166G	TMD4	83	Like WT	V	84
V188I	TMD5	84	Like WT	V	83
F196L	TMD5	89	RF, decreased cell surface expression	II	84
R213W	TMD5	82	RF, decreased cell surface expression	II	80
N281S	TMD7	76	Like WT	V	84
C289R	TMD7	76	Normal cell surface expression, no binding	III	76
D294H	TMD7	8	RF, normal cell surface expression and binding	IV	76

All *MC1R* variants currently identified are sorted by amino acid position and domain, including functional data. If functional data of a variants from different groups are conflicting, all the data and corresponding classification are presented. Abbreviations: WT, wide type; RF, reduced function; TMD, transmembrane domain; ICL, Intracellular loop, B_{max}, maximal binding value.

Additionally, F45L in TMD1 is a gain-of-function mutant with higher ligand-induced signaling than the WT MC1R.[84]

LOF or partial LOF MC1R variants have been shown to be associated with lighter skin or coat color. In humans, all *R* alleles impair signaling and are partial LOF mutants (Table 1). The weaker *r* alleles, including

Table 2 Polymorphic *MC1R* variants influencing coat color in animal species.

Species	Variants	Allele	Phenotype	Reference
Cattle	L99P	E^D	Black coat color	7
	G310delG (Y155X)	*e*	Red/white coat color	7
	c.699ins12 bp	E^1	Brown/grey coat color	93,94
Pig	L102P	E^{D1}	Black coat color	11
	D124N	E^{D2}	Black coat color	11
	nt894insCC	E^P	Red/white spots	95
	A243T	*e*	Red coat color	11
Sheep	M73K	E^D	Black coat color	12
	D121N	E^D	Black coat color	12
	R67C	*e*	Light coat color	92
Dog	G78V	E^G	Domino/grizzle phenotype	96
	M264V	E^M	Melanistic mask	96,97
	R306X	*e*	Red/yellow coat color	13,14
	c.816_817delCT(I272X)	e^2	White coat color	98
	R301C	e^A	Light hair color	99
Horse	S83F	*e*	Chestnut color	9
Red fox	C125R	E^A	Red coat color	10
Arctic fox	G5C	*e*	Blue coat color	91
	F280C	*e*	Blue coat color	91

V60L, V92M, and R163Q, have little or no signaling impairments (Table 1). LOF variant causing recessive *e* phenotype (light coat color) is also observed in several animals, including mice,[49] cattle,[7] pig,[11] dog,[13,14] arctic fox,[91] and sheep[92] (Table 2).

Constitutive activation of MC1R has been associated with a darker phenotype. This was first demonstrated in mouse.[49] Since then, constitutively active mutations have been identified in birds,[100–103] cow,[7] sheep,[12] and fox.[10] Human MC1R has very high constitutive activity.[104] Several mutants, including V38M, S41F, V51A, V60L, M128T, R163Q, F196L,

N281S, C289R, have decreased basal activities.[76,80] Similarly, in MC4R and ghrelin receptor, decreased basal signaling is also associated with human diseases (reviewed in Ref. 105–107).

Biased signaling at the MC1R may also contribute to abnormal pigmentation. cAMP signaling of MC1R has major role in regulating eumelanogenesis,[32,33] while ERK signaling is involved in proliferation and melanogenesis.[35–37] Four variants, including V60L, R151C, R160W, and D294H, have defect in cAMP signaling, but similar ERK signaling as the WT MC1R.[80,108,109] In contrast, R163Q has normal or slightly decreased cAMP production, but dramatically decreased ERK activation.[80] E92K in mouse MC1R induces constitutive cAMP production and arrestin recruitment but not ERK signaling or internalization.[110] Biased signaling with important physiological functions has been reported in several other GPCRs.[111,112]

Additionally, about 60 variants in 5'- and 3'-untranslated regions have been discovered. Numerous synonymous variants are also reported. These variants might have potential effects on *MC1R* gene expression or mRNA processing and/or stability. Further studies are needed to address these possibilities.

Interestingly, in the extinct hominid Neanderthals, R307G mutation in *MC1R* was found in both individuals sequenced.[113] R307 is located in Helix 8, a cytoplasmic helix parallel to the cell membrane and roughly perpendicular to the other seven TMDs, first identified from rhodopsin crystal structure.[114] Functional studies showed that the mutant receptor, R307G MC1R, has decreased cell surface expression, although total expression is not affected. Therefore the mutant is partially misfolded, similar to many other *MC1R* mutations as well as mutations in other GPCRs.[63,73,115] Although R307G has similar binding properties as the WT MC1R, it has decreased basal and ligand-stimulated signaling, suggesting that it is a partial LOF mutant.[113] These results suggest that Neanderthals might have pale skin and red hair, like some modern day Caucasians. The lighter skin might be advantageous in assuring adequate ultraviolet-light mediated vitamin D synthesis.

4. *MC1R* variants in large animals

Coat color is an important characteristic of breeds in domestic animals. It is one phenotype actively selected during the domestication of these animals.[116] Thus, it is not surprising that the *MC1R* gene has been extensively studied. Numerous variants have been identified in pig, dog, cattle, sheep, fox, and horse (Table 2). Cattle is the first domestic animal to have *MC1R* variants identified to be associated with different coat color phenotype.[7]

4.1 Cattle

The cattle *MC1R* gene is located on chromosome 18[117] and encodes a protein of 317 amino acids that shares high identity with human MC1R (81%) (Figs. 4 and 5). There are three common alleles (E^+, E^D, and *e*) in cattle *MC1R*, defined by two variant events (Table 2).[7,93,94,118] The E^+ allele (WT) shows different coat colors. The E^D allele (dominant allele) with the g.296T>C substitution, causing a single amino acid substitution from leucine to proline (L99P), is associated with black color. The recessive allele (*e*) is characterized by a frameshift mutation at nucleotide position 310 (310delG), resulting in a premature stop codon with only three putative TMDs (Y155X). The homozygous (*e*/*e*) animals show red/white coat color phenotypes among several cattle breeds.[7,93,94,118–124] Additionally, the E^1 allele with a 12 bp duplicated fragment at position 669 (4 amino acids) was identified in a few breeds and shows brown or grey coat colors.[93,94]

These three common alleles in cattle *MC1R* are not retrieved in several other bovids, such as yak (*Bos grunniens*),[125,126] buffalo,[127] gayal (*Bos frontalis*),[128] Chinese yakow (*Bos grunniens* × *Bos taurus*),[129] Zhoushan cattle,[130] and Sahiwal cattle.[131] There is no evidence that particular *MC1R* SNPs are associated with coat color in these Chinese and Indian cattle, suggesting that coat color could be regulated by other genes in these species. Interestingly, three major alleles are found in North American yak, from the Qinghai-Tibetan Plateau near the turn of the 20th century.[132] However, these three alleles are not involved in controlling coat color in North American yak. Subsequent studies indicate that KIT might be a major regulator in controlling coat color of these breeds.[126,132–134]

Furthermore, three variants, A191V, W223R, and I297T, were identified from Swiss cattle.[94] W223R is associated with brown coat color. Only L99P (E^D allele) was found in Karan Fries bovine (Black-white coat color),[135] and this common allele might be linked to the coat color of the Karan Fries bovine.

Other *MC1R* variants were also identified from different bovids. Six variants, including Q34X, M73L, Q114K, R142P, R229H, and A291T, were identified in yak.[125,126] G104S and M128I were discovered in buffalo.[127] Five variants, including L69P, Q114K, K238N, A291N, and

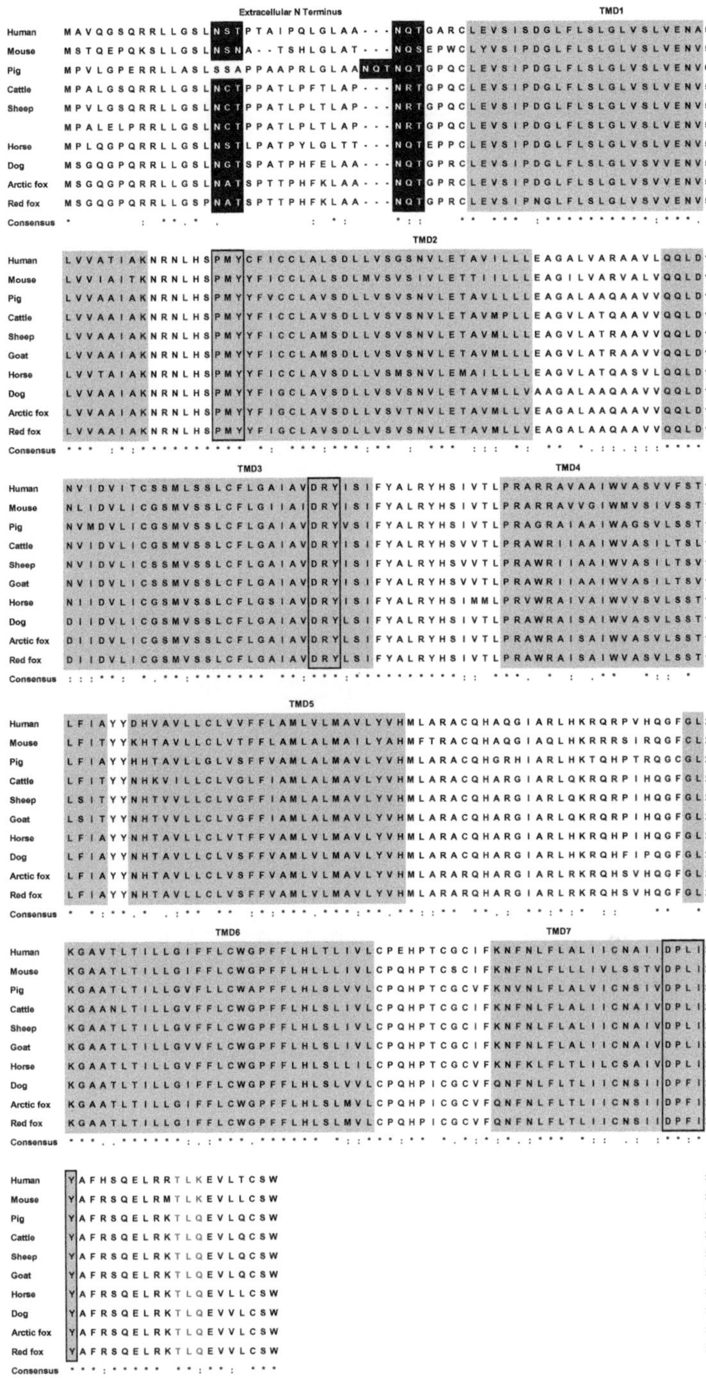

Fig. 4 Comparison of amino acid sequences of MC1R among different species. Transmembrane domains are shown in shaded boxes. PMY, DRY, DPxxY motifs are shown by open boxes. Black boxes denote the consensus sequence for *N*-linked glycosylation sites. The potential palmitoylation sites at the C termini are indicated by red color. Asterisk (*) indicates the same amino acids. MC1Rs: human (NP_002377.4), mouse (NP_032585.2), pig (NP_001008690.1), horse (NP_001108006.1), cattle (NP_776533.1), dog (NP_001014304.2), arctic fox (CAH10741.1), and red fox (XP_025860813.1).

	Human	Mouse	Pig	Cattle	Sheep	Goat	Horse	Dog	Arctic fox	Red fox
Human (317 AA)	100	76	79	81	83	83	80	81	80	80
Mouse (315 AA)		100	73	75	77	76	77	76	75	75
Pig (320 AA)			100	82	83	82	82	84	83	83
Cattle (317 AA)				100	96	95	83	84	83	83
Sheep (317 AA)					100	98	84	84	84	83
Goat (317 AA)						100	83	84	84	83
Horse (317 AA)							100	85	84	84
Dog (317 AA)								100	97	96
Arctic fox (317 AA)									100	99
Red fox (317 AA)										100

Fig. 5 Amino acid sequence identities of MC1Rs. MC1Rs: human (NP_002377.4), mouse (NP_032585.2), pig (NP_001008690.1), horse (NP_001108006.1), cattle (NP_776533.1), dog (NP_001014304.2), arctic fox (CAH10741.1), and red fox (XP_025860813.1).

I297T, were identified from Chinese yakow.[129] G114K and A291T were found in North American yak.[132] L195F and T291A were reported in gayal.[128] Zhang et al. reported seven mutations from Asian cattle, including T25I, R30Q, L95M, V113G, N184H, R301C, and T308M.[126] A291T was also identified from Japanese brown cattle.[136] Six variants, including L99P, F195L, F277S, A285T, D293N, and R302S, were identified from Karan Fries bovine.[135] L195F was also identified from Zhoushan cattle.[130]

Functional studies indicate that E^D and *e* alleles are defective in α-MSH-induced signaling; E^1 allele has higher efficacy and signaling in response to α-MSH; I297T and W223R show similar α-MSH-induced signaling with lower efficacy.[94] E^D (L99P) allele in cattle is analogous to the E^{so} allele mutation in the mouse.[49] E^{so} shares the same phenotype as $E^{so\text{-}3J}$ (dominant black), which has high basal activity (four times higher than E^+). Dark pigmentation due to high constitutive activity has also been reported in mouse[49] and pig.[11] Cattle E^D allele is also associated with dominant black color. Recessive *e* phenotype caused by LOF *MC1R* variants is common in animals.[11,13,14,49,91,92] Collectively, cattle L99P might be a constitutively active mutant, and *e* allele is a non-functional receptor.

For the other variants, Q34X is a LOF receptor and is associated with brown coat color in yaks.[137] M73K in sheep *MC1R* was identified from dominant black sheep and is constitutively active.[12] A mutation at the same position, M73L in cattle, might affect receptor function. No other studies are performed to explore the pharmacological effect of the other variants *in vitro*.

4.2 Pig

Classic genetic analyses indicates that four alleles at the *Extension* (*E*) locus are present in pigs,[138] including E^+ for WT, E^D for dominant black, E^P

for black spotting, and *e* for recessive red (Table 2). Subsequent studies revealed that the *Extension* locus corresponds to *MC1R*. Seven *MC1R* alleles (corresponding to the four phenotypically defined alleles) have been well established in pig (Table 2).[11,95,139,140] The characterization of *MC1R* polymorphism as well as mitochondrial and nuclear DNA have provided compelling evidence that pig domestication has involved both an European and an Asian subspecies of the wild boar and that some European breeds have a hybrid origin.[11,95,139] This is evident from the fact that the two *MC1R* sequences differ by five nucleotide substitutions, associated with the dominant black phenotype in European and Asian domestic pigs.[11,95,139]

Of note, pig *MC1R* encodes a protein of 320 amino acids with an insertion of three codons after codon 31 compared with *MC1R* (317 amino acids) in other mammals (Figs. 4 and 5). Early publications by Kijas et al.[11,95] reported pig MC1R as having 315 amino acids. Therefore, the numbering of codons in the pig differs from other mammals and earlier reports on pig MC1R. For example, L99P, A161V, and A240T in early reports correspond to L102P, A164V, and A243T, respectively (Table 2). No functional study has been performed on these pig mutations. The following discussion on possible causative mutations is based on the nature of the amino acid substitution and comparative data from other species.

Two variants, L102P in TMD2 and D124N in TMD3, correspond to dominant black E^{D1} and $E^{D2.}$, respectively. In Asian breeds, black coat color depends on the occurrence of the E^{D1} allele (L102P),[11,141–143] whereas the black coat is regulated by the E^{D2} allele (D124N) in European breeds.[11] Corresponding to pig L102P, L99P in cattle is associated with dominant black color.[7] L102P in pig, like L99P in cattle, might be a constitutively active mutant. For D124N in pig MC1R, the same mutation, D121N, is associated with dominant black color in sheep.[12]

E^P is a unique *MC1R* allele in pig, and it consists of two causative mutations, an insertion of two nucleotides at codon 22 (nt894insCC, numbering based on transcription initiation site) causing a frameshift and D124N in TMD3. The frameshift is expected to cause complete LOF (premature stop codon at codon 56), and thus E^P homozygotes should be completely associated with solid red coat color.[95] However, the fact is that it is usually associated with black spots on a red or white background. The insertion mutation, expanding CCCCCC to CCCCCCCC, might be unstable. Black spots might be due to somatic reversion events that restore the coding frame and receptor function.[95] The results of RT-PCR analysis from black and red areas are consistent with this hypothesis. The phenotype of E^P/E^P

homozygotes varies considerably, ranging from solid red in Tamworth pigs, red with black spots in Swedish Linderöd pigs, white with black spots in Pietrain pigs, to almost solid black in Berkshire pigs.[95] In Qingyu pig, similar mutation (g.462-463CC insertion) is also detected, and is associated with white with black spotted color.[144] Whether there is similar reversion with this mutation is not clear.

Red coat color is caused by a recessive *e* allele with two amino acid substitutions, A164V in TMD2 and A243T in TMD6. A164V is less likely to be causative because horse MC1R has valine at this position.[9] We propose that A243T is most likely the causative mutation here.

4.3 Sheep

Mutations of the *MC1R* gene causing variation in coat color have also been described in sheep (Table 2). Two major mutations (dominant black E^D allele), M73K and D121N, show complete coseggregation with dominant black coat color in several sheep breeds, including Norwegian Dala,[12] Damara sheep,[145] Massese sheep,[146] Brazilian Creole sheep,[147] Minxian Black-fur sheep,[148] Saudi indigenous sheep,[149] and Swedish Finewool sheep.[150] R67C was only detected in Valle del Belice sheep with light coat color, and represents the recessive *e* allele.[92] However, this mutation has not been identified in other sheep breeds.

M73K in sheep *MC1R* might be a constitutively active mutant. M71K mutation in mouse (corresponding to M73K in sheep) is constitutively active, with decreased maximal binding and increased affinity to α-MSH. Mouse D119N (corresponding to D121N in sheep) has decreased binding affinity (by 20 times)[151] and capacity,[12] with a low basal activity and defect in α-MSH-induced signaling.[151] The loss of function pharmacology with D121N is not consistent with the dominant black phenotype, although D124N in pig *MC1R* is also associated with black coat color.[11]

R67C mutation in sheep *MC1R* associated with light coat color might have important functional effects. The same mutation (R65C) has been identified in a beach mouse subspecies with adaptive pale coat color.[152] R65C was also reported in mammoths.[153] *In vitro* functional study indicates that R65C in bench mouse and mammoth shows a reduction in basal and ligand-stimulated cAMP generation.[152,153]

Several other variants were also identified in different sheep breeds: Q151R and S262L in Swedish Finewool,[150] K307N in thinhorn sheep,[154] D105N in Chinese Tan sheep,[155] K226E, F250, and F257Y in Mongolian sheep.[156,157] No functional study has been performed on these mutations.

Interestingly, dominant black E^D allele described in sheep are not present in goat. Nine mutations, including A81V, Q225X, K226E, Y238C, V250V, G255D, P257T, V265I, and C267W, were found in several goat breeds.[158–161] However, there is no evidence that these mutations are completely associated with different coat color in goat.

Q225X, identified from Girgentana and Derivata di Siria goats, has only four putative TMDs and is likely a LOF mutant.[158] It is not in complete association with red spots in homozygous condition in Girgentana goat. Only 33% of this allele was found in the completely red Derivata di Siria goat. These indicate that additional genes might play a major role in controlling coat color in these goats.

C267W was discovered in all Murciano-Granadina black goats, either homozygous or heterozygous,[158] as well as in black head goats (Maltese goats and crossbred goat (Boer goats x Macheng black goats)).[158,160] The conserved cysteine C267 in the third extracellular loop is essential for structural integrity of MC4R and might be involved in forming a disulfide bond.[162] Functional studies showed that C267W goat MC1R has normal cell surface expression but does not exhibit any binding and little to no response to α-MSH stimulation (Class III mutant).[160] Similar phenomenon is also found in human MC1R in that C267G shows a complete loss of binding.[163] However, LOF or partial LOF variants show light coat color in other animals. Thus, black coat color might be determined by other genes in these goat breeds.

F250V and G255D are both located at TMD6. The residue corresponding to F250 in MC1R, F254 in MC4R, is indispensable for α- and β-MSH binding.[164] This conserved residue ($Phe^{6.44}$) is also essential for TMD6 rotation.[165,166] Functional study suggests that F250V has normal cell surface expression and defects in binding (Class III mutant).[160] The defective F250V mutation might be responsible for the white body in Maltese, Boer and crossbred (Boer goats x Macheng black goats) goats.[158,160] The c.G764A allele (G255D) was found in all red Palmera goats, suggesting an association between this site and red color in this breed. This association is supported by functional study that G255D mutation (normal cell surface expression) results in a total LOF (Class III mutant), with no basal and agonist-stimulated activities in the cAMP and ERK1/2 pathway.[160]

V265I has similar pharmacological properties as the WT MC1R in ligand binding and signaling (including cAMP and ERK1/2).[160] However, K226E shows biased signaling with normal cAMP signaling and impaired ligand-stimulated ERK1/2 activation.[160] These results cannot fully explain the phenotypes. In summary, we speculate that other genes, in addition to *MC1R*, might play a major role in controlling coat color in these goat breeds.

4.4 Dog

Studies on the molecular background of coat color variation in dogs identified seven alleles: e^{1-3}, e^{A}, *E* (WT), E^{G} and E^{M} (in order of increasing black dominance) (for a comprehensive review on canine coat pigmentation, see Refs. 167, 168). Earlier studies identified ten *MC1R* polymorphic variants in different dog breeds.[13,14,169] Among them, six, including G78V, S90G, I272X, M264V, R301C, and R306X, are responsible for dog coat color (Table 2). The other four variants, including A105T, P159Q, H270D, and Q278K, may not be associated with coat color.

The dog recessive e^{1} allele, a single C>T mutation at 916 nucleotide position, causing a premature stop codon at 306 (R306X), missing 11 amino acids, was independently identified by two groups.[13,14] This allele was found in four breeds: Golden Retriever, Yellow Labrador, Labrador retrievers, and Irish Setter.[13,14,170] R306X might be a LOF receptor, resulting in red/yellow coat color in these dogs. Additionally, this allele was discovered in Akita-Inu dogs[171] and Australian Cattle dogs,[98] and is linked to white coat color. Another recessive *e* allele (e^{2}) was reported only in Alaskan and Siberian Huskies.[98] The e^{2} allele with a 2-bp deletion in the coding sequence, c.816_817delCT, results in a premature termination at codon 272 (I272X), missing TMD7 and the C-terminal tail. This LOF allele is associated with white coat color.

e^{A} allele is a mutation (C>T) changing an arginine to cysteine at position 301 of the amino acid chain (R301C).[99,172] It was first identified from over 10,000 years old Siberian and South-Eastern European dogs as well as two modern day breeds (the Alaskan Malamute and the Siberian Husky).[99] The variant might cause light coat color. R301C variant was also found in 35 dog breeds or breed varieties and was shown to be associated with partial recessive red phenotypes.[172] A mutation at the same codon (301) together with two other polymorphisms (T21A/R67C/R301S) was found in 43,000 year-old woolly mammoth, leading to an almost complete loss in basal activity and a 65% reduction in α-MSH-induced signaling, sufficient to result in substantially lighter hair color.[153] The effect of the R301C mutation alone on coat color is not well established and we speculate that R301C may have reduced function.

The E^G allele is a g.233G > T substitution, resulting in an amino acid change from a glycine to valine at residue 78 (G78V). This allele was observed only in Salukis and Afghan Hounds.[96] G78V produces the "domino" phenotype in Afghan Hound and the "grizzle" phenotype in Saluki breeds.[96] E^M allele, an A to G transition in the first position of codon 264, resulting in a methionine to valine (M264V) change, was shown to be associated with black melanistic mask in 13 dog breeds.[96,97]

S90G was discovered in several dog breeds, including Newfoundland, Flat-coated Retriever, Black Labrador, and Kunming dog.[14,173,174] S90G was detected from black dogs, but not all black dogs have S90G, indicating that dominant black coat color in dogs can be caused by *MC1R* variants or other genes. S90 in TMD2 is evolutionarily conserved across all known MC1R sequences. Alterations in the polarity and/or charge of residues in TMD2 are known to cause constitutive activation in naturally occurring *MC1R* of mice,[49] cattle,[7] and pigs.[11] Thus, S90G variant might be constitutively active.

Several mutations were also observed in other canids.[169,175] Eight *MC1R* variants were identified in Chinese raccoon dog, including G16A, E25K, V97G, V101A, A102E, N184S, F230S, and F232H. In coyotes, four mutations, F45L, G102A, P159Q, and R306X, were found.[175] Similar to dog R306X, R306X is associated with coyote white color.[175] There are no reports on the functional properties of these mutant receptors.

Different from the other mammalian species reviewed herein, there are no constitutively active *MC1R* mutations reported in dogs that cause dominant black coat color. An exciting study from Barsh and colleagues on dog coat color genetics is the dentification of the *K* locus[176] as canine β-defensin 103 (*CBD103*), the major β-defensin gene expressed in canine skin.[177] In addition to its high expression in the skin, *CBD103* is also expressed the nasal cavity, including the nares as well as respiratory and olfactory epithelia.[178] A mutation with a 3-bp deletion in coding sequence causing the loss of glycine 23 (CBD103ΔG23) was identified.[177] This allele is found in all of the Labrador retrievers (75% WT/ΔG23; 25% ΔG23/ΔG23) and Golden retrievers (54% WT/ΔG23; 46% ΔG23/ΔG23).[179] Copy number variations, with 2 to 4 copies per diploid genome, were also found in these two breeds.[179] CBD103 cannot stimulate MC1R in generating cAMP signaling. However, it can bind to the MC1R and compete with ASIP for binding to the MC1R, increasing the apparent "basal" signaling. CBD103ΔG23 binds to MC1R with higher affinity than WT CBD103 and causes increased "basal" MC1R signaling, increasing eumelanin synthesis and black coat color.[177] The same mutation is also responsible for melanism in coyote and North American gray wolf.[180]

4.5 Horse

Horse *MC1R* is located on chromosome 3 and encodes a 317 amino acid protein, sharing high identity with human MC1R (81%) (Figs. 4 and 5). A missense mutation (S83F) in horse *MC1R* is associated with chestnut color (recessive *e* phenotype) (Table 2).[9] A high percentage of horses in several breeds has shown a complete association between this mutation and the chestnut phenotype,[9,181,182] supporting the idea that it is a causative mutation. Thus, S83F might be a nonfunctional receptor. LOF variant caused recessive *e* phenotype is also observed in mice,[49] cattle,[7] pig,[11] dog,[13,14] arctic fox,[91] and sheep.[92] No other *MC1R* mutation with phenotypic effect has yet been identified in horses.

4.6 Fox

MC1R variants associated with different coat color are also reported in arctic fox (*Vulpes lagopus*) and red fox (*Vulpes vulpes*). Fox *MC1R*, with a 954-bp coding region, is predicted to encode a protein of 317 amino acids. Red fox MC1R shares 98% identity with that of arctic fox, with only 6 amino acid differences at positions 5, 14, 42, 90, 97, and 280 (Fig. 5). C125R *MC1R* mutation was found only in red fox, accounting for the darkly pigmented red fox (Table 2).[10,169,183] G5C and F280C have been reported in association with the blue coat variants in the arctic fox (Table 2).[91,169,183]

Functional studies indicates that C125R shows no ligand binding ability and might have high basal activity in cAMP signaling.[10] The same position in mouse MC1R, C123R, shows no binding or similar binding affinity and high basal activity.[10] High constitutive activity contributing to dark pigmentation has also been reported in other species.[7,11,49] G5C and F280C might be LOF mutants. Similar phenomenon is observed in several species, including mouse,[49] cattle,[7] pig,[11] dog,[13,14] and sheep.[92]

In addition, eight mutations, including G16A, A20T, E25K, A102E, C215R, H225R, F230S, and P232H, were identified in both red fox and arctic fox.[169,183] Five variants, L14P, L14T, D42N, V53I, and V97I, were only identified in red fox.[169,183] The functional properties of these mutant receptors are not known.

5. Potential therapeutic strategies and perspectives

Our view of the *MC1R* gene, its protein structure, regulation, and associated functions, have increased substantially over the past 30 years. It has been well established that this gene is a major determinant of skin and hair pigmentation, sun sensitivity, and susceptibility to skin cancer. In humans, many variants in *MC1R* with impaired protein functions have been identified. They are strongly associated with RHC (including D84E, R151C, R160W, and D294H) and increased risk of skin cancer. Thus, targeting MC1R, using agonists and antagonists to modulate the receptor signaling, might represents a therapeutic strategy.

Since most of the GPCR mutations associated with human disease are misfolded and intracellularly retained, approaches promoting increased cell surface expression could potentially be of therapeutic value. Pharmacological chaperones (pharmacoperones), small molecules that specifically bind to the target protein, stabilizing the native conformation or facilitating the folding of non-native intermediates into native conformation, could cross the cell membrane, resulting in increased cell surface expression of the receptors. Pharmacoperones have been applied to correct the trafficking and signaling of several GPCRs, including rhodopsin, calcium-sensing receptor, arginine vasopressin receptor 2, MC4R, and gonadotrophin-releasing hormone receptor,[73,184] as well as many other proteins.[185,186] The largest set of the naturally occurring *MC1R* variants is Class II (intracellularly retained mutants). Therefore, it is important and promising to explore MC1R-specific pharmacological chaperones and determine their efficacy in correcting the trafficking and signaling of Class II MC1R mutants.

New ligands, especially allosteric agonists, bind to receptor at sites different from endogenous ligands. It might be a promising strategy to correct Class III and IV mutants. Mutations with defects in binding or signaling, including I40T, V122M, M128I, R142H, R162P, R163Q, D294H, and C289R, might be rescued by new agonists or allosteric ligands. Many selective human MC1R agonists have been developed.[187–190] Importantly, one clinically successful approach to target MC1R, afamelanotide, an α-MSH analog, has been approved by Food and Drug Administration of the United States.[191] This has been shown to be an effective and safe way in the treatment of the photosensitivity related to erythropoietic protoporphyria.[192]

An alternative strategy to correct MC1R signaling is increased MC1R palmitoylation.[193–195] *MC1R* variants with RHC were shown to have decreased palmitoylation.[193] Small molecules that inhibit the depalmitoylation enzyme, acyl-protein thioesterase 2 (APT2), increase palmitoylation of RHC variants to WT levels, resulting in reduced melanoma size and prolonged survival in $Mc1r^{R151C}$-mutant mice exposed to UV.[194] Of note, palmitoylation is commonly used to regulate many signaling molecules beyond MC1R signaling.[196] Thus, how clinically useful this approach might be needs to be carefully assessed.

In animals, MC1R has a conserved function in regulating coat color and can be traced back to 43,000 years ago.[153] Reduced or LOF *MC1R* variants causing light coat color were identified in 10,000 years old dogs and 43,000 years old woolly mammoth.[99,153] Recessive *e* phenotype caused by LOF *MC1R* variants is also observed in mice,[49] cattle,[7] pig,[11] dog,[13,14] arctic fox,[91] horse,[9] and sheep.[92] In contrast, mutations with high constitutive activity contributing to dark pigmentation were found in mouse,[49] cattle,[7] red fox,[10] and pig.[11]

Given that MC1R plays a crucial role in the evolution of coat color, assessing the rate of amino acid substitution in the coding region of this gene may help to elucidate long-term evolutionary trends in coat color in a specific lineage, for example, Asian and European pig breeds.[11] In addition, MC1R could have evolved differently across taxa over time in response to environmental changes and intraspecies competition (sexual selection).[197–199] MC1R system also enables us to obtain a comprehensive view of gene evolution influenced by environmental and genomic constraints, how genetic architecture is affected by the environment through long-term adaptation, and how genetic architecture influences gene evolution in adaptive evolution.

6. Conclusion

MC1R plays an essential role in coat color determination. In humans, many *MC1R* variants are strongly associated with RHC and increased risk of melanoma. Numerous *MC1R* mutations are also identified in diverse species, including dog, cattle, horse, sheep, pig, and fox. Constitutively active mutations in *MC1R* are associated with dominant or partially dominant black/dark coat color, whereas LOF mutations in *MC1R* are associated with recessive or partially recessive light coat color phenotype. Naturally occurring *MC1R* mutations in these animals provide also an opportunity to explore receptor structure and function of human MC1R. Additionally, MC1R system enables us to gain a comprehensive understanding of gene evolution in response to environmental changes and intraspecies competition.

Acknowledgment

Ren-Lei Ji received fellowship from China Scholarship Council, People's Republic of China.

Conflict of interest

The authors declare that there is no conflict of interest that would prejudice the impartiality of this study.

References

1. Rees JL, Healy E. Melanocortin receptors, red hair, and skin cancer. *J Invest Dermatol.* 1997;2(1):94–98.
2. Gloster Jr HM, Neal K. Skin cancer in skin of color. *J Am Acad Dermatol.* 2006;55 (5):741–760. quiz 761-744.
3. Scherer D, Kumar R. Genetics of pigmentation in skin cancer—a review. *Mutat Res.* 2010;705(2):141–153.
4. Andersson L. Genetic dissection of phenotypic diversity in farm animals. *Nat Rev Genet.* 2001;2(2):130–138.
5. Sturm RA, Duffy DL. Human pigmentation genes under environmental selection. *Genome Biol.* 2012;13(9):248.
6. Pavan WJ, Sturm RA. The genetics of human skin and hair pigmentation. *Annu Rev Genomics Hum Genet.* 2019;20:41–72.
7. Klungland H, Vage DI, Gomez-Raya L, Adalsteinsson S, Lien S. The role of melanocyte-stimulating hormone (MSH) receptor in bovine coat color determination. *Mamm Genome.* 1995;6(9):636–639.
8. Valverde P, Healy E, Jackson I, Rees JL, Thody AJ. Variants of the melanocyte-stimulating hormone receptor gene are associated with red hair and fair skin in humans. *Nat Genet.* 1995;11(3):328–330.
9. Marklund L, Moller MJ, Sandberg K, Andersson L. A missense mutation in the gene for melanocyte-stimulating hormone receptor (*MC1R*) is associated with the chestnut coat color in horses. *Mamm Genome.* 1996;7(12):895–899.
10. Våge DI, Lu D, Klungland H, Lien S, Adalsteinsson S, Cone RD. A non-epistatic interaction of agouti and extension in the fox. *Nat Genet.* 1997;15(3):311–315.
11. Kijas JM, Wales R, Tornsten A, Chardon P, Moller M, Andersson L. Melanocortin receptor 1 (*MC1R*) mutations and coat color in pigs. *Genetics.* 1998;150(3):1177–1185.
12. Vage DI, Klungland H, Lu D, Cone RD. Molecular and pharmacological characterization of dominant black coat color in sheep. *Mamm Genome.* 1999;10 (1):39–43.
13. Everts RE, Rothuizen J, van Oost BA. Identification of a premature stop codon in the melanocyte-stimulating hormone receptor gene (*MC1R*) in Labrador and Golden retrievers with yellow coat colour. *Anim Genet.* 2000;31(3):194–199.
14. Newton JM, Wilkie AL, He L, et al. Melanocortin 1 receptor variation in the domestic dog. *Mamm Genome.* 2000;11(1):24–30.

15. Mountjoy KG, Robbins LS, Mortrud MT, Cone RD. The cloning of a family of genes that encode the melanocortin receptors. *Science*. 1992;257(5074):1248–1251.
16. Chhajlani V, Wikberg JE. Molecular cloning and expression of the human melanocyte stimulating hormone receptor cDNA. *FEBS Lett*. 1992;309(3):417–420.
17. Gantz I, Yamada T, Tashiro T, et al. Mapping of the gene encoding the melanocortin-1 (alpha-melanocyte stimulating hormone) receptor (MC1R) to human chromosome 16q24.3 by fluorescence *in situ* hybridization. *Genomics*. 1994;19(2):394–395.
18. Tan CP, McKee KK, Weinberg DH, et al. Molecular analysis of a new splice variant of the human melanocortin-1 receptor. *FEBS Lett*. 1999;451(2):137–141.
19. Rouzaud F, Costin GE, Yamaguchi Y, et al. Regulation of constitutive and UVR-induced skin pigmentation by melanocortin 1 receptor isoforms. *FASEB J*. 2006;20(11):1927–1929.
20. Herraiz C, Garcia-Borron JC, Jimenez-Cervantes C, Olivares C. MC1R signaling. Intracellular partners and pathophysiological implications. *Biochim Biophys Acta*. 2017;1863:2448–2461.
21. Slominski A, Wortsman J, Luger T, Paus R, Solomon S. Corticotropin releasing hormone and proopiomelanocortin involvement in the cutaneous response to stress. *Physiol Rev*. 2000;80(3):979–1020.
22. Star RA, Rajora N, Huang J, Stock RC, Catania A, Lipton JM. Evidence of autocrine modulation of macrophage nitric oxide synthase by α-melanocyte-stimulating hormone. *Proc Natl Acad Sci USA*. 1995;92(17):8016–8020.
23. Catania A, Rajora N, Capsoni F, Minonzio F, Star RA, Lipton JM. The neuropeptide α-MSH has specific receptors on neutrophils and reduces chemotaxis in vitro. *Peptides*. 1996;17(4):675–679.
24. Becher E, Mahnke K, Brzoska T, Kalden DH, Grabbe S, Luger TA. Human peripheral blood-derived dendritic cells express functional melanocortin receptor MC-1R. *Ann NY Acad Sci*. 1999;885(1):188–195.
25. Neumann Andersen G, Nagaeva O, Mandrika I, et al. MC_1 receptors are constitutively expressed on leucocyte subpopulations with antigen presenting and cytotoxic functions. *Clin Exp Immunol*. 2001;126(3):441–446.
26. Uhlen M, Fagerberg L, Hallstrom BM, et al. Proteomics. Tissue-based map of the human proteome. *Science*. 2015;347(6220):1260419.
27. Monaco G, Lee B, Xu W, et al. RNA-Seq signatures normalized by mRNA abundance allow absolute deconvolution of human immune cell types. *Cell Rep*. 2019;26 (6):1627–1640 e1627.
28. Bohm M, Luger TA, Tobin DJ, Garcia-Borron JC. Melanocortin receptor ligands: new horizons for skin biology and clinical dermatology. *J Invest Dermatol*. 2006;126 (9):1966–1975.
29. Wang W, Guo DY, Lin YJ, Tao YX. Melanocortin regulation of inflammation. *Front Endocrinol (Lausanne)*. 2019;10:683.
30. Abdel-Malek ZA, Swope VB, Starner RJ, Koikov L, Cassidy P, Leachman S. Melanocortins and the melanocortin 1 receptor, moving translationally towards melanoma prevention. *Arch Biochem Biophys*. 2014;563:4–12.
31. Harno E, Gali Ramamoorthy T, Coll AP, White A. POMC: The physiological power of hormone processing. *Physiol Rev*. 2018;98(4):2381–2430.
32. D'Orazio J, Fisher DE. Central role for cAMP signaling in pigmentation and UV resistance. *Cell Cycle*. 2011;10(1):8–9.
33. Rodriguez CI, Setaluri V. Cyclic AMP (cAMP) signaling in melanocytes and melanoma. *Arch Biochem Biophys*. 2014;563:22–27.
34. Suzuki I, Cone RD, Im S, Nordlund J, Abdel-Malek ZA. Binding of melanotropic hormones to the melanocortin receptor MC1R on human melanocytes stimulates proliferation and melanogenesis. *Endocrinology*. 1996;137(5):1627–1633.

35. Bohm M, Moellmann G, Cheng E, et al. Identification of p90RSK as the probable CREB-Ser133 kinase in human melanocytes. *Cell Growth Differ.* 1995;6 (3):291–302.
36. Cohen C, Zavala-Pompa A, Sequeira JH, et al. Mitogen-actived protein kinase activation is an early event in melanoma progression. *Clin Cancer Res.* 2002;8 (12):3728–3733.
37. Levy C, Khaled M, Fisher DE. MITF: master regulator of melanocyte development and melanoma oncogene. *Trends Mol Med.* 2006;12(9):406–414.
38. Lu D, Willard D, Patel IR, et al. Agouti protein is an antagonist of the melanocyte-stimulating-hormone receptor. *Nature.* 1994;371(6500):799–802.
39. Siegrist W, Drozdz R, Cotti R, Willard DH, Wilkison WO, Eberle AN. Interactions of α-melanotropin and agouti on B16 melanoma cells: evidence for inverse agonism of agouti. *J Recept Signal Transduct Res.* 1997;17(1-3):75–98.
40. Suzuki I, Tada A, Ollmann MM, et al. Agouti signaling protein inhibits melanogenesis and the response of human melanocytes to α-melanotropin. *J Invest Dermatol.* 1997;108 (6):838–842.
41. Ollmann MM, Lamoreux ML, Wilson BD, Barsh GS. Interaction of Agouti protein with the melanocortin 1 receptor in vitro and in vivo. *Genes Dev.* 1998;12(3):316–330.
42. Chai BX, Neubig RR, Millhauser GL, et al. Inverse agonist activity of agouti and agouti-related protein. *Peptides.* 2003;24(4):603–609.
43. Barsh GS, Ollmann MM, Wilson BD, Miller KA, Gunn TM. Molecular pharmacology of Agouti protein *in vitro* and *in vivo*. *Ann N Y Acad Sci.* 1999;885:143–152.
44. Voisey J, van Daal A. Agouti: from mouse to man, from skin to fat. *Pigment Cell Res.* 2002;15(1):10–18.
45. Lerner AB, McGuire JS. Effect of alpha- and beta-melanocyte stimulating hormones on the skin colour of man. *Nature.* 1961;189:176–179.
46. Geschwind II. Change in hair color in mice induced by injection of alpha-MSH. *Endocrinology.* 1966;79(6):1165–1167.
47. Sawyer TK, Hruby VJ, Wilkes BC, Draelos MT, Hadley ME, Bergsneider M. Comparative biological activities of highly potent active-site analogues of alpha-melanotropin. *J Med Chem.* 1982;25(9):1022–1027.
48. Tamate HB, Takeuchi T. Action of the e locus of mice in the response of phaeomelanic hair follicles to alpha-melanocyte-stimulating hormone in vitro. *Science.* 1984;224 (4654):1241–1242.
49. Robbins LS, Nadeau JH, Johnson KR, et al. Pigmentation phenotypes of variant extension locus alleles result from point mutations that alter MSH receptor function. *Cell.* 1993;72:827–834.
50. Healy E, Flannagan N, Ray A, et al. Melanocortin-1-receptor gene and sun sensitivity in individuals without red hair. *Lancet.* 2000;355(9209):1072–1073.
51. Kennedy C, ter Huurne J, Berkhout M, et al. Melanocortin 1 receptor *(MC1R)* gene variants are associated with an increased risk for cutaneous melanoma which is largely independent of skin type and hair color. *J Invest Dermatol.* 2001;117(2):294–300.
52. Rees JL. Genetics of hair and skin color. *Annu Rev Genet.* 2003;37:67–90.
53. Rees JL. The genetics of sun sensitivity in humans. *Am J Hum Genet.* 2004;75 (5):739–751.
54. Rouzaud F, Kadekaro AL, Abdel-Malek ZA, Hearing VJ. MC1R and the response of melanocytes to ultraviolet radiation. *Mutat Res.* 2005;571(1–2):133–152.
55. d'Ischia M, Wakamatsu K, Cicoira F, et al. Melanins and melanogenesis: from pigment cells to human health and technological applications. *Pigment Cell Melanoma Res.* 2015;28(5):520–544.
56. Slominski A, Tobin DJ, Shibahara S, Wortsman J. Melanin pigmentation in mammalian skin and its hormonal regulation. *Physiol Rev.* 2004;84(4):1155–1228.

57. D'Orazio JA, Nobuhisa T, Cui R, et al. Topical drug rescue strategy and skin protection based on the role of Mc1r in UV-induced tanning. *Nature*. 2006;443 (7109):340–344.
58. Cui R, Widlund HR, Feige E, et al. Central role of p53 in the suntan response and pathologic hyperpigmentation. *Cell*. 2007;128(5):853–864.
59. D'Orazio J, Jarrett S, Amaro-Ortiz A, Scott T. UV radiation and the skin. *Int J Mol Sci*. 2013;14(6):12222–12248.
60. Switonski M, Mankowska M, Salamon S. Family of melanocortin receptor (*MCR*) genes in mammals-mutations, polymorphisms and phenotypic effects. *J Appl Genet*. 2013;54(4):461–472.
61. Tao YX. Inactivating mutations of G protein-coupled receptors and diseases: structure-function insights and therapeutic implications. *Pharmacol Ther*. 2006;111(3):949–973.
62. Gerstenblith MR, Goldstein AM, Fargnoli MC, Peris K, Landi MT. Comprehensive evaluation of allele frequency differences of MC1R variants across populations. *Hum Mutat*. 2007;28(5):495–505.
63. Beaumont KA, Liu YY, Sturm RA. The melanocortin-1 receptor gene polymorphism and association with human skin cancer. *Prog Mol Biol Transl Sci*. 2009;88:85–153.
64. Garcia-Borron JC, Abdel-Malek Z, Jimenez-Cervantes C. MC1R, the cAMP pathway, and the response to solar UV: extending the horizon beyond pigmentation. *Pigment Cell Melanoma Res*. 2014;27(5):699–720.
65. Bastiaens MT, ter Huurne JA, Kielich C, et al. Melanocortin-1 receptor gene variants determine the risk of nonmelanoma skin cancer independently of fair skin and red hair. *Am J Hum Genet*. 2001;68(4):884–894.
66. Moore BS, Luo JZ, Stepanchick AN, Mirshahi T. Large scale clinical exome sequencing uncovers the scope and severity of skin disorders associated with *MC1R* genetic variants. *Genet Med*. 2021;23(12):2386–2393.
67. Lek M, Karczewski KJ, Minikel EV, et al. Analysis of protein-coding genetic variation in 60,706 humans. *Nature*. 2016;536(7616):285–297.
68. Tao YX. Molecular mechanisms of the neural melanocortin receptor dysfunction in severe early onset obesity. *Mol Cell Endocrinol*. 2005;239(1-2):1–14.
69. Hu HH, Benfodda M, Dumaz N, et al. A large French case-control study emphasizes the role of rare Mc1R variants in melanoma risk. *Biomed Res Int*. 2014;2014, 925716.
70. Duffy DL, Box NF, Chen W, et al. Interactive effects of MC1R and OCA2 on melanoma risk phenotypes. *Hum Mol Genet*. 2004;13(4):447–461.
71. Makova K, Norton H. Worldwide polymorphism at the *MC1R* locus and normal pigmentation variation in humans. *Peptides*. 2005;26(10):1901–1908.
72. Tao YX, Segaloff DL. Functional characterization of melanocortin-4 receptor mutations associated with childhood obesity. *Endocrinology*. 2003;144(10):4544–4551.
73. Tao YX, Conn PM. Chaperoning G protein-coupled receptors: from cell biology to therapeutics. *Endocr Rev*. 2014;35(4):602–647.
74. Liu T, Ji RL, Tao YX. Naturally occurring mutations in G protein-coupled receptors associated with obesity and type 2 diabetes mellitus. *Pharmacol Ther*. 2022;108044.
75. Fargnoli MC, Chimenti S, Keller G, Hofler H, Peris K. Identification of four novel melanocortin 1 receptor (MC1R) gene variants in a Mediterranean population. *Hum Mutat*. 2003;21(6):655.
76. Pérez Oliva AB, Fernéndez LP, DeTorre C, et al. Identification and functional analysis of novel variants of the human melanocortin 1 receptor found in melanoma patients. *Hum Mutat*. 2009;30(5):811–822.
77. Jimenez-Cervantes C, Germer S, Gonzalez P, Sanchez J, Sanchez CO, Garcia-Borron JC. Thr40 and Met122 are new partial loss-of-function natural mutations of the human melanocortin 1 receptor. *FEBS Lett*. 2001;508(1):44–48.

78. Box NF, Wyeth JR, O'Gorman LE, Martin NG, Sturm RA. Characterization of melanocyte stimulating hormone receptor variant alleles in twins with red hair. *Hum Mol Genet.* 1997;6(11):1891–1897.
79. Rana BK, Hewett-Emmett D, Jin L, et al. High polymorphism at the human melanocortin 1 receptor locus. *Genetics.* 1999;151(4):1547–1557.
80. Doyle JR, Fortin JP, Beinborn M, Kopin AS. Selected melanocortin 1 receptor single-nucleotide polymorphisms differentially alter multiple signaling pathways. *J Pharmacol Exp Ther.* 2012;342(2):318–326.
81. Schioth HB, Phillips SR, Rudzish R, Birch-Machin MA, Wikberg JE, Rees JL. Loss of function mutations of the human melanocortin 1 receptor are common and are associated with red hair. *Biochem Biophys Res Commun.* 1999;260(2):488–491.
82. Pastorino L, Cusano R, Bruno W, et al. Novel MC1R variants in Ligurian melanoma patients and controls. *Hum Mutat.* 2004;24(1):103.
83. Nakayama K, Soemantri A, Jin F, et al. Identification of novel functional variants of the melanocortin 1 receptor gene originated from Asians. *Hum Genet.* 2006;119 (3):322–330.
84. Ozola A, Azarjana K, Donina S, et al. Melanoma risk associated with MC1R gene variants in Latvia and the functional analysis of rare variants. *Cancer Genet.* 2013;206 (3):81–91.
85. Ringholm A, Klovins J, Rudzish R, Phillips S, Rees JL, Schioth HB. Pharmacological characterization of loss of function mutations of the human melanocortin 1 receptor that are associated with red hair. *J Invest Dermatol.* 2004;123(5):917–923.
86. Sanchez Mas J, Olivares Sanchez C, Ghanem G, et al. Loss-of-function variants of the human melanocortin-1 receptor gene in melanoma cells define structural determinants of receptor function. *Eur J Biochem.* 2002;269(24):6133–6141.
87. Garcia-Borron JC, Sanchez-Laorden BL, Jimenez-Cervantes C. Melanocortin-1 receptor structure and functional regulation. *Pigment Cell Res.* 2005;18(6):393–410.
88. Jimenez-Cervantes C, Olivares C, Gonzalez P, Morandini R, Ghanem G, Garcia-Borron JC. The Pro162 variant is a loss-of-function mutation of the human melanocortin 1 receptor gene. *J Invest Dermatol.* 2001;117(1):156–158.
89. John PR, Makova K, Li WH, Jenkins T, Ramsay M. DNA polymorphism and selection at the melanocortin-1 receptor gene in normally pigmented southern African individuals. *Ann N Y Acad Sci.* 2003;994:299–306.
90. Yang Z, Tao YX. Mutations in melanocortin-3 receptor gene and human obesity. *Prog Mol Biol Transl Sci.* 2016;140:97–129.
91. Vage DI, Fuglei E, Snipstad K, Beheim J, Landsem VM, Klungland H. Two cysteine substitutions in the *MC1R* generate the blue variant of the Arctic fox (*Alopex lagopus*) and prevent expression of the white winter coat. *Peptides.* 2005;26(10):1814–1817.
92. Fontanesi L, Beretti F, Riggio V, et al. Sequence characterization of the melanocortin 1 receptor (*MC1R*) gene in sheep with different coat colours and identification of the putative e allele at the ovine extension locus. *Small Rumin Res.* 2010;91(2–3):200–207.
93. Rouzaud F, Martin J, Gallet PF, et al. A first genotyping assay of French cattle breeds based on a new allele of the extension gene encoding the melanocortin-1 receptor (Mc1r). *Genet Sel Evol.* 2000;32(5):511–520.
94. Graphodatskaya D, Joerg H, Stranzinger G. Molecular and pharmacological characterisation of the MSH-R alleles in Swiss cattle breeds. *J Recept Signal Transduct Res.* 2002;22(1-4):421–430.
95. Kijas JM, Moller M, Plastow G, Andersson L. A frameshift mutation in *MC1R* and a high frequency of somatic reversions cause black spotting in pigs. *Genetics.* 2001;158 (2):779–785.
96. Dreger DL, Schmutz SM. A new mutation in *MC1R* explains a coat color phenotype in 2 "old" breeds: Saluki and Afghan hound. *J Hered.* 2010;101(5):644–649.

97. Schmutz SM, Berryere TG, Ellinwood NM, Kerns JA, Barsh GS. *MC1R* studies in dogs with melanistic mask or brindle patterns. *J Hered.* 2003;94(1):69–73.
98. Durig N, Letko A, Lepori V, et al. Two *MC1R* loss-of-function alleles in cream-coloured Australian cattle dogs and white Huskies. *Anim Genet.* 2018;49 (4):284–290.
99. Ollivier M, Tresset A, Hitte C, et al. Evidence of coat color variation sheds new light on ancient canids. *PLoS One.* 2013;8(10): e75110.
100. Theron E, Hawkins K, Bermingham E, Ricklefs RE, Mundy NI. The molecular basis of an avian plumage polymorphism in the wild: a *melanocortin-1-receptor* point mutation is perfectly associated with the melanic plumage morph of the bananaquit, *Coereba flaveola*. *Curr Biol.* 2001;11(8):550–557.
101. Ling MK, Lagerstrom MC, Fredriksson R, et al. Association of feather colour with constitutively active melanocortin 1 receptors in chicken. *Eur J Biochem.* 2003;270 (7):1441–1449.
102. Kerje S, Lind J, Schutz K, Jensen P, Andersson L. Melanocortin 1-receptor (*MC1R*) mutations are associated with plumage colour in chicken. *Anim Genet.* 2003;34 (4):241–248.
103. Nadeau NJ, Minvielle F, Mundy NI. Association of a Glu92Lys substitution in *MC1R* with *extended brown* in Japanese quail (*Coturnix japonica*). *Anim Genet.* 2006;37 (3):287–289.
104. Sanchez-Mas J, Hahmann C, Gerritsen I, Garcia-Borron JC, Jimenez-Cervantes C. Agonist-independent, high constitutive activity of the human melanocortin 1 receptor. *Pigment Cell Res.* 2004;17(4):386–395.
105. Tao YX. Constitutive activation of G protein-coupled receptors and diseases: Insights into mechanism of activation and therapeutics. *Pharmacol Ther.* 2008;120:129–148.
106. Tao YX. Constitutive activity in melanocortin-4 receptor: biased signaling of inverse agonists. *Adv Pharmacol.* 2014;70:135–154.
107. Wang W, Tao YX. Ghrelin receptor mutations and human obesity. *Prog Mol Biol Transl Sci.* 2016;140:131–150.
108. Herraiz C, Jimenez-Cervantes C, Zanna P, Garcia-Borron JC. Melanocortin 1 receptor mutations impact differentially on signalling to the cAMP and the ERK mitogen-activated protein kinase pathways. *FEBS Lett.* 2009;583(19):3269–3274.
109. Herraiz C, Journe F, Abdel-Malek Z, Ghanem G, Jimenez-Cervantes C, Garcia-Borron JC. Signaling from the human melanocortin 1 receptor to ERK1 and ERK2 mitogen-activated protein kinases involves transactivation of cKIT. *Mol Endocrinol.* 2011;25(1):138–156.
110. Benned-Jensen T, Mokrosinski J, Rosenkilde MM. The E92K melanocortin 1 receptor mutant induces cAMP production and arrestin recruitment but not ERK activity indicating biased constitutive signaling. *PLoS One.* 2011;6(9): e24644.
111. Yang LK, Tao YX. Biased signaling at neural melanocortin receptors in regulation of energy homeostasis. *Biochim Biophys Acta Mol Basis Dis.* 2017;1863:2486–2495.
112. Yang LK, Hou ZS, Tao YX. Biased signaling in naturally occurring mutations of G protein-coupled receptors associated with diverse human diseases. *Biochim Biophys Acta Mol Basis Dis.* 2021;1867(1): 165973.
113. Lalueza-Fox C, Rompler H, Caramelli D, et al. A melanocortin 1 receptor allele suggests varying pigmentation among Neanderthals. *Science.* 2007;318(5855):1453–1455.
114. Palczewski K, Kumasaka T, Hori T, et al. Crystal structure of rhodopsin: a G protein-coupled receptor. *Science.* 2000;289(5480):739–745.
115. Beaumont KA, Newton RA, Smit DJ, Leonard JH, Stow JL, Sturm RA. Altered cell surface expression of human MC1R variant receptor alleles associated with red hair and skin cancer risk. *Hum Mol Genet.* 2005;14(15):2145–2154.

116. Linderholm A, Larson G. The role of humans in facilitating and sustaining coat colour variation in domestic animals. *Semin Cell Dev Biol.* 2013;24(6-7):587–593.
117. Werth LA, Hawkins GA, Eggen A, et al. Rapid communication: *melanocyte stimulating hormone receptor* (*MC1R*) maps to bovine chromosome 18. *J Anim Sci.* 1996;74(1):262.
118. Go H-A, Lee M-J, Oh D-R, et al. Fast identification of Korean cattle based on simultaneous detection of various single nucleotide polymorphisms markers by capillary electrophoresis. *Bull Korean Chem Soc.* 2009;30(9):2141–2144.
119. Joerg H, Fries HR, Meijerink E, Stranzinger GF. Red coat color in Holstein cattle is associated with a deletion in the MSHR gene. *Mamm Genome.* 1996;7(4):317–318.
120. Sasazaki J, Usui M, Mannen H, Huira C, Tsuji S. Allele frequencies of the extension locus encoding the melanocortin-1 receptor in Japanese and Korean cattle. *Anim Sci J.* 2005;76(2):129–132.
121. Guastella AM, Sorbolini S, Zuccaro A, et al. Melanocortin 1 receptor (*MC1R*) gene polymorphisms in three Italian cattle breeds. *Anim Prod Sci.* 2011;51(11):1039–1043.
122. Niemi M, Sajantila A, Vilkki J. Temporal variation in coat colour (genotypes) supports major changes in the nordic cattle population after iron age. *Anim Genet.* 2016;47 (4):495–498.
123. Kasprzak-Filipek K, Sawicka-Zugaj W, Litwinczuk Z, Chabuz W, Sveistiene R, Bulla J. Polymorphism of the melanocortin 1 receptor (*MC1R*) gene and its role in determining the coat colour of central European cattle breeds. *Animals (Basel).* 2020;10(10).
124. Santana CJS, Alves JS, Pinto LFB, Costa RB, de Camargo GMF. A deletion in the *MC1R* gene alters coat color in Guzerat cattle. *Anim Genet.* 2021;52(6):896–897.
125. Chen SY, Huang Y, Zhu Q, Fontanesi L, Yao YG, Liu YP. Sequence characterization of the *MC1R* gene in yak (*Poephagus grunniens*) breeds with different coat colors. *J Biomed Biotechnol.* 2009;2009: 861046.
126. Zhang Y, Li Q, Ye S, et al. New variants in the melanocortin 1 receptor gene (*MC1R*) in Asian cattle. *Anim Genet.* 2014;45(4):609–610.
127. Miao Y, Wu G, Wang L, et al. The role of *MC1R* gene in buffalo coat color. *Sci China Life Sci.* 2010;53(2):267–272.
128. Xi D, Liu Q, Huo Y, et al. Nucleotide diversity of the melanocortin 1 receptor gene (*MC1R*) in the gayal (*Bos frontalis*). *Mol Biol Rep.* 2012;39(7):7293–7301.
129. Xi D, Wu M, Fan Y, et al. Isolation and characteristics of the melanocortin 1 receptor gene (*MC1R*) in the Chinese yakow (*Bos grunniensxBos taurus*). *Gene.* 2012;498 (2):259–263.
130. Jiang L, Kon T, Chen C, et al. Whole-genome sequencing of endangered Zhoushan cattle suggests its origin and the association of *MC1R* with black coat colour. *Sci Rep.* 2021;11(1):17359.
131. Goud TS, Upadhyay RC, Pichili VBR, Onteru SK, Chadipiralla K. Molecular characterization of coat color gene in Sahiwal versus Karan Fries bovine. *J Genet Eng Biotechnol.* 2021;19(1):22.
132. Petersen JL, Kalbfleisch TS, Parris M, Tietze SM, Cruickshank J. MC1R and KIT haplotypes associate with pigmentation phenotypes of north American Yak (Bos grunniens). *J Hered.* 2020;111(2):182–193.
133. Durkin K, Coppieters W, Drogemuller C, et al. Serial translocation by means of circular intermediates underlies colour sidedness in cattle. *Nature.* 2012;482 (7383):81–84.
134. Brenig B, Beck J, Floren C, et al. Molecular genetics of coat colour variations in White Galloway and White Park cattle. *Anim Genet.* 2013;44(4):450–453.
135. Goud TS, Upadhyay RC, Onteru SK, Pichili VBR, Chadipiralla K. Identification and sequence characterization of melanocortin 1 receptor gene (*MC1R*) in *Bos indicus* versus (*Bos taurus* X *Bos indicus*). *Anim Biotechnol.* 2020;31(4):283–294.

136. Matsumoto H, Kojya M, Takamuku H, et al. *MC1R* c.310G >- and c.871G > A determine the coat color of Kumamoto sub-breed of Japanese Brown cattle. *Anim Sci J*. 2020;91(1): e13367.
137. Zhang MQ, Xu X, Luo SJ. The genetics of brown coat color and white spotting in domestic yaks (*Bos grunniens*). *Anim Genet*. 2014;45(5):652–659.
138. Ollivier L, Sellier P. Pig genetics: a review. *Ann Genet Sel Anim*. 1982;14(4):481–544.
139. Giuffra E, Kijas JM, Amarger V, Carlborg O, Jeon JT, Andersson L. The origin of the domestic pig: independent domestication and subsequent introgression. *Genetics*. 2000;154(4):1785–1791.
140. Gustafsson AC, Kijas JM, Alderborn A, Uhlen M, Andersson L, Lundeberg J. Screening and scanning of single nucleotide polymorphisms in the pig melanocortin 1 receptor gene (*MC1R*) by pyrosequencing. *Anim Biotechnol*. 2001;12(2):145–153.
141. Shi K, Wang A, Li N, Deng X. Single nucleotide polymorphism analysis on melanocortin receptor 1 (*MC1R*) of Chinese native pig. *Sci China C Life Sci*. 2004;47(3):287–292.
142. Lai F, Ren J, Ai H, et al. Chinese white Rongchang pig does not have the dominant white allele of *KIT* but has the dominant black allele of *MC1R*. *J Hered*. 2007;98 (1):84–87.
143. Mao H, Ren J, Ding N, Xiao S, Huang L. Genetic variation within coat color genes of *MC1R* and *ASIP* in Chinese brownish red Tibetan pigs. *Anim Sci J*. 2010;81 (6):630–634.
144. Wu X, Tan Z, Shen L, et al. Coat colour phenotype of Qingyu pig is associated with polymorphisms of melanocortin receptor 1 gene. *Asian-Australas J Anim Sci*. 2017;30 (7):938–943.
145. Vage DI, Fleet MR, Ponz R, et al. Mapping and characterization of the dominant black colour locus in sheep. *Pigment Cell Res*. 2003;16(6):693–697.
146. Fontanesi L, Dall'Olio S, Beretti F, Portolano B, Russo V. Coat colours in the Massese sheep breed are associated with mutations in the agouti signalling protein (*ASIP*) and melanocortin 1 receptor (*MC1R*) genes. *Animal*. 2011;5(1):8–17.
147. Hepp D, Goncalves GL, Moreira GR, et al. Identification of the e allele at the extension locus (*MC1R*) in Brazilian Creole sheep and its role in wool color variation. *Genet Mol Res*. 2012;11(3):2997–3006.
148. Yang GL, Fu DL, Lang X, et al. Mutations in *MC1R* gene determine black coat color phenotype in Chinese sheep. *Scientific World J*. 2013;2013: 675382.
149. Mahmoud AH, Mashaly AM, Rady AM, Al-Anazi KM, Saleh AA. Allelic variation of melanocortin-1 receptor locus in Saudi indigenous sheep exhibiting different color coats. *Asian-Australas J Anim Sci*. 2017;30(2):154–159.
150. Rochus CM, Westberg Sunesson K, Jonas E, Mikko S, Johansson AM. Mutations in *ASIP* and *MC1R*: dominant black and recessive black alleles segregate in native Swedish sheep populations. *Anim Genet*. 2019;50(6):712–717.
151. Lu D, Vage DI, Cone RD. A ligand-mimetic model for constitutive activation of the melanocortin-1 receptor. *Mol Endocrinol*. 1998;12(4):592–604.
152. Hoekstra HE, Hirschmann RJ, Bundey RA, Insel PA, Crossland JP. A single amino acid mutation contributes to adaptive beach mouse color pattern. *Science*. 2006;313 (5783):101–104.
153. Rompler H, Rohland N, Lalueza-Fox C, et al. Nuclear gene indicates coat-color polymorphism in mammoths. *Science*. 2006;313(5783):62.
154. Loehr J, Worley K, Moe J, Carey J, Coltman D. *MC1R* variants correlate with thinhorn sheep colour cline but not individual colour. *Can J Zool*. 2008;86(2):147–150.
155. Gebreselassie G, Liang B, Berihulay H, et al. Genomic mapping identifies two genetic variants in the *MC1R* gene for coat colour variation in Chinese Tan sheep. *PLoS One*. 2020;15(8): e0235426.

156. Ganbold O, Manjula P, Lee SH, et al. Sequence characterization and polymorphism of melanocortin 1 receptor gene in some goat breeds with different coat color of Mongolia. *Asian-Australas J Anim Sci.* 2019;32(7):939–948.
157. Ganbold O, Lee S, Seo D, et al. Genetic diversity and the origin of Mongolian native sheep. *Livest Sci.* 2019;220:17–25.
158. Fontanesi L, Beretti F, Riggio V, et al. Missense and nonsense mutations in melanocortin 1 receptor (*MC1R*) gene of different goat breeds: association with red and black coat colour phenotypes but with unexpected evidences. *BMC Genet.* 2009;10:47.
159. Badaoui B, Manunza A, Castello A, et al. Technical note: advantages and limitations of authenticating Palmera goat dairy products by pyrosequencing the melanocortin 1 receptor (*MC1R)* gene. *J Dairy Sci.* 2014;97(11):7293–7297.
160. Xiong Q, Chai J, Chen M, Tao YX. Identification and pharmacological analyses of eight naturally occurring caprine melanocortin-1 receptor mutations in three different goat breeds. *Gen Comp Endocrinol.* 2016;235:1–10.
161. Feng T, Zhang Z, Qu M, et al. Three SNPs Identified in the *MC1R* Gene of Nubian goats may relate to the wool color. *Pak J Zool.* 2020;52(3):843.
162. Tarnow P, Schoneberg T, Krude H, Gruters A, Biebermann H. Mutationally induced disulfide bond formation within the third extracellular loop causes melanocortin 4 receptor inactivation in patients with obesity. *J Biol Chem.* 2003;278 (49):48666–48673.
163. Frandberg PA, Doufexis M, Kapas S, Chhajlani V. Cysteine residues are involved in structure and function of melanocortin 1 receptor: substitution of a cysteine residue in transmembrane segment two converts an agonist to antagonist. *Biochem Biophys Res Commun.* 2001;281(4):851–857.
164. Huang H, Tao YX. Pleiotropic functions of the transmembrane domain 6 of human melanocortin-4 receptor. *J Mol Endocrinol.* 2012;49(3):237–248.
165. Lebon G, Warne T, Edwards PC, et al. Agonist-bound adenosine A2A receptor structures reveal common features of GPCR activation. *Nature.* 2011;474(7352): 521–525.
166. Rasmussen SG, Choi HJ, Fung JJ, et al. Structure of a nanobody-stabilized active state of the β_2 adrenoceptor. *Nature.* 2011;469(7329):175–180.
167. Kaelin CB, Barsh GS. Genetics of pigmentation in dogs and cats. *Annu Rev Anim Biosci.* 2013;1:125–156.
168. Brancalion L, Haase B, Wade CM. Canine coat pigmentation genetics: a review. *Anim Genet.* 2022;53(1):3–34.
169. Nowacka-Woszuk J, Salamon S, Gorna A, Switonski M. Missense polymorphisms in the *MC1R* gene of the dog, red fox, arctic fox and Chinese raccoon dog. *J Anim Breed Genet.* 2013;130(2):136–141.
170. Schmutz SM, Berryere TG, Goldfinch AD. *TYRP1* and *MC1R* genotypes and their effects on coat color in dogs. *Mamm Genome.* 2002;13(7):380–387.
171. Oguro-Okano M, Honda M, Yamazaki K, Okano K. Mutations in the melanocortin 1 receptor, beta-defensin103 and agouti signaling protein genes, and their association with coat color phenotypes in Akita-inu dogs. *J Vet Med Sci.* 2011;73(7):853–858.
172. Anderson H, Honkanen L, Ruotanen P, Mathlin J, Donner J. Comprehensive genetic testing combined with citizen science reveals a recently characterized ancient *MC1R* mutation associated with partial recessive red phenotypes in dog. *Canine Med Genet.* 2020;7(1):16.
173. Kerns JA, Olivier M, Lust G, Barsh GS. Exclusion of melanocortin-1 receptor (*Mc1r*) and *agouti* as candidates for dominant black in dogs. *J Hered.* 2003;94(1):75–79.
174. Wang GD, Cheng LG, Fan RX, et al. Signature of balancing selection at the *MC1R* gene in Kunming dog populations. *PLoS One.* 2013;8(2): e55469.

175. Brockerville RM, McGrath MJ, Pilgrim BL, Marshall HD. Sequence analysis of three pigmentation genes in the Newfoundland population of *Canis latrans* links the Golden Retriever *Mc1r* variant to white coat color in coyotes. *Mamm Genome*. 2013;24 (3–4):134–141.
176. Kerns JA, Cargill EJ, Clark LA, et al. Linkage and segregation analysis of black and brindle coat color in domestic dogs. *Genetics*. 2007;176(3):1679–1689.
177. Candille SI, Kaelin CB, Cattanach BM, et al. A β-defensin mutation causes black coat color in domestic dogs. *Science*. 2007;318:1418–1423.
178. Aono S, Dennis JC, He S, Wang W, Tao YX, Morrison EE. Exploring pleiotropic functions of canine β-defensin 103: nasal cavity expression, antimicrobial activity, and melanocortin receptor activity. *Anat Rec (Hoboken)*. 2021;304:210–221.
179. Leonard BC, Marks SL, Outerbridge CA, et al. Activity, expression and genetic variation of canine β-defensin 103: a multifunctional antimicrobial peptide in the skin of domestic dogs. *J Innate Immun*. 2012;4(3):248–259.
180. Anderson TM, vonHoldt BM, Candille SI, et al. Molecular and evolutionary history of melanism in North American gray wolves. *Science*. 2009;323(5919):1339–1343.
181. Rieder S, Taourit S, Mariat D, Langlois B, Guerin G. Mutations in the agouti (*ASIP*), the extension (*MC1R*), and the brown (*TYRP1*) loci and their association to coat color phenotypes in horses (*Equus caballus*). *Mamm Genome*. 2001;12(6):450–455.
182. Shang S, Yu Y, Zhao Y, et al. Synergy between *MC1R* and *ASIP* for coat color in horses (*Equus caballus*). *J Anim Sci*. 2019;97(4):1578–1585.
183. Liu Z, Gong Y, Feng M, Duan L, Li Y, Li X. Genetic variations of the coding region of the melanocortin receptor 1 (*MC1R*) gene in the fox. *Vet Dermatol*. 2016;27 (3):135–e136.
184. Wang W, Guo DY, Tao YX. Therapeutic strategies for diseases caused by loss-of-function mutations in G protein-coupled receptors. *Prog Mol Biol Transl Sci*. 2019;161:181–210.
185. Tao YX, Conn PM. Pharmacoperones as novel therapeutics for diverse protein conformational diseases. *Physiol Rev*. 2018;98(2):697–725.
186. Hou ZS, Ulloa-Aguirre A, Tao YX. Pharmacoperone drugs: targeting misfolded proteins causing lysosomal storage-, ion channels-, and G protein-coupled receptors-associated conformational disorders. *Expert Rev Clin Pharmacol*. 2018;11(6):611–624.
187. Almeida Scalvino S, Chapelle A, Hajem N, et al. Efficacy of an agonist of alpha-MSH, the palmitoyl tetrapeptide-20, in hair pigmentation. *Int J Cosmet Sci*. 2018;40 (5):516–524.
188. Mowlazadeh Haghighi S, Zhou Y, Dai J, Sawyer JR, Hruby VJ, Cai M. Replacement of Arg with Nle and modified D-Phe in the core sequence of MSHs, Ac-His-D-Phe-Arg-Trp-NH2, leads to hMC1R selectivity and pigmentation. *Eur J Med Chem*. 2018;151:815–823.
189. Jackson E, Heidl M, Imfeld D, Meeus L, Schuetz R, Campiche R. Discovery of a highly selective MC1R agonists pentapeptide to be used as a skin pigmentation enhancer and with potential anti-aging properties. *Int J Mol Sci*. 2019;20(24).
190. Koikov L, Starner RJ, Swope VB, et al. Development of hMC1R selective small agonists for sunless tanning and prevention of genotoxicity of UV in melanocytes. *J Invest Dermatol*. 2021;141(7):1819–1829.
191. Ratner M. Patients with porphyria bask in sunlight of FDA approval. *Nat Biotechnol*. 2019;37(12):1390–1391.
192. Langendonk JG, Balwani M, Anderson KE, et al. Afamelanotide for erythropoietic protoporphyria. *N Engl J Med*. 2015;373(1):48–59.
193. Chen S, Zhu B, Yin C, et al. Palmitoylation-dependent activation of MC1R prevents melanomagenesis. *Nature*. 2017;549(7672):399–403.

194. Chen S, Han C, Miao X, et al. Targeting MC1R depalmitoylation to prevent melanomagenesis in redheads. *Nat Commun.* 2019;10(1):877.
195. Guida S, Guida G, Goding CR. MC1R functions, expression, and implications for targeted therapy. *J Invest Dermatol.* 2022;142(2):293–302 e291.
196. Wang Y, Lu H, Fang C, Xu J. Palmitoylation as a signal for delivery. *Adv Exp Med Biol.* 2020;1248:399–424.
197. Majerus ME, Mundy NI. Mammalian melanism: natural selection in black and white. *Trends Genet.* 2003;19(11):585–588.
198. Nadeau NJ, Burke T, Mundy NI. Evolution of an avian pigmentation gene correlates with a measure of sexual selection. *Proc Biol Sci.* 2007;274(1620):1807–1813.
199. Suzuki H. Evolutionary and phylogeographic views on *Mc1r* and *Asip* variation in mammals. *Genes Genet Syst.* 2013;88(3):155–164.

CHAPTER SEVEN

Mutations in melanocortin-4 receptor: From fish to men

Ya-Xiong Tao*
Department of Anatomy, Physiology and Pharmacology, College of Veterinary Medicine, Auburn University, Auburn, AL, United States
*Corresponding author: e-mail address: taoyaxi@auburn.edu

Contents

Abstract

Melanocortin-4 receptor (MC4R), expressed abundantly in the hypothalamus, is a critical regulator of energy homeostasis, including both food intake and energy expenditure. Shortly after the publication in 1997 of the *Mc4r* knockout phenotypes in mice, including increased food intake and severe obesity, the first mutations in *MC4R* were reported in humans in 1998. Studies in the subsequent two decades have established *MC4R* mutation as the most common monogenic form of obesity, especially in early-onset severe obesity. Studies in animals, from fish to mammals, have established the conserved physiological roles of MC4R in all vertebrates in regulating energy balance. Drug targeting MC4R has been recently approved for treating morbid genetic obesity. How the MC4R can be exploited for animal production is highly worthy of active investigation.

Progress in Molecular Biology and Translational Science, Volume 189
ISSN 1877-1173
https://doi.org/10.1016/bs.pmbts.2022.03.003

Abbreviations

ACTH	adrenocorticotropin
AgRP	Agouti-related protein
ASIP	Agouti-signaling protein
ERK1/2	extracellular signal regulated kinase 1/2
GPCR	G protein-coupled receptor
MC4R	melanocortin-4 receptor
MCR	melanocortin receptor
MSH	melanocyte-stimulating hormone
NDP-MSH	[Nle^4,$DPhe^7$]-α-MSH
POMC	proopiomelanocortin
TMD	transmembrane domain
WT	wild-type

1. Introduction

Melanocortin, first appeared in a publication in 1979,[1] is in fact an active area of research since the 1940s (for examples, see Refs. [2,3]). Earlier studies focused on two hormones, adrenocorticotropin (ACTH) and α-melanocyte-stimulating hormone (α-MSH). ACTH is a hormone produced in the anterior pituitary gland that regulates adrenal gland hormone secretion as well as cell proliferation, whereas α-MSH is a hormone made in the skin that controls skin pigmentation. If we consider even earlier studies of Allen[4] and Smith[5] exploring the relationship between pituitary and pigmentation, the history of melanocortin studies is more than 100 years old. In 1932, Harvey Cushing reported cases of basophil adenomas in the pituitary, which we now call Cushing syndrome.[6] Another landmark study is that of Edward Kendall and Philip Hench on ACTH and cortisol[7,8] that earned them a Nobel Prize in Physiology or Medicine in 1950.

With the use of modern molecular biology techniques, the field has exploded, partly due to the cloning of additional melanocortin receptors (MCRs) in addition to the well-recognized MSH receptor (now called MC1R) and ACTH-R (now called MC2R). Indeed, a search in PubMed on March 22, 2022, using the term "melanocortin," resulted in almost 60,000 hits.

Melanocortins consist of four major peptides, including α-, β-, and γ-MSHs, and ACTH, derived from a precursor, proopiomelanocortin (POMC), by tissue-specific posttranslational processing (see Refs. [9, 10] for excellent comprehensive reviews on POMC). Another peptide hormone derived from POMC, β-endorphin, exerts its actions through μ-opioid receptor and is primarily involved in regulation of pain and not considered as a melanocortin. Among the jawed vertebrates, POMC is expressed in corticotropes in the anterior pituitary, melanotrophs in the intermediate lobe of the pituitary, and neurons in the central nervous system and brainstem as well as in the melanocytes in the skin. Extensive studies were performed, starting in the 1950s, with these hormones, resulting in the cloning of *POMC* cDNA, first reported in bovine[11] (for a history of the early studies on the melanocortin peptides, see Ref. [12]).

The cloning of the MC1R[13,14] and MC2R[13] in 1992 represented a major milestone in the field of melanocortin research. Shortly, three additional MCRs, named based on the sequence of their cloning, MC3R,[15,16] MC4R,[17,18] and MC5R,[19–23] were identified, mostly by degenerate PCR technique. MC3R and MC4R were shown to be highly expressed in the central nervous system.[16,18] All five MCRs are G protein-coupled receptors (GPCRs) that are primarily coupled to the stimulatory G protein, Gs, with receptor activation increasing intracellular cAMP level.[24,25] Other G proteins (such as Gi and Gq) and signaling pathways can also be activated by some of these MCRs. Activation of Gq and extracellular signal regulated kinase 1/2 (ERK1/2) by the MC4R will be discussed in this article.

Although the primary ligand for the MC4R was identified as α-MSH, the physiological function of MC4R was not known when first cloned. A series of seminal papers published in the next few years quickly established the critical importance of MC4R in regulating energy homeostasis. Although earlier studies demonstrated that α-MSH decreases food intake,[26–28] the receptor mediating this effect was not definitely known until the knockout of *Mc4r* in mice.[29] The homozygous *Mc4r*$^{-/-}$ mice have increased food intake and decreased energy expenditure and severe obesity, together with hyperinsulinemia and increased linear growth. In addition, central administration of MC4R agonist, melanotan II, decreases food intake whereas co-administration of the MC4R antagonist, SHU9119, blocks melanotan II's action.[30] These studies published in 1997 were very quickly followed by human genetics studies reporting *MC4R* mutations in obese humans in 1998–1999.[31–33] The studies that followed in the next two decades are discussed in this chapter.

2. Mutations in human *MC4R*

2.1 Leptin-regulated melanocortin circuit

As demonstrated in Fig. 1, leptin produced by white adipocytes can cross the blood-brain barrier, binds to its receptors in two subsets of neurons in the

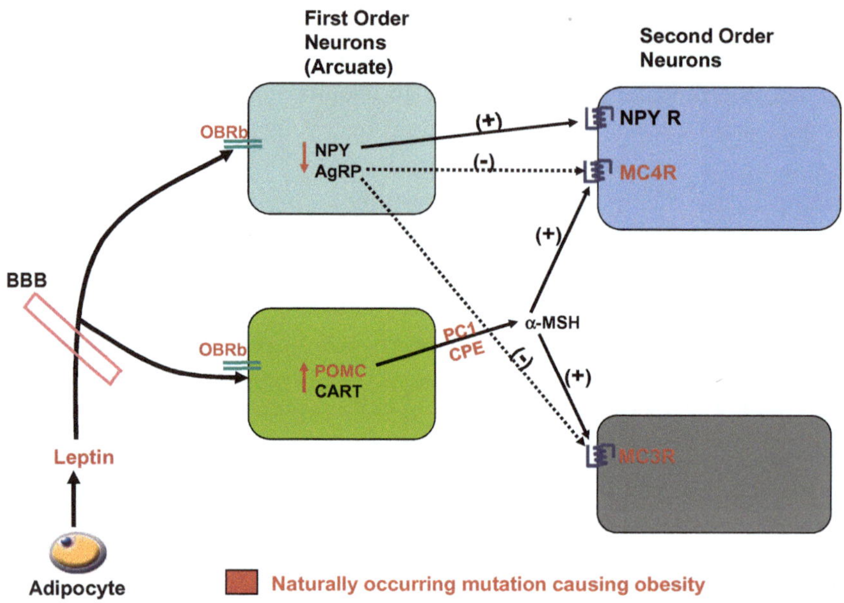

Fig. 1 Leptin-regulated melanocortin circuit. Mutations in genes encoding the proteins highlighted in red cause monogenic obesity in humans. *Modified from Spiegelman BM, Flier JS. Obesity and the regulation of energy balance.* Cell. *2001;104(4):531–543.*

arcuate nucleus. One subset of neurons express POMC that can be cleaved into the MSHs, especially α-MSH. Another subset of neurons express Agouti-related protein (AgRP) and neuropeptide Y. The former subset is anorexigenic and the latter is orexigenic. The second order neurons express MC4R and neuropeptide Y receptors. α-MSH is an agonist for the MC4R and AgRP is an inverse agonist (or negative antagonist) for the MC4R.[34–36]

The critical importance of leptin-regulated melanocortin circuit in maintaining energy homeostasis is demonstrated by monogenic obesity when a single gene, including leptin, leptin receptor, *POMC*, *PCSK1* (the gene encoding the enzyme, prohormone convertase 1/3, cleaving POMC into MSHs), and *MC4R*, is mutated (see Ref. [37] for a recent review). Using advanced neuroscience and genetic techniques, the MC4R circuity regulating energy homeostasis is being mapped in detail.[38]

It should be highlighted that recent data showed that in addition to the central nervous system, the *MC4R* is also expressed in the retina and Fallopian tube at high levels, even higher than the brain (Fig. 2). It is also

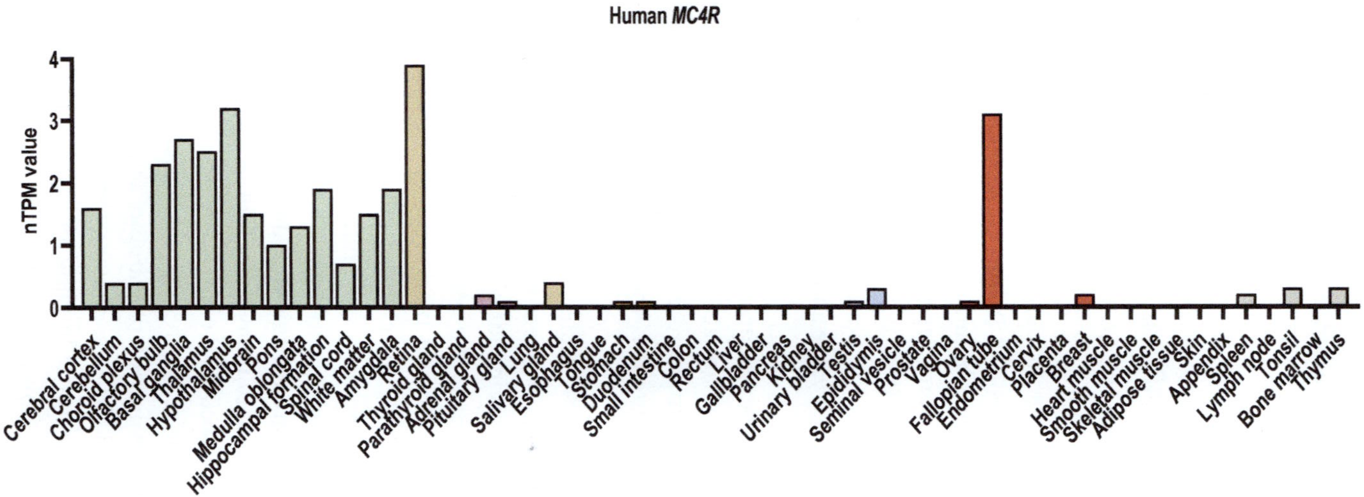

Fig 2 *MC4R* expression in human tissues. nTPM indicates normalized protein-coding transcripts per million. Color-coding is based on tissue groups with common functions. *The data were extracted from https://www.proteinatlas.org/ (Ref. Uhlen M, Fagerberg L, Hallstrom BM, et al. Proteomics. Tissue-based map of the human proteome.* Science. *2015;347(6220):1260419.*

expressed at low levels in salivary gland, epididymis, tonsil, and thymus, and very low levels in adrenal gland, breast, and spleen (Fig. 2). The potential physiological roles of the MC4R in these tissues have not been investigated yet.

2.2 Classification of MC4R mutants

As indicted above, the first *MC4R* mutations were reported in 1998 and 1999 in morbidly obese children.[31–33] The first two mutations were frameshift mutations, at the transmembrane domain (TMD) 5 and between TMDs 6 and 7, respectively, that can be predicted to cause a total loss of function.[31,32] However, the effects of missense mutations subsequently reported cannot be easily predicted. Some reports used software to predict whether a missense mutation might be defective or not. From extensive experimental validation and inspection of the literature, it can be concluded that these predictions have very limited value and is not a viable alternative for experimental verification.

When I was trying to establish my independent line of research, with my training in studying GPCR pharmacology and interest in diseases associated with mutations in gonadotropin receptors, I asked two simple questions: (1) Do the mutations reported in the clinical studies result in defective mutant receptors? (2) And if they do, what are the molecular defects? Although the pace of clinical studies was accelerating, detailed functional studies on the mutant receptors were relatively few.[39–41] In our first study, we generated eight mutations identified from obese patients and studied the cell surface and total expression by confocal microscopy using epitope tag antibody (the receptor cDNA was modified by insertion of c-myc epitope tag after the initiating codon), as well as ligand binding and signaling.[42] We showed that the major defect in loss-of-function mutations is intracellular retention, likely due to protein misfolding.[42] Similar conclusion was reached independently by several groups studying different mutations.[43–46] This is consistent with observations in many other GPCRs and membrane proteins as well as enzymes (reviewed in Refs. [47–52]).

In another collaborative study, we identified an inframe deletion mutation resulting in the loss of five amino acids in TMD2, Δ88-92.[53] Interestingly, we showed that the mutant receptor is expressed well on the cell surface but has no ligand binding and signaling.[53] In another study on the related MC3R, we showed that a mutant MC3R, I183N, has normal ligand binding (by inference, expressed on the cell surface since the peptide ligand cannot pass through the cell membrane) but is devoid of ligand-stimulated signaling.[54]

Taken into account data from other groups on the MC4R as well as extensive studies in other GPCRs associated with numerous human diseases (such as our own on I183N MC3R), we proposed a classification scheme to catalog the mutant receptors into five different classes: Class I, defective protein production, mostly due to frameshift and nonsense mutations; Class II, intracellular retention, mostly trapped in the endoplasmic reticulum with a few in Golgi apparatus; Class III, defective binding with mutant receptor expressed on the cell surface: Class IV, defective signaling with mutant receptor expressed on the cell surface that can bind to the ligand; and Class V, mutants with no overt functional defect[42,55] (Fig. 3). We extend this classification scheme to other GPCRs.[47,49]

Numerous studies have followed since the first reports of *MC4R* mutations causing monogenic obesity.[31–33] Indeed, although mutations in several other genes, including leptin, leptin receptor, *POMC*, *PCSK1*, and single-minded 1, among others, can cause monogenic obesity, mutations in *MC4R* are the most common monogenic form of obesity, accounting

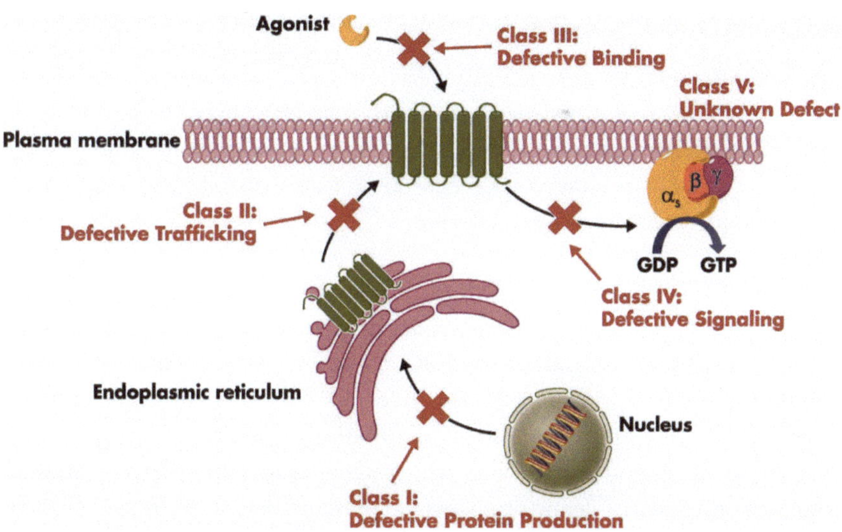

Fig. 3 Classification of naturally occurring *MC4R* mutations. *Reprinted with permission from Tao YX, Conn PM. Chaperoning G protein-coupled receptors: from cell biology to therapeutics.* Endocr Rev. *2014;35(4):602–647. Copyright 2014 The Endocrine Society.*

for up to 5% of early-onset morbid obesity cases.[56] We have summarized some of these earlier studies in several review articles.[55,57–59] Hinney and colleagues further updated this field of research, especially their own extensive contribution to identification of numerous *MC4R* mutations.[60] Several recent articles further reviewed different aspects of this exciting field; interested readers are encouraged to consult these excellent articles.[61–67] Fig. 4 listed many new *MC4R* mutations identified from large scale genome sequencing.

The recent structural elucidation of the MC4R represents a major milestone in MC4R research.[69–72] These data will significantly facilitate studies on naturally occurring mutations by adding a structural perspective. Before the structures were reported, homology modeling and molecular dynamic simulations, using other GPCRs as templates, have been used to gain insights into the structure-function perturbations with the naturally occurring mutations.[63,65,73–77]

2.3 Constitutive activity and biased signaling of MC4R mutants

An important pharmacological property of GPCRs, including the MC4R, is the basal or constitutive activity, signaling in the absence of any agonist.[78] In our own experiments, of the four human MCRs (with the exception of MC2R), MC1R has the highest constitutive activity, followed by MC4R, with MC3R and MC5R having very little constitutive activities.[42,79,80] It was suggested that decreased constitutive activity may be one cause of monogenic obesity caused by *MC4R* mutations,[81] similar to mutations in ghrelin receptor[82] (reviewed in Ref. [83]). We also observed several mutant MC4Rs with decreased basal activities[54,84,85] (see Refs. [59, 74] for focused reviews).

Of special interest is the identification of six mutations that increase constitutive activity but were identified from obese patients, including H76R,[86] S127L,[87] D146N,[86] P230L,[87] L250Q,[40] and F280L.[88] We and others showed that they are constitutively active.[74,84,89,90] Constitutive activation of the MC4R, as one type of gain-of-function mutation,[83] is expected to tonically inhibit food intake and stimulate energy expenditure, resulting in a constitutionally lean phenotype, perhaps even anorexia nervosa. Increased basal signaling is not consistent with the obesity phenotype in these patients.

In addition to the classical Gs-adenylate cyclase-cAMP pathway, MC4R is also coupled to other G proteins as well as activation of ERK1/2 (for a

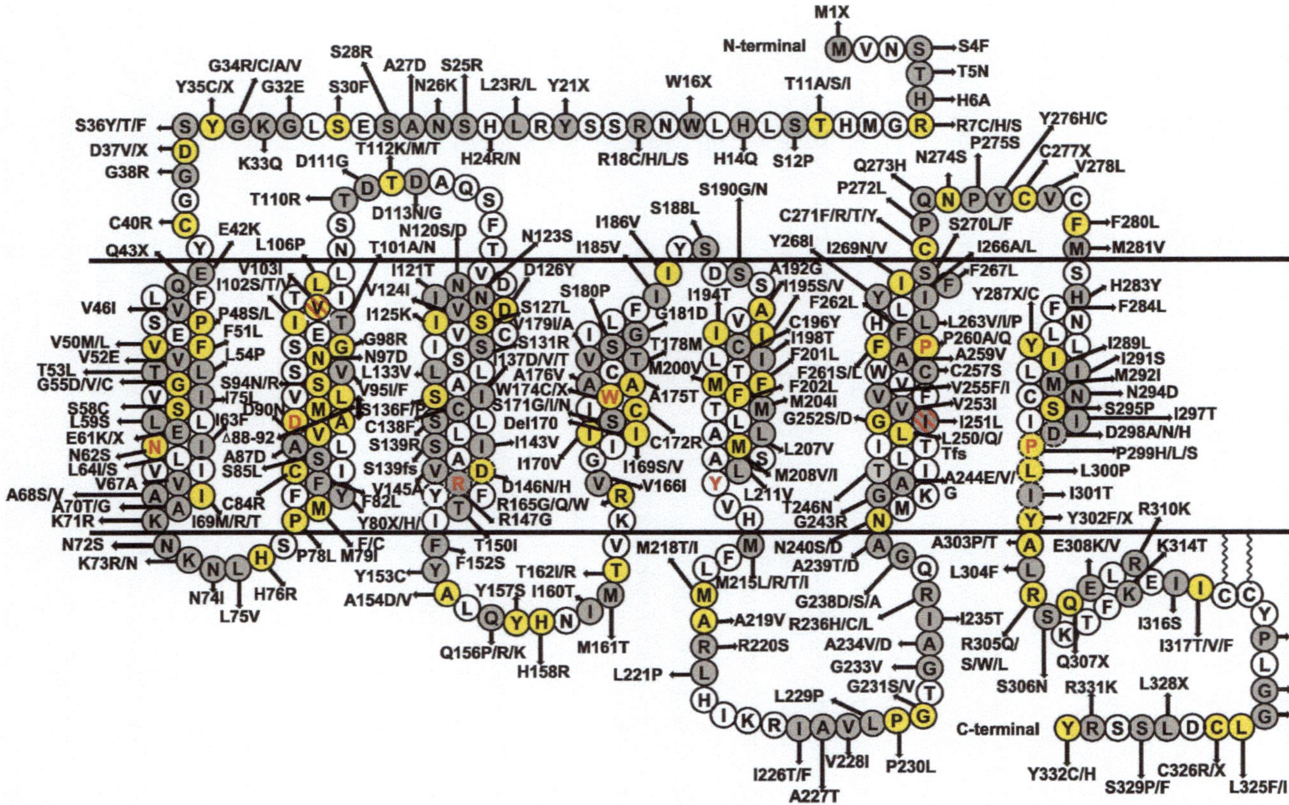

Fig. 4 Naturally occurring *MC4R* mutations. Mutations registered in gnomAD database v2.1.1 (https://gnomad.broadinstitute.org/) (Ref. [68]) and previously reported are listed. Frameshift mutations were not included in the figure. Residues in red font are the most conserved residues in each transmembrane domain (TMD). E/DRY and N/DPxxY motifs are labeled in dashed line circles. The residues filled with yellow highlight the mutations that our lab have investigated their pharmacology. *Reprinted with permission from Liu T, Ji RL, Tao YX. Naturally occurring mutations in G protein-coupled receptors associated with obesity and type 2 diabetes mellitus.* Pharmacol Ther. *2022:108044. Copyright 2022 Elsevier.*

comprehensive review, see Ref. [58]). We asked whether differential signaling might explain the paradox.[91] We showed that inverse agonists at the Gs-cAMP pathway (decreasing basal cAMP signaling) can serve as agonists in the ERK1/2 pathway (activating the ERK1/2 signaling). Different mutants can also respond differently to these ligands.[91,92] This study represents one of the earliest reports on biased signaling in naturally occurring *MC4R* mutations[91] (reviewed in Refs. [93–95]).

Studies of Class V mutants, those with similar Gs-cAMP signaling as the wild-type (WT) MC4R (as reported previously by us[96]), showed that some mutants have defects in ERK1/2 signaling, including decreased basal and ligand-stimulated signaling.[97] We suggest that defect in ERK1/2 signaling may be another cause of obesity due to *MC4R* mutations,[97] expanding the mutations that are pathogenic for obesity.

Subsequently, a series of investigations have compared multiple aspects of MC4R pharmacology and reported biased signaling and its physiological relevance[98–103] (reviewed in Refs. [93,95]). Importantly, the recently approved MC4R agonist used for treatment of genetic obesity, setmelanotide, was shown to have biased signaling[104–113] (see Refs. [114,115] for reviews). These important studies revitalized the field of MC4R for obesity treatment, after the earlier disappointing results with another MC4R agonist that has significant side effects, including increased blood pressure and heart rate.[116]

2.4 Correction of MC4R mutants

With the classification of mutants as described above, different strategies can be devised to correct the defects.[47] For Class I mutants due to premature stop codons, antibiotics can be used to decrease the codon-anticodon pairing fidelity and allow the translation to proceed through the premature termination codon to the natural termination codon. This has been studied both *in vitro* and *in vivo* in the MC4R with variable success.[117,118] It was surprising that functional rescue can be achieved with W16X and Y35X located in the N-terminus but less successful for Q307X at the C-terminus, considering that Q307X lacks a much shorter stretch of amino acids than the first two mutations.[117] *In vivo* experiments with W16X knockin mice result in no correction of obesity.[118] Much more further studies are needed to address the potential therapeutic strategy for these nonsense mutations.

For Class III and IV mutants that cannot bind or respond to endogenous ligands, novel ligands might be able to use different residues on the receptor for binding and signaling. There are also some reports along this line. For example, Haskell-Luevano's group reported that several peptidic and small molecule ligands can rescue several MC4R mutants.[119,120] IPSEN developed two novel MC4R agonists, IRC-022493 and IRC-022511, that can efficiently activate MC4R mutants with impaired response to the endogenous agonist, α-MSH.[121]

Our lab has been focusing on correcting Class II mutants using small molecule ligands as pharmacological chaperones or pharmacoperones ("pharmacoperone" was first coined by the late P. Michael Conn[122]). This strategy, first explored with enzymes causing lysosomal storage disease,[123] was then tested by Bouvier and colleagues in GPCRs, with misfolded V2 vasopressin receptor that cause nephrogenic diabetes insipidus.[124] Subsequently, this strategy has been explored widely in several other GPCRs, including gonadotropin-releasing hormone receptor, luteinizing hormone receptor, follicle-stimulating hormone receptor, glucagon receptor, and melanin-concentrating hormone receptor 1, among others (reviewed in Refs. [48–52,125,126]).

We were the first to report that a small molecule MC4R inverse agonist, ML00253764, first developed at Millennium Pharmaceuticals,[127] can function as a pharmacoperone, rescuing two mutant MC4Rs, C84R and W174C, to the cell surface and the rescued mutants can generate signaling with increased cAMP production in response to agonist stimulation.[90] Subsequently, I showed that multiple additional mutants can be rescued to the cell surface.[58] We have since identified another antagonist and a small molecule agonist that can serve as pharmacological chaperones.[128,129] Fig. 5 shows the dramatic rescue of most of the mutants to the cell surface by the high affinity antagonist, Ipsen 5i.[129] Other groups also identified several pharmacological chaperones for misfolded MC4Rs.[130,131] In a recent review article, we comprehensively summarized these *in vitro* studies.[132] Of special interest is a study from Bouvier's group extending these findings to *in vivo* knockin mouse experiments, demonstrating the therapeutic potential of pharmacological chaperones *in vivo*.[133]

Lifestyle intervention for treatment of obesity in *MC4R* mutation carriers has met with limited success.[134,135] Activation of molecules independent of or downstream of MC4R and bariatric surgery has been used to treat obese patients with *MC4R* mutations. For example, glucagon-like peptide-1 receptor agonist, liraglutide, has been used successfully to treat *MC4R* mutation-carrying obese patients, including patients who are heterozygous or homozygous for *MC4R* mutations.[136–138] Improved body weight and glucose handling are observed. However, gastrointestinal side

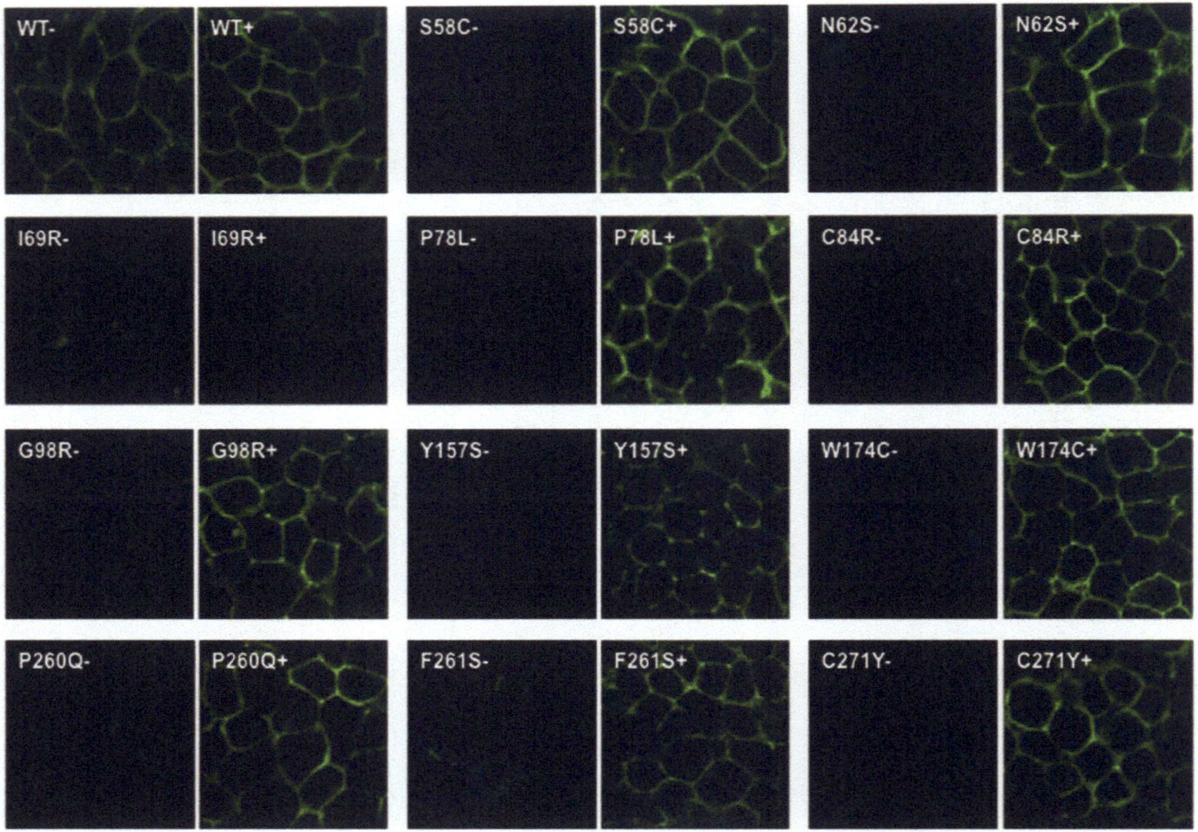

Fig. 5 Ipsen 5i as a MC4R pharmacological chaperone. Ipsen 5i rescues the cell surface expression of most of the intracellularly retained mutant MC4Rs stably expressed in HEK293 cells. HEK293 cells stably expressing WT or mutant MC4Rs were treated with buffer (−) or 10^{-6} M Ipsen 5i (+) for 24 h. Confocal microscopy was used to visualize the cell surface receptor in non-permeabilized cells. *Reproduced with permission from Ref. Tao YX, Huang H. Ipsen 5i is a novel potent pharmacoperone for intracellularly retained melanocortin-4 receptor mutants.* Front Endocrinol (Lausanne). *2014;5:131.*

effect can limit the long-term outcome.[138] The potential of semaglutide for treating obesity in *MC4R* mutation carriers has not been reported. Pharmacotherapy together with lifestyle intervention, such as diet and exercise, is another line of investigation worth pursuing for obese patients with *MC4R* mutations.

Many studies reported bariatric surgery in *MC4R* mutation-carrying patients with varying degrees of success, with some reporting better weight loss in *MC4R* mutation carriers than non-carriers.[139–151] Some of these studies reported outcomes of long-term follow-up, with significant weight regain. One type of surgery might achieve better outcome than another. Obesity surgery is reserved as the last resort for morbidly obese patients with metabolic complications.[152]

2.5 Protective effect of *MC4R* variants

Two common polymorphisms, V103I (minor allele frequency 4%) and I251L (minor allele frequency 2%), first identified by two groups,[33,153] confer protection against obesity (by 21% and 50%, respectively),[101,139,154–158] although there are also studies reporting opposite findings (see Refs. [96,159,160] for examples). The effect of V103I is more pronounced in East Asian people.[157]

Functional studies showed that V103I has similar cell surface expression, binding, and signaling properties to endogenous agonists but has two-fold lower affinity for the antagonist, AgRP.[161] Decreased antagonism by AgRP might lead to decreased orexigenic signaling. V251L has increased basal activity,[161] therefore likely exerting a tonic inhibition on food intake and stimulation on energy expenditure, resulting in decreased body weight. Another study reported that these two variants do not have any functional difference compared with the WT MC4R.[162] Other aspects of receptor pharmacology, including homodimerization, internalization, and trafficking of internalized receptor (recycling or degradation) might also be affected.[103]

When these variants were knocked into mice using CRISPR/Cas9, it was shown that the knockin mice are shorter with decreased fat mass and enhanced glucose handling.[163] However, they are still susceptible to the detrimental effect of high fat diet,[163] demonstrating gene-environment interaction in obesity pathogenesis: a better genetics can still be corrupted by a bad environment. Similar interaction between common *MC4R* polymorphisms and dietary intake on body weight has also been reported.[164,165]

2.6 Polymorphisms near *MC4R*

It should also be pointed out that there are extensive genetic studies identifying a polymorphism, rs17782313 (located at 188 kb downstream of the *MC4R* gene, T>C), that is strongly associated with a predisposition to polygenic obesity and type 2 diabetes mellitus, with higher total energy intake and dietary fat intake.[166–169] Another study identified another polymorphism, rs129070134, at 150 kb downstream of *MC4R*, that is in high linkage disequilibrium with rs17782313, and is associated with waist circumference and insulin resistance in Indians.[170] Later studies replicated the significant associations of these polymorphisms with body mass index, waist circumference, and obesity risk.[171–176] About 14% increased risk of obesity is conferred by rs17782313.[158]

In Caucasian children, the near-*MC4R* locus has been consistently shown to be associated with increased body mass index by 6.7% standard deviation.[177] However, inconsistent observations were made in children of other racial/ethnic origins. In African American children and adolescents, no association of the near-*MC4R* locus with risk of obesity was observed.[178] In Chinese children and adolescents, inconsistent results were obtained.[179,180] Recent systematic reviews and meta-analysis suggested that depending on the ethnicity of the population studied, rs17782313 can be associated with obesity in children, especially in Caucasian and Asian children.[181,182]

Despite the strong reproducibility of the association of these near-*MC4R* polymorphisms and increased risk of obesity, the causal variant and the biological pathway involved remain to be elucidated.[158] Detailed clinical observations, including increased adult height and obesity in children, are consistent with MC4R being involved.[158] Further studies are needed.

3. Mutations in pig *MC4R*

Pig is rapidly being developed as a model of obesity, type 2 diabetes mellitus, and metabolic syndrome.[183,184] In growing pigs, administration of [Nle^4,DPhe^7]-α-MSH (NDP-MSH), a superpotent analog of α-MSH, decreases feed intake but the MC4R antagonists, SHU-9119 and AgRP, fail to increase feed intake, questioning whether endogenous melanocortin system is also critical in regulating feed intake in the pig.[185] Signaling experiments confirm that SHU-9119 is an antagonist of pig MC4R.[185] CRISP/Cas technology was used to knockout *MC4R* in miniature pig.[186] The *MC4R*$^{-/-}$ pigs are hyperphagic with increased body fat that develop into hepatic steatosis without the need of feeding a high-fat diet,[186] demonstrating that the MC4R indeed has similar roles in regulating energy homeostasis in the pig.

In 2000, using a candidate gene approach, Rothschild and colleagues identified a missense mutation in the porcine *MC4R* gene, D298N, that is associated with fatness, growth, and feed intake.[187] However, the effect is relatively modest, with the average backfat thickness of homozygotes differing by 0.9 mm.[188] Subsequent genotype-phenotype association studies on this locus resulted in divergent conclusions, with some studies supporting the original association[189–197] while other studies failed to find an association.[188,198–202]

D298, designated as residue 7.49 based on the numbering scheme of Ballesteros and Weinstein,[203] lies in the highly conserved N/DPxxY motif. The majority of Family A (rhodopsin-like) GPCRs indeed has an Asn at this position, suggesting that Asn at 7.49 confers normal functions in many family A GPCRs. MCRs, together with a few other GPCRs such as gonadotropin-releasing hormone receptor and prostanoid receptors, have an Asp at this position. Therefore, we speculated that either Asn or Asp at this position might produce a functional receptor. However, it should be noted that all the MCRs cloned to date have Asp at this position, suggesting that Asp(7.49) might be functionally important. Based on extensive studies on other GPCRs, we hypothesized that D298N porcine MC4R has normal function. We therefore generated the mutant porcine MC4R and analyzed its function.

It was first reported that D298N porcine MC4R is totally devoid of signaling in response to agonist stimulation.[204] However, in our structure-function studies on the human MC4R, we found that D298N human MC4R has relatively normal functions.[205] Therefore we sought to confirm whether indeed porcine D298N MC4R is defective in signaling. In our binding experiments, similar binding capacities and affinities for the natural agonist α-MSH and the natural antagonist AgRP are observed for the WT and D298N porcine MC4Rs. Similarly, in signaling assays measuring intracellular cAMP levels after agonist stimulation, we showed that the two receptors exhibit similar potencies and maximal responses to α-MSH stimulation.[205] Therefore, we conclude that D298N has no overt functional defect.

Subsequent *in vivo* studies were performed by Barb, Tuggle, and colleagues. Injection of the superpotent agonist, NDP-MSH, results in similar feed intake in pigs harboring WT or D298N *MC4R*.[206] In addition to no difference in body weight, backfat depth, or concentrations of serum metabolites measured between pig with WT or D298N genotype, detailed transcriptional profiling also reveals no significant difference in gene expression that can be correlated to the *MC4R* genotype.[200] These results provide convincing *in vivo* data suggesting that D298N has intact signaling.

Recently, Wang and colleagues reported that, using luciferase reporter assay, D298N lacks constitutive activity[207] whereas we report that D298N has decreased constitutive activity.[205] In our hands using radioimmunoassay to measure cAMP levels directly, the constitutive activities in mammalian MC4Rs, including human[42,54,84,85,90,96,208–210] (reviewed extensively in Ref.[59]), giant panda,[211] and pig[205] MC4Rs, are modest, higher than cells transfected with human MC3R[79] or empty vector, but much lower than fish MC4Rs.[212–218] The use of reporter gene assay amplifies the signal. They also reported decreased cell surface expression.[207] We did not measure cell surface expression because we observed similar maximal binding.[205] Binding can only be measured with cell surface receptor since peptide ligand cannot cross the cell membrane.

Another *MC4R* mutation, R236H, was identified in a Piétrain founder pig, a relatively slow-growing pig.[192] It was not found in the Mangalitsa pigs in the same breeding program. R236 is located in the third intracellular loop that could be important for G protein coupling. However, the functional properties of this mutant are not different from the WT pig MC4R.[205] It was suggested that D298N and R236H have different effects, with D298N primarily affecting growth rate and R236H affecting back fat and growth in one pig population, and the two variants interact with each other, a potential explanation of some studies failing to demonstrating an effect of D298N,[219] although another group suggested that R236H is not associated with any production trait whereas D298N is associated with average daily gain and feed gain ration.[197,220] Therefore, no consensus can be reached regarding the relevance of these two variants with production traits.

4. Mutations in dog *MC4R*

Dog, human's best friend,[221] has contributed significantly to our understanding of physiology and pathophysiology, such as glucose homeostasis and the discovery of insulin,[222] coat color genetics,[223,224] olfaction,[225–227] retinitis pigmentosa,[228] narcolepsy,[229] muscular dystrophy,[230] body size and shape,[231] autism,[232] urologic diseases,[233] and cancer,[234,235] just to cite a few prominent examples. Dog is also extensively used in preclinical tests for toxicity of potential new therapeutics.[236] Dog fanciers created numerous breeds using a small number of founders with desired traits. The ensuing dramatic differences in the different breeds of dogs provide an excellent resource for genetic studies into these difference as well as breed-specific diseases and other complex traits, including behavior, especially with the availability of a high quality genome sequence[237] (reviewed in Refs. [238,239]).

With the increasing prevalence of obesity in humans, dog obesity is also increasing.[240] Therefore, a better understanding of canine obesity can be beneficial to both veterinary and human medicine. With the knowledge of *MC4R* mutations and human obesity, two studies identified the same mutation, V213F, in canine *MC4R*.[241,242] Since no functional studies were performed in the original reports, we generated the mutant receptor and studied its pharmacology using endogenous as well as synthetic superpotent agonists.[243] Ligand binding and signaling experiments showed that indeed V213F has similar pharmacological properties as the WT canine MC4R,[243] suggesting that this mutant allele is likely not to cause obesity in dogs. Subsequently, N101T was shown to be significantly associated with body weight traits in Beagle dogs.[244] Whether N101T has any functional effect has not been studied yet.

A recently study sequenced *MC4R* gene, including the entire coding region as well as the 5′- and 3′-flanking regions in 270 dogs from four breeds predisposed to obesity, including Labrador Retriever, Golden Retriever, Beagle, and Cocker Spaniel.[245] Although several polymorphic variants were identified, the authors concluded that *MC4R* gene polymorphism is unlikely to be associated with dog obesity or body weight.[245]

In studies with Chinese raccoon dog and Arctic fox, no significant *MC4R* mutations were identified, with only silent mutations identified in the coding region and polymorphisms in the 5′-untranslated region not associated with body weight.[246,247] In blue fox, one missense mutation, A244E, together with silent mutations, were identified to be significantly associated with body weight.[248] Interestingly, combined polymorphisms in *MC3R* and *MC4R* have significant effects on body weight and abdominal circumference,[248] similar to the observations in mice with both *Mc3r* and *Mc4r* knockout.[249] In human MC4R, A244E was shown to have decreased basal and ligand-stimulated signaling,[44,87,161] therefore a loss-of-function mutation, consistent with it having an effect on energy balance in blue fox.

It should be highlighted that although no conclusive evidence of *MC4R* mutation causing dog obesity is present in the literature so far, mutation in *POMC* that cause dog obesity has been presented.[250] This *POMC* mutation, disrupting the production of β-MSH, is common in obesity-prone Labrador retriever and associated with body weight, adiposity, and food motivation.[250] Similar mutation that disrupts β-MSH generation has also been identified in obese humans.[251,252] These studies highlight the importance of β-MSH in regulating energy homeostasis in both humans and dogs. However, another study reported no association between this *POMC* deletion mutation and diabetes mellitus in a population of 118 Labrador retriever dogs.[253] These authors suggest that their data add to the literature demonstrating the lack of a direct link between obesity and diabetes mellitus in dogs.[253]

In summary, current literature highlights the similarity in melanocortin regulation of body weight between human and dog. Further studies might identify pathogenic mutations in canine *MC4R* in obese dogs.

5. Mutations in other mammalian *MC4R*

Cow, a ruminant, is an excellent model for studying regulation of food intake and energy balance, especially interactions of short-term signals.[254] Dairy cows have tremendous energy requirement, providing a powerful model to study interactions of short- and long-term signals in long-term energy status to meet the needs of the particular lactation cycle.[254] The availability of bovine genome sequence, at the Year of Ox, is beneficial for rapid advancement in this area of research, to gain a better understanding of ruminant biology and improve animal agriculture.[255]

Bovine *MC4R* is mapped to bovine chromosome 24,[256] with high synteny to human *MC4R* chromosome location, 18q21.[257] Several variants have been identified in bovine *MC4R* from different cattle breeds, including T11A, A20T, V145A, V166M, A172T, L286V,[256,258–260] as well as in 5′-untransated regions,[261] that might be associated with various production traits, such as live weight, carcass weight, and backfat thickness. No functional studies on any of these variants have been performed. In water buffalo, although only silent mutations and polymorphisms were identified, g.1104C > T (silent mutation) is significantly associated with milk yield, suggesting that the *MC4R* gene might affect milk production.[262]

In human MC4R, we and others showed that T11A had no overt functional defects.[54,97,161] In our structure-function studies of human MC4R, we showed that V145A has higher signaling than WT receptor, with no other defects.[210] There are no functional data in corresponding mutations in human MC4R for the other mutations identified in cattle. Further studies are needed to establish a potential causal relationship of *MC4R* variants with the production traits measured.

Domestic yak is a species living at high altitude that is closely related to the low-altitude cattle.[263] Sequencing of 354 yaks identified five polymorphisms, three synonymous and two non-synonymous (V286L and R331S).[264] V286L is associated with increased feed intake, body weight, and average daily gain.[264] L286V mutation is found in cattle that is associated with backfat thickness.[259] The functional implications of V286L and R331S in yak MC4R have not been studied yet.

Sheep is another ruminant that is continuously grazing (instead of intermittent feeding as in most of the other species, including humans). Sheep can serve as a large animal model of obesity and insulin resistance.[265] The availability of sheep genome greatly facilitates studies on sheep physiology.[266] Earlier studies demonstrated that fasting increases AgRP protein expression and neuronal activity in AgRP neurons; injection of AgRP stimulates food intake in *ad libitum*-fed sheep.[267,268] AgRP also blocks endotoxin-induced inhibition of food intake.[269] These results suggest that MC4R is also a critical regulator of food intake in sheep.[270]

In German Merino sheep, two polymorphisms in the 3′-untranslated region and two in the coding region (one synonymous and one missense, G98R) were identified.[271] The 1016G/A mutation in the 3′-untranslated region is significantly associated with body weight, average daily gain, back fat thickness, and loin-eye area whereas G98R mutation is associated with significantly higher loin-eye area.[271] G98R mutations has also been identified in human *MC4R*[272] and shown to cause a total loss of function due to lack of cell surface expression of the mutant receptor.[42] We can use pharmacological chaperones to rescue the mutant receptor to the cell surface; however, it still cannot generate ligand-stimulated signaling, suggesting that G98R is also defective in ligand binding or G protein activation.[58,128,129,132] In Hu sheep, eight polymorphisms in the *MC4R* promoter region are associated with one or more production traits.[273] *MC4R* polymorphisms were also identified in Hu sheep and East Friesian x Hu crossbred sheep that are associated with birth weight, weaning weight, and body size.[274,275]

In a population of domestic shorthaired cats from UK, a missense mutation, L31P, was identified.[276] There is no significant difference in genotype frequencies between nondiabetic lean and overweight cats. However, there is significant difference in allele frequency between overweight diabetic and nondiabetic cats.[276] Functional significance of this mutation is not clear. No corresponding mutation was identified in human MC4R.

In the rabbit, there is one *MC4R* mutation reported, G34D, that is associated with body weight.[277] Mutations at the homologous position have been identified in human *MC4R*, with G34R, G34C, G34A, and G34V mutations reported.[95] No functional data exist for these mutations in either human or rabbit MC4R.

6. Mutations in avian *MC4R*

Chicken is widely studied for academic and agricultural purposes. Nobel Prize-winning work of Peyton Rous on cancer-causing virus and Rita Levi-Montalcini's work on nerve growth factor used chickens as the materials for study.[278,279] Chicken is also instrumental in advancing our understanding of limb development and genetic diseases such as muscular dystrophy and epilepsy. The chicken genome sequence has been published in 2004.[280]

Avian melanocortin and melanocortin receptor genes are conserved between the birds and mammals. Avian MC4R is expressed similarly as the mammalian MC4Rs, therefore it was suggested that avian MC4R may have similar functions as the mammalian counterparts.[281–285] *In vivo* experiments with Japanese quail and broiler breeder hens confirmed the roles of the melanocortin system in appetite control,[283,285] mediated by activation of the MC4R in the ventromedial hypothalamus and paraventricular nucleus of the hypothalamus.[286] A comprehensive study on the chicken MC4R pharmacology has been published by Yajun Wang and his colleagues.[287]

Several chicken *MC4R* polymorphisms were identified to be associated with different production traits, such as higher body weight or feed conversion ratio.[288–291] For example, one study identified a missense mutation, G21R, that is closely related to body weight, eviscerated weight, and weight of thigh meat.[288] Sharma and colleagues identified two missense variants, Q18H and S76L, with the S76L heterozygote associated with greater body weight in female broilers.[290]

We performed detailed functional characterization of these mutations, including expression, ligand binding, and signaling, to demonstrate a potential causality between *MC4R* mutations and production traits in chicken.[292] Flow cytometry experiments demonstrated that compared with WT chicken MC4R, G21R has decreased cell surface and total expression whereas Q18H and S76L have similar total and cell surface expression levels as the WT receptor. All three mutant chicken MC4Rs have decreased binding. Interestingly, we showed that Q18H is constitutively active, which should be associated with decreased appetite and consequently decreased body weight. All three mutants have decreased signaling in response to α-MSH or β-MSH stimulation, suggesting that they are loss-of-function mutations, with G21R as Class I, Q18H and S76L as Class III mutations. In summary, these functional data are supportive of potential association of these *MC4R* variants with production traits.[292]

Of special interest is our data on β-MSH. We showed that similar binding and signaling properties are observed for α-MSH and β-MSH at the chicken MC4R. Because rodents lack the N-terminal cleavage site for β-MSH and are β-MSH deficient,[293] studies on the *in vivo* regulation of energy homeostasis focused on α-MSH. Human genetic studies demonstrated the critical importance of β-MSH in energy balance in humans.[251,252] Our data showed that both α-MSH and β-MSH can activate chicken MC4R, consistent with *in vivo* experiments.[294]

Silent polymorphisms in *MC4R* have also been identified that have effects on carcass traits in Landes geese.[295] Independent confirmation studies are needed and the mechanism of the association is also unclear. In laying ducks, *POMC* and *MC4R* are expressed lower in ducks with high residual feed intake compared to the ducks with low residual feed intake.[296] To the best of my knowledge, no genetic study on duck *MC4R* has been published.

In summary, it is likely that mutations in *MC4R* are relevant for chicken production traits, since the mutations identified have clear functional defects. Whether mutations in *MC4R* in other avian species such as duck and geese are also relevant for production traits remain to be further investigated.

7. Mutations in fish *mc4r*

Similar to mammals, in fishes, α-Msh and Agrp immunoreactivity is expressed in neurons homologous to the arcuate nucleus and responds to metabolic signal such as fasting.[297,298] In teleosts, sequences of *pomc* revealed that there is no γ-Msh sequence.[299] However, in cartilaginous fish, there is an additional melanocortin peptide, δ-Msh, suggesting that MSHs diverged at cartilaginous and bony fishes.[299] Consistent with the absence of γ-MSH, many teleosts lack the Mc3r (comprehensively listed in Ref. [300]), the MCR with relatively high affinity for γ-MSH,[16] serving as another example of co-evolution of ligand and receptor.[300]

Different from the Mc3r, Mc4r has been found in all fishes studied. The late eminent comparative endocrinologist, Richard Peter, was the first to comprehensively study fish Mc4r.[301] Since then, Mc4r has been characterized in teleosts[212–218,302–309] as well as in cartilaginous fish.[310–312]

In fish, Mc4r has been shown to act as a regulator of energy balance. A series of experiments demonstrate that Mc4r is an important regulator of food intake and energy homeostasis in goldfish and rainbow trout, with Mc4r agonists decreasing food intake and antagonists increasing food intake.[301,313–315] In barfin flounder, *mc4r* is expressed in the liver and this expression is modulated by fasting.[316] In common carp, *agrp1* and *mc4r* expression in the brain is decreased after fasting and increased after refeeding, when compared with normal fed controls.[317] However, in Atlantic salmon, expression of *mc4r*, *agrp*, or *pomc* is not changed by fasting and feeding.[318] Common carp overexpressing growth hormone has increased food intake and decreased energy expenditure, which may be explained by increased *agrp* expression.[319] In "cobalt" rainbow trout, lack of α-Msh causes obesity and liver fat dyshomeostasis.[320] In zebrafish, knockout of *mc4r* or genetic manipulation to delete α-Msh sequence results in hyperphagia and increased body weight[20,321]; administration of Mc4r agonists to fish lacking α-Msh ameliorates these phenotypes.[321] The inhibition of melanin-concentrating hormone and octadecaneuropeptide on food intake is also mediated by the Mc4r[322,323] (see Ref. [324] for a comprehensive review on neuroendocrine regulation of food intake in fishes).

In our pharmacology studies of fish Mc4rs, we revealed a unique property that is observed in all the species we have studied so far, i.e. dramatically increased constitutive activities compared to human MC4R. The human MC4R has been shown to have some constitutive activity in Gs-cAMP signaling,[59] and N-terminus is an important modulator of constitutive activity in human MC4R.[81] Mutations leading to decreased constitutive activities are associated with obesity pathogenesis.[55,81,83,84] Compared to human MC4R, teleost Mc4rs have much higher constitutive activity in cAMP signaling.[212–218,325] Constitutive activation of Erk1/2 signaling in some fish but not others has also been observed.[217,325] The potential relevance of constitutive activity in teleost Mc4rs remains to be studied more extensively.

In many fishes, *mc4r* is expressed more widely in different tissues (reviewed in Ref. [326]), including in the gonads, although dimorphic expression has also been reported (for example, expressed in the testis but not in the ovary[217]). Therefore, Mc4rs might play an important role in regulating fish reproductive function.[213,215,327] Indeed, several studies reported the roles of Mc4r in regulating sexual behavior and reproductive function.[327–332] In mammals, although there are some studies on MC4R involvement in reproduction-related hormone release (reviewed in Ref. [58]), lack of MC4R has very modest effect on reproduction.[333,334] However, in zebrafish, both Pomc and Agrp neurons project to the pituitary to regulate secretion of several pituitary hormones, affecting growth and reproduction.[335]

Especially relevant for the present discussion, two studies reporting mutations in the *mc4r* in fishes are of special interest. Clifford Tabin at Harvard Medical School is a pioneer in the study of Mexican cavefish. To adapt to the nutrient-poor environment, cave fish have increased appetite when food is available, even in the fed state, increased feed efficiency and body fat storage, and resistance to weight loss when fasted.[336] When Tabin and his colleagues sequenced the *mc4r* from surface and Tinaja cavefish, three missense mutations in highly conserved residues were identified: G145S in the second intracellular loop, V162I in TMD4, and M259T in TMD6.[336] Of these mutations, G145S might be the most important mutation because the corresponding mutation in human *MC4R* is associated with obesity.[337] Functional studies revealed that the mutant Mc4r has decreased constitutive activity (similar to our results on the analogous mutation in human MC4R, A154D[54]) and signaling in response to NDP-MSH stimulation.[336] Indeed, we showed that A154D human MC4R is also defective in signaling to ERK1/2.[97] These results strongly suggest that defective Mc4r signaling in the cave fish is the likely cause of hyperphagia and obesity phenotype of cave fish, which is the thrifty genotype beneficial in evolution.[336]

Fish of the genus *Xiphophorus*, also called either platyfish or swordtails, in the family Poeciliidae of order Cyprinodontiformes, are live-bearers widely used in genetic studies (https://www.xiphophorus.txstate.edu), especially in the genetics of pigmentation and melanoma research since the 1920s.[338] These fishes also provide an excellent case study of genetic control of fecundity in females, puberty onset in both sexes, and body size and reproductive success in males. Males that become sexually mature earlier are smaller in adult size than those that mature later.[339]

Of special interest is the *P* (Pituitary or Puberty) locus first identified in the 1970s.[339–341] It was not until 2010 that Manfred Schartl and colleagues identified the gene in the *P* locus responsible for the timing of puberty onset and male adult size.[328] In *X. nigrensis* and *X. multilineatus*, puberty onset and body length are determined by *mc4r* mutations and copy number variations of the mutant allele. Unlike the WT (A allele) Mc4r, the mutant alleles (B1 and B2) have no signaling and exert dominant negative activity on the WT Mc4r.[328] The dominant negative activity is different from most of the human *MC4R* mutations, where even intracellularly retained mutant receptors do not have dominant negative activity) (reviewed in Ref. [55]).

In addition, the male body size is directly proportional to the copy number of nonfunctional *mc4r* B alleles.[328] In *X. helleri*, only WT *mc4r* is identified and it is expressed at higher level in larger males than the smaller ones.[332] Higher expression of the Mc4r might cause increased appetite and larger adult size. In *Xiphophorus maculatus*, dramatic increase in *mc4r* copy number (up to 10 copies on both the X and Y chromosomes) and different kinds of mutations, including in-frame insertions/deletions and nonsense and missense mutations in the coding region and various mutations in the promoter region, have occurred.[329] Mutant receptors have changes in the basal activity as well as in ligand binding and hormone-induced signaling.[329]

Agouti-signaling protein (ASIP, also simply called Agouti) is an inverse agonist of the MC1R,[342,343] decreasing the high basal activity of MC1R.[344] It also antagonizes the MC4R and ubiquitously overexpressed *Asip*, including in the hypothalamus, inhibits MC4R action and causes severe obesity in the Agouti/yellow mouse.[30] However, transgenic overexpression of *asip* in zebrafish, *Danio rerio*, significantly delays early growth but does not change the timing of puberty.[345] Similar observations were made in medaka (*Oryzias latipes*), where *mc4r* knockout changes growth rate but not puberty timing,[346] different from the results obtained in swordtail fish described above.

It should be noted that although Agrp has been shown to be an inverse agonist for teleost Mc4rs,[325,347] as far as I am aware, there is no report of Asip as an antagonist for teleost Mc4rs. Potential different effects of ASIP on mammalian and teleost MC4Rs need to be investigated. Indeed, we observed dramatic difference in mammalian and teleost MC4Rs in their interactions with small molecule ligands, where the small molecule ligands serve as orthosteric ligands in mammalian MC4Rs but allosteric ligands in fish Mc4rs.[212,214,216,325] One ligand can even have different effect on Mc4rs from different fishes.[325]

8. Conclusions

In summary, the melanocortin system is an ancient system highly conserved in the vertebrates. The function of the MC4R in regulating energy homeostasis is also highly conserved in vertebrates. *MC4R* mutations or variants are associated with traits that are related to energy balance, for example, surviving and thriving in a nutrient-poor environment by increased appetite in the fed state, conferring a reproductive advantage by growing to a larger size, or causing obesity in humans. *MC4R* mutations in large domestic animals are relatively rare, especially when functional relevance is considered, raising questions regarding their causal relationship. Many variants identified in these animals do not cause an overt functional defect. In contrast, *mc4r* mutations identified in fishes have functional defects, more convincingly demonstrate a causal relationship.

Conflict of interest

The author declares that there is no conflict of interest that would prejudice the impartiality of this study.

References

1. Chrétien M, Benjannet S, Gossard F, et al. From β-lipotropin to β-endorphin and 'pro-opio-melanocortin'. *Can J Biochem.* 1979;57(9):1111–1121.
2. Sayers G, Sayers MA. Regulatory effect of adrenal cortical extract on elaboration of pituitary adrenotropic hormone. *Proc Soc Exp Biol Med.* 1945;60:162.
3. Simpson ME, Li CH, Evans HM. Effect of adrenocorticotropic hormone (ACTH) on hypophysectomized adrenal-demedullated rats. *Proc Soc Exp Biol Med.* 1945;60:329–332.
4. Allen BM. The results of extirpation of the anterior lobe of the hypophysis and of the thyroid of *Rana pipiens* larvae. *Science.* 1916;44(1143):755–758.
5. Smith PE. Experimental ablation of the hypophysis in the frog embryo. *Science.* 1916;44 (1130):280–282.
6. Cushing H. The basophil adenomas of the pituitary body and their clinical manifestations (pituitary basophilism). 1932. *Obes Res.* 1994;2(5):486–508.
7. Ingle DJ, Kendall EC. Atrophy of the adrenal cortex of the rat produced by the administration of large amounts of cortin. *Science.* 1937;86(2228):245.
8. Hench PS, Kendall EC, Slocumb CH, Polley HF. Effects of cortisone acetate and pituitary ACTH on rheumatoid arthritis, rheumatic fever and certain other conditions. *Arch Intern Med (Chic).* 1950;85(4):545–666.
9. Smith AI, Funder JW. Proopiomelanocortin processing in the pituitary, central nervous system, and peripheral tissues. *Endocr Rev.* 1988;9(1):159–179.
10. Harno E, Gali Ramamoorthy T, Coll AP, White A. POMC: the physiological power of hormone processing. *Physiol Rev.* 2018;98(4):2381–2430.
11. Nakanishi S, Inoue A, Kita T, et al. Nucleotide sequence of cloned cDNA for bovine corticotropin-beta-lipotropin precursor. *Nature.* 1979;278(5703):423–427.
12. Lerner AB. The discovery of the melanotropins. A history of pituitary endocrinology. *Ann N Y Acad Sci.* 1993;680:1–12.
13. Mountjoy KG, Robbins LS, Mortrud MT, Cone RD. The cloning of a family of genes that encode the melanocortin receptors. *Science.* 1992;257(5074):1248–1251.
14. Chhajlani V, Wikberg JE. Molecular cloning and expression of the human melanocyte stimulating hormone receptor cDNA. *FEBS Lett.* 1992;309(3):417–420.
15. Gantz I, Konda Y, Tashiro T, et al. Molecular cloning of a novel melanocortin receptor. *J Biol Chem.* 1993;268(11):8246–8250.

16. Roselli-Rehfuss L, Mountjoy KG, Robbins LS, et al. Identification of a receptor for γ melanotropin and other proopiomelanocortin peptides in the hypothalamus and limbic system. *Proc Natl Acad Sci U S A.* 1993;90(19):8856–8860.
17. Gantz I, Miwa H, Konda Y, et al. Molecular cloning, expression, and gene localization of a fourth melanocortin receptor. *J Biol Chem.* 1993;268(20):15174–15179.
18. Mountjoy KG, Mortrud MT, Low MJ, Simerly RB, Cone RD. Localization of the melanocortin-4 receptor (MC4-R) in neuroendocrine and autonomic control circuits in the brain. *Mol Endocrinol.* 1994;8(10):1298–1308.
19. Chhajlani V, Muceniece R, Wikberg JE. Molecular cloning of a novel human melanocortin receptor. *Biochem Biophys Res Commun.* 1993;195(2):866–873.
20. Gantz I, Shimoto Y, Konda Y, Miwa H, Dickinson CJ, Yamada T. Molecular cloning, expression, and characterization of a fifth melanocortin receptor. *Biochem Biophys Res Commun.* 1994;200(3):1214–1220.
21. Barrett P, MacDonald A, Helliwell R, Davidson G, Morgan P. Cloning and expression of a new member of the melanocyte-stimulating hormone receptor family. *J Mol Endocrinol.* 1994;12(2):203–213.
22. Griffon N, Mignon V, Facchinetti P, Diaz J, Schwartz JC, Sokoloff P. Molecular cloning and characterization of the rat fifth melanocortin receptor. *Biochem Biophys Res Commun.* 1994;200(2):1007–1014.
23. Labbe O, Desarnaud F, Eggerickx D, Vassart G, Parmentier M. Molecular cloning of a mouse melanocortin 5 receptor gene widely expressed in peripheral tissues. *Biochemistry.* 1994;33(15):4539–4543.
24. Cone RD. Studies on the physiological functions of the melanocortin system. *Endocr Rev.* 2006;27(7):736–749.
25. Tao YX. Melanocortin receptors. *Biochim Biophys Acta.* 2017;1863(10 Pt A):2411–2413.
26. Panskepp J, Reilly P, Bishop P, Meeker RB, Vilberg TR, Kastin AJ. Effects of α-MSH on motivation, vigilance and brain respiration. *Pharmacol Biochem Behav.* 1976;5 (suppl 1):59–64.
27. Vergoni AV, Poggioli R, Bertolini A. Corticotropin inhibits food intake in rats. *Neuropeptides.* 1986;7(2):153–158.
28. Poggioli R, Vergoni AV, Bertolini A. ACTH-(1-24) and alpha-MSH antagonize feeding behavior stimulated by kappa opiate agonists. *Peptides.* 1986;7(5):843–848.
29. Huszar D, Lynch CA, Fairchild-Huntress V, et al. Targeted disruption of the melanocortin-4 receptor results in obesity in mice. *Cell.* 1997;88(1):131–141.
30. Fan W, Boston BA, Kesterson RA, Hruby VJ, Cone RD. Role of melanocortinergic neurons in feeding and the agouti obesity syndrome. *Nature.* 1997;385(6612):165–168.
31. Vaisse C, Clement K, Guy-Grand B, Froguel P. A frameshift mutation in human *MC4R* is associated with a dominant form of obesity. *Nat Genet.* 1998;20(2):113–114.
32. Yeo GS, Farooqi IS, Aminian S, Halsall DJ, Stanhope RG, O'Rahilly S. A frameshift mutation in *MC4R* associated with dominantly inherited human obesity. *Nat Genet.* 1998;20(2):111–112.
33. Hinney A, Schmidt A, Nottebom K, et al. Several mutations in the melanocortin-4 receptor gene including a nonsense and a frameshift mutation associated with dominantly inherited obesity in humans. *J Clin Endocrinol Metab.* 1999;84(4):1483–1486.
34. Ollmann MM, Wilson BD, Yang YK, et al. Antagonism of central melanocortin receptors in vitro and in vivo by agouti-related protein. *Science.* 1997;278(5335):135–138.
35. Nijenhuis WA, Oosterom J, Adan RA. AgRP(83-132) acts as an inverse agonist on the human melanocortin-4 receptor. *Mol Endocrinol.* 2001;15(1):164–171.
36. Haskell-Luevano C, Monck EK. Agouti-related protein functions as an inverse agonist at a constitutively active brain melanocortin-4 receptor. *Regul Pept.* 2001;99(1):1–7.
37. Loos RJF, Yeo GSH. The genetics of obesity: from discovery to biology. *Nat Rev Genet.* 2022;23(2):120–133.

38. Krashes MJ, Lowell BB, Garfield AS. Melanocortin-4 receptor-regulated energy homeostasis. *Nat Neurosci*. 2016;19(2):206–219.
39. Ho G, MacKenzie RG. Functional characterization of mutations in melanocortin-4 receptor associated with human obesity. *J Biol Chem*. 1999;274(50):35816–35822.
40. Vaisse C, Clement K, Durand E, Hercberg S, Guy-Grand B, Froguel P. Melanocortin-4 receptor mutations are a frequent and heterogeneous cause of morbid obesity. *J Clin Invest*. 2000;106(2):253–262.
41. Farooqi IS, Yeo GS, Keogh JM, et al. Dominant and recessive inheritance of morbid obesity associated with melanocortin 4 receptor deficiency. *J Clin Invest*. 2000;106 (2):271–279.
42. Tao YX, Segaloff DL. Functional characterization of melanocortin-4 receptor mutations associated with childhood obesity. *Endocrinology*. 2003;144(10):4544–4551.
43. Yeo GSH, Lank EJ, Farooqi IS, Keogh J, Challis BG, O'Rahilly S. Mutations in the human melanocortin-4 receptor gene associated with severe familial obesity disrupts receptor function through multiple molecular mechanisms. *Hum Mol Genet*. 2003;12 (5):561–574.
44. Lubrano-Berthelier C, Durand E, Dubern B, et al. Intracellular retention is a common characteristic of childhood obesity-associated MC4R mutations. *Hum Mol Genet*. 2003;12(2):145–153.
45. VanLeeuwen D, Steffey ME, Donahue C, Ho G, MacKenzie RG. Cell surface expression of the melanocortin-4 receptor is dependent on a C-terminal di-isoleucine sequence at codons 316/317. *J Biol Chem*. 2003;278(18):15935–15940.
46. Nijenhuis WA, Garner KM, VanRozen RJ, Adan RA. Poor cell surface expression of human melanocortin-4 receptor mutations associated with obesity. *J Biol Chem*. 2003;278:22939–22945.
47. Tao YX. Inactivating mutations of G protein-coupled receptors and diseases: structure-function insights and therapeutic implications. *Pharmacol Ther*. 2006;111 (3):949–973.
48. Conn PM, Ulloa-Aguirre A, Ito J, Janovick JA. G protein-coupled receptor trafficking in health and disease: lessons learned to prepare for therapeutic mutant rescue in vivo. *Pharmacol Rev*. 2007;59(3):225–250.
49. Tao YX, Conn PM. Chaperoning G protein-coupled receptors: from cell biology to therapeutics. *Endocr Rev*. 2014;35(4):602–647.
50. Hou ZS, Ulloa-Aguirre A, Tao YX. Pharmacoperone drugs: targeting misfolded proteins causing lysosomal storage-, ion channels-, and G protein-coupled receptors-associated conformational disorders. *Expert Rev Clin Pharmacol*. 2018;11(6):611–624.
51. Tao YX, Conn PM. Pharmacoperones as novel therapeutics for diverse protein conformational diseases. *Physiol Rev*. 2018;98(2):697–725.
52. Marinko JT, Huang H, Penn WD, Capra JA, Schlebach JP, Sanders CR. Folding and misfolding of human membrane proteins in health and disease: from single molecules to cellular proteostasis. *Chem Rev*. 2019;119(9):5537–5606.
53. Donohoue PA, Tao YX, Collins M, Yeo GSH, O'Rahilly S, Segaloff DL. Deletion of codons 88-92 of the melanocortin-4 receptor gene: a novel deleterious mutation in an obese female. *J Clin Endocrinol Metab*. 2003;88(12):5841–5845.
54. Tao YX, Segaloff DL. Functional analyses of melanocortin-4 receptor mutations identified from patients with binge eating disorder and nonobese or obese subjects. *J Clin Endocrinol Metab*. 2005;90(10):5632–5638.
55. Tao YX. Molecular mechanisms of the neural melanocortin receptor dysfunction in severe early onset obesity. *Mol Cell Endocrinol*. 2005;239(1-2):1–14.
56. Farooqi IS, Keogh JM, Yeo GS, Lank EJ, Cheetham T, O'Rahilly S. Clinical spectrum of obesity and mutations in the melanocortin 4 receptor gene. *N Engl J Med*. 2003;348 (12):1085–1095.

57. Tao YX. Mutations in melanocortin-4 receptor and human obesity. *Prog Mol Biol Transl Sci*. 2009;88:173–204.
58. Tao YX. The melanocortin-4 receptor: physiology, pharmacology, and pathophysiology. *Endocr Rev*. 2010;31(4):506–543.
59. Tao YX. Constitutive activity in melanocortin-4 receptor: biased signaling of inverse agonists. *Adv Pharmacol*. 2014;70:135–154.
60. Hinney A, Volckmar AL, Knoll N. Melanocortin-4 receptor in energy homeostasis and obesity pathogenesis. *Prog Mol Biol Transl Sci*. 2013;114:147–191.
61. Fani L, Bak S, Delhanty P, van Rossum EF, van den Akker EL. The melanocortin-4 receptor as target for obesity treatment: a systematic review of emerging pharmacological therapeutic options. *Int J Obes*. 2014;38(2):163–169.
62. Kühnen P, Krude H, Biebermann H. Melanocortin-4 receptor signalling: importance for weight regulation and obesity treatment. *Trends Mol Med*. 2019;25(2):136–148.
63. Sharma S, Garfield AS, Shah B, et al. Current mechanistic and pharmacodynamic understanding of melanocortin-4 receptor activation. *Molecules*. 2019;24(10):1892.
64. Heyder N, Kleinau G, Szczepek M, et al. Signal transduction and pathogenic modifications at the melanocortin-4 receptor: a structural perspective. *Front Endocrinol (Lausanne)*. 2019;10:515.
65. Kleinau G, Heyder NA, Tao YX, Scheerer P. Structural complexity and plasticity of signaling regulation at the melanocortin-4 receptor. *Int J Mol Sci*. 2020;21(16):5728.
66. Yeo GSH, Chao DHM, Siegert AM, et al. The melanocortin pathway and energy homeostasis: from discovery to obesity therapy. *Mol Metab*. 2021;48, 101206.
67. Fatima MT, Ahmed I, Fakhro KA, Akil ASA. Melanocortin-4 receptor complexity in energy homeostasis, obesity and drug development strategies. *Diabetes Obes Metab*. 2022;24:583–598.
68. Lek M, Karczewski KJ, Minikel EV, et al. Analysis of protein-coding genetic variation in 60,706 humans. *Nature*. 2016;536(7616):285–297.
69. Yu J, Gimenez LE, Hernandez CC, et al. Determination of the melanocortin-4 receptor structure identifies Ca^{2+} as a cofactor for ligand binding. *Science*. 2020;368 (6489):428–433.
70. Israeli H, Degtjarik O, Fierro F, et al. Structure reveals the activation mechanism of the MC4 receptor to initiate satiation signaling. *Science*. 2021;372(6544):808–814.
71. Heyder NA, Kleinau G, Speck D, et al. Structures of active melanocortin-4 receptor-Gs-protein complexes with NDP-α-MSH and setmelanotide. *Cell Res*. 2021;31 (11):1176–1189.
72. Zhang H, Chen LN, Yang D, et al. Structural insights into ligand recognition and activation of the melanocortin-4 receptor. *Cell Res*. 2021;31(11):1163–1175.
73. Tan K, Pogozheva ID, Yeo GS, et al. Functional characterization and structural modeling of obesity associated mutations in the melanocortin 4 receptor. *Endocrinology*. 2009;150(1):114–125.
74. Tao YX, Huang H, Wang ZQ, Yang F, Williams JN, Nikiforovich GV. Constitutive activity of neural melanocortin receptors. *Methods Enzymol*. 2010;484:267–279.
75. Shahlaei M, Mousavi A. A conformational analysis study on the melanocortin 4 receptor using multiple molecular dynamics simulations. *Chem Biol Drug Des*. 2015;86 (3):309–321.
76. Saleh N, Kleinau G, Heyder N, Clark T, Hildebrand PW, Scheerer P. Binding, thermodynamics, and selectivity of a non-peptide antagonist to the melanocortin-4 receptor. *Front Pharmacol*. 2018;9:560.
77. Falls BA, Zhang Y. Insights into the allosteric mechanism of setmelanotide (RM-493) as a potent and first-in-class melanocortin-4 receptor (MC4R) agonist to treat rare genetic disorders of obesity through an in silico approach. *ACS Chem Neurosci*. 2018;10(3):1055–1065.

78. Lefkowitz RJ, Cotecchia S, Samama P, Costa T. Constitutive activity of receptors coupled to guanine nucleotide regulatory proteins. *Trends Pharmacol Sci.* 1993;14(8):303–307.
79. Tao YX. Functional characterization of novel melanocortin-3 receptor mutations identified from obese subjects. *Biochim Biophys Acta.* 2007;1772(10):1167–1174.
80. Gimenez LE, Noblin TA, Williams SY, et al. Demonstration of a common DPhe[7] to DNal(2')[7] peptide ligand antagonist switch for the melanocortin-3 and melanocortin-4 receptors identifies systematic mischaracterization of the pharmacological properties of melanocortin peptides. *J Med Chem.* 2022. https://doi.org/10.1021/acs.jmedchem.1c01295. in press.
81. Srinivasan S, Lubrano-Berthelier C, Govaerts C, et al. Constitutive activity of the melanocortin-4 receptor is maintained by its N-terminal domain and plays a role in energy homeostasis in humans. *J Clin Invest.* 2004;114(8):1158–1164.
82. Pantel J, Legendre M, Cabrol S, et al. Loss of constitutive activity of the growth hormone secretagogue receptor in familial short stature. *J Clin Invest.* 2006;116(3):760–768.
83. Tao YX. Constitutive activation of G protein-coupled receptors and diseases: insights into mechanism of activation and therapeutics. *Pharmacol Ther.* 2008;120:129–148.
84. Wang ZQ, Tao YX. Functional studies on twenty novel naturally occurring melanocortin-4 receptor mutations. *Biochim Biophys Acta.* 2011;1812:1190–1199.
85. Hohenadel MG, Thearle MS, Grice BA, et al. Brain-derived neurotrophic factor in human subjects with function-altering melanocortin-4 receptor variants. *Int J Obes.* 2014;38(8):1068–1074.
86. Stutzmann F, Tan K, Vatin V, et al. Prevalence of melanocortin-4 receptor deficiency in Europeans and their age-dependent penetrance in multigenerational pedigrees. *Diabetes.* 2008;57(9):2511–2518.
87. Hinney A, Hohmann S, Geller F, et al. Melanocortin-4 receptor gene: case-control study and transmission disequilibrium test confirm that functionally relevant mutations are compatible with a major gene effect for extreme obesity. *J Clin Endocrinol Metab.* 2003;88(9):4258–4267.
88. Beckers S, Zegers D, de Freitas F, et al. Identification and functional characterization of novel mutations in the melanocortin-4 receptor. *Obes Facts.* 2010;3(5):304–311.
89. Proneth B, Xiang Z, Pogozheva ID, et al. Molecular mechanism of the constitutive activation of the L250Q human melanocortin-4 receptor polymorphism. *Chem Biol Drug Des.* 2006;67(3):215–229.
90. Fan ZC, Tao YX. Functional characterization and pharmacological rescue of melanocortin-4 receptor mutations identified from obese patients. *J Cell Mol Med.* 2009;13:3268–3282.
91. Mo XL, Tao YX. Activation of MAPK by inverse agonists in six naturally occurring constitutively active mutant human melanocortin-4 receptors. *Biochim Biophys Acta.* 2013;1832(12):1939–1948.
92. Yang Z, Tao YX. Biased signaling initiated by agouti-related peptide through human melanocortin-3 and -4 receptors. *Biochim Biophys Acta.* 2016;1862(9):1485–1494.
93. Yang LK, Tao YX. Biased signaling at neural melanocortin receptors in regulation of energy homeostasis. *Biochim Biophys Acta Mol basis Dis.* 2017;1863:2486–2495.
94. Yang LK, Hou ZS, Tao YX. Biased signaling in naturally occurring mutations of G protein-coupled receptors associated with diverse human diseases. *Biochim Biophys Acta Mol basis Dis.* 2021;1867(1), 165973.
95. Liu T, Ji RL, Tao YX. Naturally occurring mutations in G protein-coupled receptors associated with obesity and type 2 diabetes mellitus. *Pharmacol Ther.* 2022;108044. https://doi.org/10.1016/j.pharmthera.2021.108044.
96. Rong R, Tao YX, Cheung BM, Xu A, Cheung GC, Lam KS. Identification and functional characterization of three novel human melanocortin-4 receptor gene variants in an obese Chinese population. *Clin Endocrinol.* 2006;65(2):198–205.

97. He S, Tao YX. Defect in MAPK signaling as a cause for monogenic obesity caused by inactivating mutations in the melanocortin-4 receptor gene. *Int J Biol Sci*. 2014;10 (10):1128–1137.
98. Li YQ, Shrestha Y, Pandey M, et al. $G_{q/11}\alpha$ and $G_s\alpha$ mediate distinct physiological responses to central melanocortins. *J Clin Invest*. 2016;126(1):40–49.
99. Podyma B, Sun H, Wilson EA, et al. The stimulatory G protein $G_s\alpha$ is required in melanocortin 4 receptor-expressing cells for normal energy balance, thermogenesis, and glucose metabolism. *J Biol Chem*. 2018;293(28):10993–11005.
100. Gillyard T, Fowler K, Williams SY, Cone RD. Obesity-associated mutant melanocortin-4 receptors with normal Gαs coupling frequently exhibit other discoverable pharmacological and biochemical defects. *J Neuroendocrinol*. 2019;31(10), e12795.
101. Lotta LA, Mokrosinski J, Mendes de Oliveira E, et al. Human gain-of-function *MC4R* variants show signaling bias and protect against obesity. *Cell*. 2019;177 (3):597–607 e599.
102. Sharma S, Thibodeau S, Lytton J. Signal pathway analysis of selected obesity-associated melanocortin-4 receptor class V mutants. *Biochim Biophys Acta Mol basis Dis*. 2020;1866 (8), 165835.
103. Brouwers B, de Oliveira EM, Marti-Solano M, et al. Human MC4R variants affect endocytosis, trafficking and dimerization revealing multiple cellular mechanisms involved in weight regulation. *Cell Rep*. 2021;34(12), 108862.
104. Kumar KG, Sutton GM, Dong JZ, et al. Analysis of the therapeutic functions of novel melanocortin receptor agonists in MC3R- and MC4R-deficient C57BL/6J mice. *Peptides*. 2009;30(10):1892–1900.
105. Kievit P, Halem H, Marks DL, et al. Chronic treatment with a melanocortin-4 receptor agonist causes weight loss, reduces insulin resistance, and improves cardiovascular function in diet-induced obese rhesus macaques. *Diabetes*. 2013;62(2):490–497.
106. Chen KY, Muniyappa R, Abel BS, et al. RM-493, a melanocortin-4 receptor (MC4R) agonist, increases resting energy expenditure in obese individuals. *J Clin Endocrinol Metab*. 2015;100(4):1639–1645.
107. Molden BM, Cooney KA, West K, Van Der Ploeg LH, Baldini G. Temporal cAMP signaling selectivity by natural and synthetic MC4R agonists. *Mol Endocrinol*. 2015;29 (11):1619–1633.
108. Kühnen P, Clement K, Wiegand S, et al. Proopiomelanocortin deficiency treated with a melanocortin-4 receptor agonist. *N Engl J Med*. 2016;375(3):240–246.
109. Collet TH, Dubern B, Mokrosinski J, et al. Evaluation of a melanocortin-4 receptor (MC4R) agonist (Setmelanotide) in MC4R deficiency. *Mol Metab*. 2017;6(10): 1321–1329.
110. Clément K, Biebermann H, Farooqi IS, et al. MC4R agonism promotes durable weight loss in patients with leptin receptor deficiency. *Nat Med*. 2018;24(5):551–555.
111. Clément K, van den Akker E, Argente J, et al. Efficacy and safety of setmelanotide, an MC4R agonist, in individuals with severe obesity due to LEPR or POMC deficiency: single-arm, open-label, multicentre, phase 3 trials. *Lancet Diabetes Endocrinal*. 2020; 8:960–970.
112. Haws R, Brady S, Davis E, et al. Effect of setmelanotide, a melanocortin-4 receptor agonist, on obesity in Bardet-Biedl syndrome. *Diabetes Obes Metab*. 2020;22(11):2133–2140.
113. Kühnen P, Wabitsch M, von Schnurbein J, et al. Quality of life outcomes in two phase 3 trials of setmelanotide in patients with obesity due to LEPR or POMC deficiency. *Orphanet J Rare Dis*. 2022;17(1):38.
114. Ryan DH. Setmelanotide: what does it mean for clinical care of patients with obesity? *Lancet Diabetes Endocrinol*. 2020;8(12):933–935.
115. Markham A. Setmelanotide: first approval. *Drugs*. 2021;81(3):397–403.

116. Greenfield JR, Miller JW, Keogh JM, et al. Modulation of blood pressure by central melanocortinergic pathways. *N Engl J Med.* 2009;360(1):44–52.
117. Brumm H, Muhlhaus J, Bolze F, et al. Rescue of melanocortin 4 receptor (MC4R) nonsense mutations by aminoglycoside-mediated read-through. *Obesity (Silver Spring).* 2012;20(5):1074–1081.
118. Bolze F, Rink N, Brumm H, et al. Characterization of the melanocortin-4-receptor nonsense mutation W16X *in vitro* and *in vivo*. *Pharm J.* 2013;13(1):80–93.
119. Xiang Z, Pogozheva ID, Sorenson NB, et al. Peptide and small molecules rescue the functional activity and agonist potency of dysfunctional human melanocortin-4 receptor polymorphisms. *Biochemistry.* 2007;46(28):8273–8287.
120. Haslach EM, Huang H, Dirain M, et al. Identification of tetrapeptides from a mixture based positional scanning library that can restore nM full agonist function of the L106P, I69T, I102S, A219V, C271Y, and C271R human melanocortin-4 polymorphic receptors (hMC4Rs). *J Med Chem.* 2014;57(11):4615–4628.
121. Roubert P, Dubern B, Plas P, et al. Novel pharmacological MC4R agonists can efficiently activate mutated MC4R from obese patient with impaired endogenous agonist response. *J Endocrinol.* 2010;207(2):177–183.
122. Conn PM, Ulloa-Aguirre A, Janovick JA. "Pharmacoperone": what's in a word? *Pharmacol Res.* 2014;83:1–2.
123. Fan JQ, Ishii S, Asano N, Suzuki Y. Accelerated transport and maturation of lysosomal α-galactosidase A in Fabry lymphoblasts by an enzyme inhibitor. *Nat Med.* 1999;5(1):112–115.
124. Morello JP, Salahpour A, Laperrière A, et al. Pharmacological chaperones rescue cell-surface expression and function of misfolded V2 vasopressin receptor mutants. *J Clin Invest.* 2000;105(7):887–895.
125. Ulloa-Aguirre A, Janovick JA, Brothers SP, Conn PM. Pharmacologic rescue of conformationally-defective proteins: implications for the treatment of human disease. *Traffic.* 2004;5(11):821–837.
126. Maya-Nunez G, Ulloa-Aguirre A, Janovick JA, Conn PM. Pharmacological chaperones correct misfolded GPCRs and rescue function: protein trafficking as a therapeutic target. *Subcell Biochem.* 2012;63:263–289.
127. Vos TJ, Caracoti A, Che JL, et al. Identification of 2-[2-[2-(5-bromo-2-methoxyphenyl)-ethyl]-3-fluorophenyl]-4,5-dihydro-1H-imidazole (ML00253764), a small molecule melanocortin 4 receptor antagonist that effectively reduces tumor-induced weight loss in a mouse model. *J Med Chem.* 2004;47(7):1602–1604.
128. Huang H, Tao YX. A small molecule agonist THIQ as a novel pharmacoperone for intracellularly retained melanocortin-4 receptor mutants. *Int J Biol Sci.* 2014;10(8):817–824.
129. Tao YX, Huang H. Ipsen 5i is a novel potent pharmacoperone for intracellularly retained melanocortin-4 receptor mutants. *Front Endocrinol (Lausanne).* 2014;5:131.
130. Rene P, Le Gouill C, Pogozheva ID, et al. Pharmacological chaperones restore function to MC4R mutants responsible for severe early-onset obesity. *J Pharmacol Exp Ther.* 2010;335(3):520–532.
131. Wang XH, Wang HM, Zhao BL, Yu P, Fan ZC. Rescue of defective MC4R cell-surface expression and signaling by a novel pharmacoperone Ipsen 17. *J Mol Endocrinol.* 2014;53(1):17–29.
132. Huang H, Wang W, Tao YX. Pharmacological chaperones for the misfolded melanocortin-4 receptor associated with human obesity. *Biochim Biophys Acta Mol Basis Dis.* 2017;1863(10 Pt A):2496–2507.
133. Rene P, Lanfray D, Richard D, Bouvier M. Pharmacological chaperone action in humanized mouse models of MC4R-linked obesity. *JCI Insight.* 2021;6(4):e132778.

134. Kühnen P, Wiegand S, Biebermann H. Pharmacological treatment strategies for patients with monogenic obesity. *J Pediatr Endocrinol Metab*. 2020. https://doi.org/10.1515/jpem-2020-0129.
135. Trier C, Hollensted M, Schnurr TM, et al. Obesity treatment effect in Danish children and adolescents carrying *Melanocortin-4 Receptor* mutations. *Int J Obes*. 2021;45(1):66–76.
136. Iepsen EW, Zhang J, Thomsen HS, et al. Patients with obesity caused by melanocortin-4 receptor mutations can be treated with a glucagon-like peptide-1 receptor agonist. *Cell Metab*. 2018;28(1):23–32 e23.
137. Iepsen EW, Have CT, Veedfald S, et al. GLP-1 receptor agonist treatment in morbid obesity and type 2 diabetes due to pathogenic homozygous melanocortin-4 receptor mutation: a case report. *Cell Rep Med*. 2020;1(1), 100006.
138. Camtosun E, Akinci A, Kayas L, Ciftci N, Tekedereli I. Liraglutide treatment experience in morbid obese adolescent with a *MC4R* gene variant: side effects reduce success. *J Clin Res Pediatr Endocrinol*. 2021.
139. Mirshahi UL, Still CD, Masker KK, Gerhard GS, Carey DJ, Mirshahi T. The *MC4R (I251L)* allele is associated with better metabolic status and more weight loss after gastric bypass surgery. *J Clin Endocrinol Metab*. 2011;96(12):E2088–E2096.
140. Aslan IR, Ranadive SA, Ersoy BA, Rogers SJ, Lustig RH, Vaisse C. Bariatric surgery in a patient with complete MC4R deficiency. *Int J Obes*. 2011;35(3):457–461.
141. Aslan IR, Campos GM, Calton MA, Evans DS, Merriman RB, Vaisse C. Weight loss after Roux-en-Y gastric bypass in obese patients heterozygous for MC4R mutations. *Obes Surg*. 2011;21(7):930–934.
142. Valette M, Poitou C, Le Beyec J, Bouillot JL, Clement K, Czernichow S. Melanocortin-4 receptor mutations and polymorphisms do not affect weight loss after bariatric surgery. *PLoS One*. 2012;7(11), e48221.
143. Hatoum IJ, Stylopoulos N, Vanhoose AM, et al. Melanocortin-4 receptor signaling is required for weight loss after gastric bypass surgery. *J Clin Endocrinol Metab*. 2012;97(6):E1023–E1031.
144. Censani M, Conroy R, Deng L, et al. Weight loss after bariatric surgery in morbidly obese adolescents with *MC4R* mutations. *Obesity*. 2014;22(1):225–231.
145. Elkhenini HF, New JP, Syed AA. Five-year outcome of bariatric surgery in a patient with melanocortin-4 receptor mutation. *Clin Obes*. 2014;4(2):121–124.
146. Jelin EB, Daggag H, Speer AL, et al. Melanocortin-4 receptor signaling is not required for short-term weight loss after sleeve gastrectomy in pediatric patients. *Int J Obes*. 2016;40(3):550–553.
147. Li YY, Zhang H, Tu YF, et al. Monogenic obesity mutations lead to less weight loss aafter bariatric surgery: a 6-year follow-up study. *Obes Surg*. 2019;29(4):1169–1173.
148. Cooiman MI, Kleinendorst L, Aarts EO, et al. Genetic obesity and bariatric surgery outcome in 1014 patients with morbid obesity. *Obes Surg*. 2020;30(2):470–477.
149. Poitou C, Puder L, Dubern B, et al. Long-term outcomes of bariatric surgery in patients with bi-allelic mutations in the *POMC, LEPR*, and *MC4R* genes. *Surg Obes Relat Dis*. 2021;17(8):1449–1456.
150. Cooiman MI, Alsters SIM, Duquesnoy M, et al. Long-term weight outcome after bariatric surgery in patients with melanocortin-4 receptor gene variants: a case-control study of 105 patients. *Obes Surg*. 2022;32(3):837–844.
151. Fojas EGF, Radha SK, Ali T, Nadler EP, Lessan N. Weight and glycemic control outcomes of bariatric surgery and pharmacotherapy in patients with melanocortin-4 receptor deficiency. *Front Endocrinol (Lausanne)*. 2021;12, 792354.
152. Vos N, Oussaada SM, Cooiman MI, et al. Bariatric surgery for monogenic non-syndromic and syndromic obesity disorders. *Curr Diab Rep*. 2020;20(9):44.

153. Gotoda T, Scott J, Aitman TJ. Molecular screening of the human melanocortin-4 receptor gene: identification of a missense variant showing no association with obesity, plasma glucose, or insulin. *Diabetologia.* 1997;40(8):976–979.
154. Geller F, Reichwald K, Dempfle A, et al. Melanocortin-4 receptor gene variant I103 is negatively associated with obesity. *Am J Hum Genet.* 2004;74(3):572–581.
155. Young EH, Wareham NJ, Farooqi S, et al. The V103I polymorphism of the MC4R gene and obesity: population based studies and meta-analysis of 29 563 individuals. *Int J Obes.* 2007;31(9):1437–1441.
156. Stutzmann F, Vatin V, Cauchi S, et al. Non-synonymous polymorphisms in melanocortin-4 receptor protect against obesity: the two facets of a Janus obesity gene. *Hum Mol Genet.* 2007;16(15):1837–1844.
157. Wang D, Ma J, Zhang S, et al. Association of the *MC4R* V103I polymorphism with obesity: a Chinese case-control study and meta-analysis in 55,195 individuals. *Obesity (Silver Spring).* 2010;18(3):573–579.
158. Loos RJ. The genetic epidemiology of *melanocortin 4 receptor* variants. *Eur J Pharmacol.* 2011;660(1):156–164.
159. Marti A, Corbalan MS, Forga L, Martinez JA, Hinney A, Hebebrand J. A novel nonsense mutation in the melanocortin-4 receptor associated with obesity in a Spanish population. *Int J Obes Relat Metab Disord.* 2003;27(3):385–388.
160. Yurtcu E, Yilmaz A, Ozkurt Z, et al. Melanocortin-4 receptor gene polymorphisms in obese patients. *Biochem Genet.* 2009;47(3-4):295–300.
161. Xiang Z, Litherland SA, Sorensen NB, et al. Pharmacological characterization of 40 human melanocortin-4 receptor polymorphisms with the endogenous proopiomelanocortin-derived agonists and the agouti-related protein (AGRP) antagonist. *Biochemistry.* 2006; 45(23):7277–7288.
162. Ochoa MC, Azcona C, Biebermann H, et al. A novel mutation Thr162Arg of the melanocortin 4 receptor gene in a Spanish children and adolescent population. *Clin Endocrinol.* 2007;66(5):652–658.
163. Rojo D, McCarthy C, Raingo J, Rubinstein M. Mouse models for V103I and I251L gain of function variants of the human MC4R display decreased adiposity but are not protected against a hypercaloric diet. *Mol Metab.* 2020;42, 101077.
164. Marti A, Moreno-Aliaga MJ, Hebebrand J, Martinez JA. Genes, lifestyles and obesity. *Int J Obes Relat Metab Disord.* 2004;28(suppl 3):S29–S36.
165. Adamska-Patruno E, Bauer W, Bielska D, et al. An association between diet and *MC4R* genetic polymorphism, in relation to obesity and metabolic parameters—a cross sectional population-based study. *Int J Mol Sci.* 2021;22(21):12044.
166. Loos RJ, Lindgren CM, Li S, et al. Common variants near *MC4R* are associated with fat mass, weight and risk of obesity. *Nat Genet.* 2008;40(6):768–775.
167. Qi L, Kraft P, Hunter DJ, Hu FB. The common obesity variant near *MC4R* gene is associated with higher intakes of total energy and dietary fat, weight change and diabetes risk in women. *Hum Mol Genet.* 2008;17(22):3502–3508.
168. Xi B, Chandak GR, Shen Y, Wang Q, Zhou D. Association between common polymorphism near the *MC4R* gene and obesity risk: a systematic review and meta-analysis. *PLoS One.* 2012;7(9), e45731.
169. Xi B, Takeuchi F, Chandak GR, et al. Common polymorphism near the *MC4R* gene is associated with type 2 diabetes: data from a meta-analysis of 123,373 individuals. *Diabetologia.* 2012;55(10):2660–2666.
170. Chambers JC, Elliott P, Zabaneh D, et al. Common genetic variation near *MC4R* is associated with waist circumference and insulin resistance. *Nat Genet.* 2008;40 (6):716–718.
171. Willer CJ, Speliotes EK, Loos RJ, et al. Six new loci associated with body mass index highlight a neuronal influence on body weight regulation. *Nat Genet.* 2009;41(1):25–34.

172. Thorleifsson G, Walters GB, Gudbjartsson DF, et al. Genome-wide association yields new sequence variants at seven loci that associate with measures of obesity. *Nat Genet.* 2009;41(1):18–24.
173. Lindgren CM, Heid IM, Randall JC, et al. Genome-wide association scan meta-analysis identifies three Loci influencing adiposity and fat distribution. *PLoS Genet.* 2009;5(6), e1000508.
174. Heard-Costa NL, Zillikens MC, Monda KL, et al. *NRXN3* is a novel locus for waist circumference: a genome-wide association study from the CHARGE Consortium. *PLoS Genet.* 2009;5(6), e1000539.
175. Meyre D, Delplanque J, Chevre JC, et al. Genome-wide association study for early-onset and morbid adult obesity identifies three new risk loci in European populations. *Nat Genet.* 2009;41(2):157–159.
176. Scherag A, Dina C, Hinney A, et al. Two new Loci for body-weight regulation identified in a joint analysis of genome-wide association studies for early-onset extreme obesity in French and german study groups. *PLoS Genet.* 2010;6(4), e1000916.
177. den Hoed M, Ekelund U, Brage S, et al. Genetic susceptibility to obesity and related traits in childhood and adolescence: influence of loci identified by genome-wide association studies. *Diabetes.* 2010;59(11):2980–2988.
178. Grant SF, Bradfield JP, Zhang H, et al. Investigation of the locus near *MC4R* with childhood obesity in Americans of European and African ancestry. *Obesity (Silver Spring).* 2009;17(7):1461–1465.
179. Ng MC, Tam CH, So WY, et al. Implication of genetic variants near *NEGR1*, *SEC16B*, *TMEM18*, *ETV5/DGKG*, *GNPDA2*, *LIN7C/BDNF*, *MTCH2*, *BCDIN3D/FAIM2*, *SH2B1*, *FTO*, *MC4R*, and *KCTD15* with obesity and type 2 diabetes in 7705 Chinese. *J Clin Endocrinol Metab.* 2010;95(5):2418–2425.
180. Wu L, Xi B, Zhang M, et al. Associations of six single nucleotide polymorphisms in obesity-related genes with BMI and risk of obesity in Chinese children. *Diabetes.* 2010;59(12):3085–3089.
181. Dastgheib SA, Bahrami R, Setayesh S, et al. Evidence from a meta-analysis for association of MC4R rs17782313 and FTO rs9939609 polymorphisms with susceptibility to obesity in children. *Diabetes Metab Syndr.* 2021;15(5), 102234.
182. Resende CMM, Silva H, Campello CP, et al. Polymorphisms on rs9939609 *FTO* and rs17782313 *MC4R* genes in children and adolescent obesity: a systematic review. *Nutrition.* 2021;91-92, 111474.
183. Spurlock ME, Gabler NK. The development of porcine models of obesity and the metabolic syndrome. *J Nutr.* 2008;138(2):397–402.
184. Koopmans SJ, Schuurman T. Considerations on pig models for appetite, metabolic syndrome and obese type 2 diabetes: from food intake to metabolic disease. *Eur J Pharmacol.* 2015;759:231–239.
185. Barb CR, Robertson AS, Barrett JB, Kraeling RR, Houseknecht KL. The role of melanocortin-3 and -4 receptor in regulating appetite, energy homeostasis and neuroendocrine function in the pig. *J Endocrinol.* 2004;181(1):39–52.
186. Hao H, Lin R, Li Z, et al. MC4R deficiency in pigs results in hyperphagia and ultimately hepatic steatosis without high-fat diet. *Biochem Biophys Res Commun.* 2019;520(3):651–656.
187. Kim KS, Larsen N, Short T, Plastow G, Rothschild MF. A missense variant of the porcine melanocortin-4 receptor (MC4R) gene is associated with fatness, growth, and feed intake traits. *Mamm Genome.* 2000;11(2):131–135.
188. Andersson L. Melanocortin receptor variants with phenotypic effects in horse, pig, and chicken. *Ann N Y Acad Sci.* 2003;994:313–318.
189. Liu GL, Jiang SW, Xiong YZ, Zheng R, Qu YC. Molecular screening of MC4R gene and association with fat traits in pig resource family. *Yi Chuan Xue Bao.* 2002;29(6):497–501.

190. Hernandez-Sanchez J, Visscher P, Plastow G, Haley C. Candidate gene analysis for quantitative traits using the transmission disequilibrium test: the example of the melanocortin 4-receptor in pigs. *Genetics*. 2003;164(2):637–644.
191. Houston RD, Cameron ND, Rance KA. A melanocortin-4 receptor (MC4R) polymorphism is associated with performance traits in divergently selected Large White pig populations. *Anim Genet*. 2004;35(5):386–390.
192. Meidtner K, Wermter AK, Hinney A, Remschmidt H, Hebebrand J, Fries R. Association of the melanocortin 4 receptor with feed intake and daily gain in F2 Mangalitsa x Pietrain pigs. *Anim Genet*. 2006;37(3):245–247.
193. Piorkowska K, Tyra M, Rogoz M, Ropka-Molik K, Oczkowicz M, Rozycki M. Association of the melanocortin-4 receptor (MC4R) with feed intake, growth, fatness and carcass composition in pigs raised in Poland. *Meat Sci*. 2010;85(2):297–301.
194. Van den Maagdenberg K, Stinckens A, Claeys E, et al. The *Asp298Asn* missense mutation in the porcine melanocortin-4 receptor (*MC4R*) gene can be used to affect growth and carcass traits without an effect on meat quality. *Animal*. 2007;1(8):1089–1098.
195. Fan B, Onteru SK, Du ZQ, Garrick DJ, Stalder KJ, Rothschild MF. Genome-wide association study identifies Loci for body composition and structural soundness traits in pigs. *PLoS One*. 2011;6(2), e14726.
196. Davoli R, Braglia S, Valastro V, et al. Analysis of *MC4R* polymorphism in Italian Large White and Italian Duroc pigs: association with carcass traits. *Meat Sci*. 2012;90(4):887–892.
197. Fontanesi L, Galimberti G, Calo DG, et al. Identification and association analysis of several hundred single nucleotide polymorphisms within candidate genes for back fat thickness in Italian Large White pigs using a selective genotyping approach. *J Anim Sci*. 2012;90(8):2450–2464.
198. Park HB, Carlborg O, Marklund S, Andersson L. Melanocortin-4 receptor (MC4R) genotypes have no major effect on fatness in a Large White x Wild Boar intercross. *Anim Genet*. 2002;33(2):155–157.
199. Stachowiak M, Szydlowski M, Obarzanek-Fojt M, Switonski M. An effect of a missense mutation in the porcine melanocortin-4 receptor (MC4R) gene on production traits in Polish pig breeds is doubtful. *Anim Genet*. 2006;37(1):55–57.
200. Lkhagvadorj S, Qu L, Cai W, et al. Microarray gene expression profiles of fasting induced changes in liver and adipose tissues of pigs expressing the melanocortin-4 receptor D298N variant. *Physiol Genomics*. 2009;38(1):98–111.
201. Munoz G, Alcazar E, Fernandez A, et al. Effects of porcine *MC4R* and *LEPR* polymorphisms, gender and Duroc sire line on economic traits in Duroc x Iberian crossbred pigs. *Meat Sci*. 2011;88(1):169–173.
202. Ai H, Ren J, Zhang Z, et al. Detection of quantitative trait loci for growth- and fatness-related traits in a large-scale White Duroc x Erhualian intercross pig population. *Anim Genet*. 2012;43(4):383–391.
203. Ballesteros JA, Weinstein H. Integrated methods for the construction of three-dimensional models and computational probing of structure-function relations in G protein-coupled receptors. *Methods Neurosci*. 1995;25:366–428.
204. Kim KS, Reecy JM, Hsu WH, Anderson LL, Rothschild MF. Functional and phylogenetic analyses of a melanocortin-4 receptor mutation in domestic pigs. *Domest Anim Endocrinol*. 2004;26(1):75–86.
205. Fan ZC, Sartin JL, Tao YX. Pharmacological analyses of two naturally occurring porcine melanocortin-4 receptor mutations in domestic pigs. *Domest Anim Endocrinol*. 2008;34(4):383–390.
206. Barb CR, Hausman GJ, Rekaya R, et al. Gene expression in hypothalamus, liver, and adipose tissues and food intake response to melanocortin-4 receptor agonist in pigs expressing melanocortin-4 receptor mutations. *Physiol Genomics*. 2010;41(3):254–268.

207. Zhang J, Li J, Wu C, et al. The Asp298Asn polymorphism of melanocortin-4 receptor (MC4R) in pigs: evidence for its potential effects on MC4R constitutive activity and cell surface expression. *Anim Genet.* 2020;51(5):694–706.
208. Roth CL, Ludwig M, Woelfle J, et al. A novel melanocortin-4 receptor gene mutation in a female patient with severe childhood obesity. *Endocrine.* 2009;36(1):52–59.
209. Huang H, Tao YX. Pleiotropic functions of the transmembrane domain 6 of human melanocortin-4 receptor. *J Mol Endocrinol.* 2012;49(3):237–248.
210. Mo XL, Yang R, Tao YX. Functions of transmembrane domain 3 of human melanocortin-4 receptor. *J Mol Endocrinol.* 2012;49(3):221–235.
211. Wang ZQ, Wang W, Shi L, Chai JT, Zhang XJ, Tao YX. Molecular cloning and pharmacological characterization of giant panda (*Ailuropoda melanoleuca*) melanocortin-4 receptor. *Gen Comp Endocrinol.* 2016;229:32–40.
212. Li JT, Yang Z, Chen HP, et al. Molecular cloning, tissue distribution, and pharmacological characterization of melanocortin-4 receptor in spotted scat, *Scatophagus argus. Gen Comp Endocrinol.* 2016;230-231:143–152.
213. Li L, Yang Z, Zhang YP, He S, Liang XF, Tao YX. Molecular cloning, tissue distribution, and pharmacological characterization of melanocortin-4 receptor in grass carp (*Ctenopharyngodon idella*). *Domest Anim Endocrinol.* 2017;59:140–151.
214. Yi TL, Yang LK, Ruan GL, Yang DQ, Tao YX. Melanocortin-4 receptor in swamp eel (*Monopterus albus*): cloning, tissue distribution, and pharmacology. *Gene.* 2018; 678:79–89.
215. Rao YZ, Chen R, Zhang Y, Tao YX. Orange-spotted grouper melanocortin-4 receptor: modulation of signaling by MRAP2. *Gen Comp Endocrinol.* 2019;284, 113234.
216. Zhang KQ, Hou ZS, Wen HS, et al. Melanocortin-4 receptor in spotted sea bass, *Lateolabrax maculatus*: cloning, tissue distribution, physiology, and pharmacology. *Front Endocrinol (Lausanne).* 2019;10:705.
217. Tao M, Ji RL, Huang L, et al. Regulation of melanocortin-4 receptor pharmacology by two isoforms of melanocortin receptor accessory protein 2 in topmouth culter (*Culter alburnus*). *Front Endocrinol (Lausanne).* 2020;11:538.
218. Wen ZY, Liu T, Qin CJ, et al. MRAP2 interaction with melanocortin-4 receptor in snakehead (*Channa argus*). *Biomolecules.* 2021;11(3):481.
219. Fan B, Onteru SK, Plastow GS, Rothschild MF. Detailed characterization of the porcine *MC4R* gene in relation to fatness and growth. *Anim Genet.* 2009;40(4): 401–409.
220. Fontanesi L, Buttazzoni L, Galimberti G, Calò DG, Scotti E, Russo V. Association between melanocortin 4 receptor (*MC4R*) gene haplotypes and carcass and production traits in Italian Large White pigs evaluated with a selective genotyping approach. *Livest Sci.* 2013;157:48–56.
221. Smith TD, Van Valkenburgh B. The dog-human connection. *Anat Rec (Hoboken).* 2021;304(1):10–18.
222. Lewis GF, Brubaker PL. The discovery of insulin revisited: lessons for the modern era. *J Clin Invest.* 2021;131(1).
223. Candille SI, Kaelin CB, Cattanach BM, et al. A β-defensin mutation causes black coat color in domestic dogs. *Science.* 2007;318:1418–1423.
224. Brancalion L, Haase B, Wade CM. Canine coat pigmentation genetics: a review. *Anim Genet.* 2022;53(1):3–34.
225. Quignon P, Kirkness E, Cadieu E, et al. Comparison of the canine and human olfactory receptor gene repertoires. *Genome Biol.* 2003;4(12):R80.
226. Ache BW, Young JM. Olfaction: diverse species, conserved principles. *Neuron.* 2005;48(3):417–430.
227. Aono S, Dennis JC, He S, Wang W, Tao YX, Morrison EE. Exploring pleiotropic functions of canine β-defensin 103: nasal cavity expression, antimicrobial activity, and melanocortin receptor activity. *Anat Rec (Hoboken).* 2021;304:210–221.

228. Kijas JW, Cideciyan AV, Aleman TS, et al. Naturally occurring *rhodopsin* mutation in the dog causes retinal dysfunction and degeneration mimicking human dominant retinitis pigmentosa. *Proc Natl Acad Sci U S A*. 2002;99(9):6328–6333.
229. Lin L, Faraco J, Li R, et al. The sleep disorder canine narcolepsy is caused by a mutation in the hypocretin (orexin) receptor 2 gene. *Cell*. 1999;98(3):365–376.
230. Kornegay JN, Bogan JR, Bogan DJ, et al. Canine models of Duchenne muscular dystrophy and their use in therapeutic strategies. *Mamm Genome*. 2012;23(1-2):85–108.
231. Sutter NB, Bustamante CD, Chase K, et al. A single *IGF1* allele is a major determinant of small size in dogs. *Science*. 2007;316(5821):112–115.
232. Topal J, Roman V, Turcsan B. The dog (*Canis familiaris*) as a translational model of autism: it is high time we move from promise to reality. *Wiley Interdiscip Rev Cogn Sci*. 2019;10(4), e1495.
233. Ruetten H, Vezina CM. Relevance of dog as an animal model for urologic diseases. *Prog Mol Biol Transl Sci*. 2022;189. https://doi.org/10.1016/bs.pmbts.2021.1012.1002.
234. Ostrander EA, Dreger DL, Evans JM. Canine cancer genomics: lessons for canine and human health. *Annu Rev Anim Biosci*. 2019;7:449–472.
235. Nance R, Sajib AB, ASmith BF. Canine models of human cancer: bridging the gap to improve precision medicine. *Prog Mol Biol Transl Sci*. 2022;189. https://doi.org/10.1016/bs.pmbts.2021.1012.1003.
236. Parkinson C, Grasso P. The use of the dog in toxicity tests on pharmaceutical compounds. *Hum Exp Toxicol*. 1993;12(2):99–109.
237. Lindblad-Toh K, Wade CM, Mikkelsen TS, et al. Genome sequence, comparative analysis and haplotype structure of the domestic dog. *Nature*. 2005;438(7069):803–819.
238. Sutter NB, Ostrander EA. Dog star rising: the canine genetic system. *Nat Rev Genet*. 2004;5(12):900–910.
239. Schoenebeck JJ, Ostrander EA. Insights into morphology and disease from the dog genome project. *Annu Rev Cell Dev Biol*. 2014;30:535–560.
240. German AJ. The growing problem of obesity in dogs and cats. *J Nutr*. 2006;136 (suppl 7):1940S–1946S.
241. Skorczyk A, Stachowiak M, Szczerbal I, et al. Polymorphism and chromosomal location of the MC4R (melanocortin-4 receptor) gene in the dog and red fox. *Gene*. 2007;392(1-2):247–252.
242. van den Berg L, van den Berg SM, Martens EECP, et al. Analysis of variation in the melanocortin-4 receptor gene (mc4r) in golden retriever dogs. *Anim Genet*. 2010; 41:557.
243. Yan J, Tao YX. Pharmacological characterization of canine melancortin-4 receptor and its natural variant V213F. *Domest Anim Endocrinol*. 2011;41(2):91–97.
244. Zeng RX, Zhang YB, Du P. SNPs of melanocortin 4 receptor (*MC4R*) associated with body weight in Beagle dogs. *Exp Anim*. 2014;63(1):73–78.
245. Mankowska M, Nowacka-Woszuk J, Graczyk A, Ciazynska P, Stachowiak M, Switonski M. Polymorphism and methylation of the *MC4R* gene in obese and non-obese dogs. *Mol Biol Rep*. 2017;44(4):333–339.
246. Nowacka-Woszuk J, Skorczyk A, Flisikowski K, Szydlowski M, Switonski M. Polymorphic variants within a putative upstream open reading frame of the *MC4R* gene do not affect body weight of farmed red foxes. *Anim Genet*. 2012;43(4):480–481.
247. Skorczyk A, Flisikowski K, Switonski M. A comparative analysis of *MC4R* gene sequence, polymorphism, and chromosomal localization in Chinese raccoon dog and Arctic fox. *DNA Cell Biol*. 2012;31(5):732–738.
248. Yu DY, Wu RZ, Zhao Y, et al. Polymorphisms of four candidate genes and their correlations with growth traits in blue fox (*Alopex lagopus*). *Gene*. 2019;717, 143987.
249. Chen AS, Marsh DJ, Trumbauer ME, et al. Inactivation of the mouse melanocortin-3 receptor results in increased fat mass and reduced lean body mass. *Nat Genet*. 2000;26 (1):97–102.

250. Raffan E, Dennis RJ, O'Donovan CJ, et al. A deletion in the canine *POMC* gene is associated with weight and appetite in obesity-prone Labrador retriever dogs. *Cell Metab*. 2016;23(5):893–900.
251. Lee YS, Challis BG, Thompson DA, et al. A POMC variant implicates β-melanocyte-stimulating hormone in the control of human energy balance. *Cell Metab*. 2006;3 (2):135–140.
252. Biebermann H, Castaneda TR, van Landeghem F, et al. A role for β-melanocyte-stimulating hormone in human body-weight regulation. *Cell Metab*. 2006;3 (2):141–146.
253. Davison LJ, Holder A, Catchpole B, O'Callaghan CA. The canine *POMC* gene, obesity in Labrador retrievers and susceptibility to diabetes mellitus. *J Vet Intern Med*. 2017;31(2):343–348.
254. Allen MS, Bradford BJ, Harvatine KJ. The cow as a model to study food intake regulation. *Annu Rev Nutr*. 2005;25:523–547.
255. Bovine Genome Sequencing and Analysis Consortium, Elsik CG, Tellam RL, et al. The genome sequence of taurine cattle: a window to ruminant biology and evolution. *Science*. 2009;324(5926):522–528.
256. Haegeman A, Coopman F, Jacobs K, Mattheeuws M, Van Zeveren A, Peelman L. Bovine melanocortin receptor 4: cDNA sequence, polymorphisms and mapping. *Anim Genet*. 2001;32(4):189–192.
257. Magenis RE, Smith L, Nadeau JH, Johnson KR, Mountjoy KG, Cone RD. Mapping of the ACTH, MSH, and neural (MC3 and MC4) melanocortin receptors in the mouse and human. *Mamm Genome*. 1994;5(8):503–508.
258. Valle E, Habermann FA, Moore SS, Crews DH, Benkel BF. Genomic localization and SNP discovery in the bovine melanocortin receptor 4 gene (*MC4R*). *Anim Genet*. 2004;35(4):351–352.
259. Huang M, Gao X, Li JY, Ren HY, Chen JB, Xu SZ. Polymorphisms in MC4R gene and correlations with economic traits in cattle. *Mol Biol Rep*. 2010;37(8):3941–3944.
260. Seong J, Suh DS, Park KD, Lee HK, Kong HS. Identification and analysis of *MC4R* polymorphisms and their association with economic traits of Korean cattle (Hanwoo). *Mol Biol Rep*. 2012;39(4):3597–3601.
261. Liu GY, Raza SHA, Zhou L, et al. The genetic polymorphisms of melanocortin-4 receptor gene are associated with carcass quality traits in a Chinese indigenous beef cattle breed. *Res Vet Sci*. 2020;132:202–206.
262. Deng TX, Pang CY, Liu MQ, Zhang C, Liang XW. Synonymous single nucleotide polymorphisms in the *MC4R* gene that are significantly associated with milk production traits in water buffaloes. *Genet Mol Res*. 2016;15(2).
263. Qiu Q, Zhang G, Ma T, et al. The yak genome and adaptation to life at high altitude. *Nat Genet*. 2012;44(8):946–949.
264. Cai X, Mipam TD, Zhao FF, Sun L. SNPs detected in the yak *MC4R* gene and their association with growth traits. *Animal*. 2015;9(7):1097–1103.
265. Adam CL, Findlay PA, Aitken RP, Milne JS, Wallace JM. *In vivo* changes in central and peripheral insulin sensitivity in a large animal model of obesity. *Endocrinology*. 2012;153 (7):3147–3157.
266. Jiang Y, Xie M, Chen W, et al. The sheep genome illuminates biology of the rumen and lipid metabolism. *Science*. 2014;344(6188):1168–1173.
267. Adam CL, Archer ZA, Findlay PA, Thomas L, Marie M. Hypothalamic gene expression in sheep for cocaine- and amphetamine-regulated transcript, pro-opiomelanocortin, neuropeptide Y, agouti-related peptide and leptin receptor and responses to negative energy balance. *Neuroendocrinology*. 2002;75(4):250–256.
268. Wagner CG, McMahon CD, Marks DL, Daniel JA, Steele B, Sartin JL. A role for agouti-related protein in appetite regulation in a species with continuous nutrient delivery. *Neuroendocrinology*. 2004;80(4):210–218.

269. Sartin JL, Marks DL, McMahon CD, et al. Central role of the melanocortin-4 receptors in appetite regulation after endotoxin. *J Anim Sci.* 2008;86(10):2557–2567.
270. Sartin JL, Wagner CG, Marks DL, et al. Melanocortin-4 receptor in sheep: a potential site for therapeutic intervention in disease models. *Domest Anim Endocrinol.* 2005;29 (2):446–455.
271. Zuo B, Liu G, Peng Y, et al. Melanocortin-4 receptor (*MC4R*) polymorphisms are associated with growth and meat quality traits in sheep. *Mol Biol Rep.* 2014;41 (10):6967–6974.
272. Kobayashi H, Ogawa Y, Shintani M, et al. A novel homozygous missense mutation of melanocortin-4 receptor (MC4R) in a Japanese woman with severe obesity. *Diabetes.* 2002;51(1):243–246.
273. Shishay G, Liu G, Jiang X, et al. Variation in the promoter region of the *MC4R* gene elucidates the association of body measurement traits in Hu sheep. *Int J Mol Sci.* 2019;20(2).
274. Song XM, Jiang JF, Zhang GZ, Shi FX, Jiang YQ. DNA polymorphisms of the Hu sheep melanocortin-4 receptor gene associated with birth weight and 45-day weaning weight. *Genet Mol Res.* 2012;11(4):4432–4441.
275. Wang YL, Wang CL, Zhang J, et al. Three novel *MC4R* SNPs associated with growth traits in Hu sheep and East Friesian × Hu crossbred sheep. *Small Rumin Res.* 2015;125:26–33.
276. Forcada Y, Holder A, Church DB, Catchpole B. A polymorphism in the melanocortin 4 receptor gene (*MC4R:c.92C> T*) is associated with diabetes mellitus in overweight domestic shorthaired cats. *J Vet Intern Med.* 2014;28(2):458–464.
277. Fontanesi L, Scotti E, Cisarova K, et al. A missense mutation in the rabbit melanocortin 4 receptor (MC4R) gene is associated with finishing weight in a meat rabbit line. *Anim Biotechnol.* 2013;24(4):268–277.
278. Rous P. The challenge to man of the neoplastic cell. *Science.* 1967;157(3784):24–28.
279. Levi-Montalcini R. The nerve growth factor 35 years later. *Science.* 1987;237 (4819):1154–1162.
280. International Chicken Genome Sequencing C. Sequence and comparative analysis of the chicken genome provide unique perspectives on vertebrate evolution. *Nature.* 2004;432(7018):695–716.
281. Takeuchi S, Teshigawara K, Takahashi S. Widespread expression of Agouti-related protein (AGRP) in the chicken: a possible involvement of AGRP in regulating peripheral melanocortin systems in the chicken. *Biochim Biophys Acta.* 2000;1496(2-3):261–269.
282. Richards MP. Genetic regulation of feed intake and energy balance in poultry. *Poult Sci.* 2003;82(6):907–916.
283. Takeuchi S, Takahashi S, Okimoto R, Schioth HB, Boswell T. Avian melanocortin system: α-MSH may act as an autocrine/paracrine hormone: a minireview. *Ann N Y Acad Sci.* 2003;994:366–372.
284. Boswell T, Takeuchi S. Recent developments in our understanding of the avian melanocortin system: its involvement in the regulation of pigmentation and energy homeostasis. *Peptides.* 2005;26(10):1733–1743.
285. Boswell T, Dunn IC. Regulation of the avian central melanocortin system and the role of leptin. *Gen Comp Endocrinol.* 2015;vol: 221:278–283. in press.
286. Lear T, Liu L, O'Donnell M, et al. Alpha-melanocyte stimulating hormone-induced anorexia in Japanese quail (*Coturnix japonica*) likely involves the ventromedial hypothalamus and paraventricular nucleus of the hypothalamus. *Gen Comp Endocrinol.* 2017;252:97–102.
287. Zhang J, Li X, Zhou Y, et al. The interaction of MC3R and MC4R with MRAP2, ACTH, α-MSH and AgRP in chickens. *J Endocrinol.* 2017;234:155–174.
288. Qiu X, Li N, Deng X, Zhao X, Meng Q, Wang X. The single nucleotide polymorphisms of chicken melanocortin-4 receptor (MC4R) gene and their association analysis with carcass traits. *Sci China C Life Sci.* 2006;49(6):560–566.

289. Li CY, Li H. Association of MC4R gene polymorphisms with growth and body composition traits in chicken. *Asian-Aust J Anim Sci.* 2006;19:763–768.
290. Sharma P, Bottje W, Okimoto R. Polymorphisms in uncoupling protein, melanocortin 3 receptor, melanocortin 4 receptor, and pro-opiomelanocortin genes and association with production traits in a commercial broiler line. *Poult Sci.* 2008;87(10):2073–2086.
291. Kubota S, Vandee A, Keawnakient P, Molee W, Yongsawatdikul J, Molee A. Effects of the MC4R, CAPN1, and ADSL genes on body weight and purine content in slow-growing chickens. *Poult Sci.* 2019;98(10):4327–4337.
292. Wang ZQ, Huang JS, Zhou JH, Shi L, Jiang XF, Tao YX. Pharmacologic analyses of four chicken melanocortin-4 receptor mutations. *Domest Anim Endocrinol.* 2016;54:68–75.
293. Bennett HP. Biosynthetic fate of the amino-terminal fragment of pro-opiomelanocortin within the intermediate lobe of the mouse pituitary. *Peptides.* 1986;7(4):615–622.
294. Smith ML, Prall B, Nandar W, Cline MA. β-Melanocyte-stimulating hormone potently reduces appetite via the hypothalamus in chicks. *J Neuroendocrinol.* 2008;20 (2):220–226.
295. Huang Y, Wang Y, He D, Liu Y. Genetic diversity of the melanocortin 4 receptor (MC4R) gene and its association with slaughter traits in the Landes goose. *Biochem Genet.* 2010;48(11-12):944–953.
296. Zeng T, Chen L, Du X, et al. Association analysis between feed efficiency studies and expression of hypothalamic neuropeptide genes in laying ducks. *Anim Genet.* 2016;47 (5):606–609.
297. Song Y, Golling G, Thacker TL, Cone RD. Agouti-related protein (AGRP) is conserved and regulated by metabolic state in the zebrafish, *Danio rerio*. *Endocrine.* 2003;22 (3):257–265.
298. Forlano PM, Cone RD. Conserved neurochemical pathways involved in hypothalamic control of energy homeostasis. *J Comp Neurol.* 2007;505(3):235–248.
299. Navarro S, Soletto L, Puchol S, Rotllant J, Soengas JL, Cerda-Reverter JM. 60 YEARS OF POMC: POMC: an evolutionary perspective. *J Mol Endocrinol.* 2016;56(4):T113–T118.
300. Ji RL, Huang L, Wang Y, et al. Topmouth culter melanocortin-3 receptor: regulation by two isoforms of melanocortin-2 receptor accessory protein 2. *Endocr Connect.* 2021;10:1489–1501.
301. Cerdá-Reverter JM, Ringholm A, Schioth HB, Peter RE. Molecular cloning, pharmacological characterization, and brain mapping of the melanocortin 4 receptor in the goldfish: involvement in the control of food intake. *Endocrinology.* 2003;144 (6):2336–2349.
302. Ringholm A, Fredriksson R, Poliakova N, et al. One melanocortin 4 and two melanocortin 5 receptors from zebrafish show remarkable conservation in structure and pharmacology. *J Neurochem.* 2002;82(1):6–18.
303. Haitina T, Klovins J, Andersson J, et al. Cloning, tissue distribution, pharmacology and three-dimensional modelling of melanocortin receptors 4 and 5 in rainbow trout suggest close evolutionary relationship of these subtypes. *Biochem J.* 2004;380 (Pt 2):475–486.
304. Klovins J, Haitina T, Fridmanis D, et al. The melanocortin system in Fugu: determination of POMC/AGRP/MCR gene repertoire and synteny, as well as pharmacology and anatomical distribution of the MCRs. *Mol Biol Evol.* 2004;21(3):563–579.
305. Song Y, Cone RD. Creation of a genetic model of obesity in a teleost. *FASEB J.* 2007;21(9):2042–2049.
306. Sebag JA, Zhang C, Hinkle PM, Bradshaw AM, Cone RD. Developmental control of the melanocortin-4 receptor by MRAP2 proteins in zebrafish. *Science.* 2013;341 (6143):278–281.

307. Wei R, Yuan D, Zhou C, et al. Cloning, distribution and effects of fasting status of melanocortin 4 receptor (MC4R) in *Schizothorax prenanti*. *Gene*. 2013;532(1): 100–107.
308. Wolverton EA, Wong MK, Davis PE, Hoglin B, Braasch I, Dores RM. Analyzing the signaling properties of gar (*Lepisosteus oculatus*) melanocortin receptors: evaluating interactions with MRAP1 and MRAP2. *Gen Comp Endocrinol*. 2019;282, 113215.
309. Wang M, Chen YJ, Zhu M, et al. Pharmacological modulation of melanocortin-4 receptor by melanocortin receptor accessory protein 2 in Nile tilapia. *Gen Comp Endocrinol*. 2019;282, 113219.
310. Ringholm A, Klovins J, Fredriksson R, et al. Presence of melanocortin (MC4) receptor in spiny dogfish suggests an ancient vertebrate origin of central melanocortin system. *Eur J Biochem*. 2003;270(2):213–221.
311. Takahashi A, Davis P, Reinick C, Mizusawa K, Sakamoto T, Dores RM. Characterization of melanocortin receptors from stingray *Dasyatis akajei*, a cartilaginous fish. *Gen Comp Endocrinol*. 2016;232:115–124.
312. Barney E, Dores MR, McAvoy D, et al. Elephant shark melanocortin receptors: novel interactions with MRAP1 and implication for the HPI axis. *Gen Comp Endocrinol*. 2019;272:42–51.
313. Cerdá-Reverter JM, Schioth HB, Peter RE. The central melanocortin system regulates food intake in goldfish. *Regul Pept*. 2003;115(2):101–113.
314. Cerdá-Reverter JM, Peter RE. Endogenous melanocortin antagonist in fish: structure, brain mapping, and regulation by fasting of the goldfish agouti-related protein gene. *Endocrinology*. 2003;144(10):4552–4561.
315. Schjolden J, Schioth HB, Larhammar D, Winberg S, Larson ET. Melanocortin peptides affect the motivation to feed in rainbow trout (*Oncorhynchus mykiss*). *Gen Comp Endocrinol*. 2009;160(2):134–138.
316. Kobayashi Y, Tsuchiya K, Yamanome T, Schioth HB, Kawauchi H, Takahashi A. Food deprivation increases the expression of melanocortin-4 receptor in the liver of barfin flounder, *Verasper moseri*. *Gen Comp Endocrinol*. 2008;155(2):280–287.
317. Wan YM, Zhang Y, Ji PF, Li Y, Xu P, Sun XW. Molecular characterization of CART, AgRP, and MC4R genes and their expression with fasting and re-feeding in common carp (*Cyprinus carpio*). *Mol Biol Rep*. 2012;39(3):2215–2223.
318. Kalananthan T, Lai F, Gomes AS, Murashita K, Handeland S, Ronnestad I. The melanocortin system in Atlantic salmon (*Salmo salar* L.) and its role in appetite control. *Front Neuroanat*. 2020;14:48.
319. Zhong CR, Song YL, Wang YP, et al. Increased food intake in growth hormone-transgenic common carp (*Cyprinus carpio* L.) may be mediated by upregulating Agouti-related protein (AgRP). *Gen Comp Endocrinol*. 2013;192:81–88.
320. Yada T, Moriyama S, Suzuki Y, et al. Relationships between obesity and metabolic hormones in the "cobalt" variant of rainbow trout. *Gen Comp Endocrinol*. 2002;128 (1):36–43.
321. Hsieh YW, Tsai YW, Lai HH, Lai CY, Lin CY, Her GM. Depletion of alpha-melanocyte-stimulating hormone induces insatiable appetite and gains in energy reserves and body weight in zebrafish. *Biomedicine*. 2021;9(8):941.
322. Shimakura S, Miura T, Maruyama K, et al. α-Melanocyte-stimulating hormone mediates melanin-concentrating hormone-induced anorexigenic action in goldfish. *Horm Behav*. 2008;53(2):323–328.
323. Matsuda K, Kojima K, Wada K, et al. The anorexigenic action of the octadecaneuropeptide (ODN) in goldfish is mediated through the MC4R- and subsequently the CRH receptor-signaling pathways. *J Mol Neurosci*. 2010;42(1):74–79.
324. Volkoff H. The neuroendocrine regulation of food intake in fish: a review of current knowledge. *Front Neurosci*. 2016;10:540.

325. Yang Z, Liang XF, Li GL, Tao YX. Biased signaling in fish melanocortin-4 receptors (MC4Rs): divergent pharmacology of four ligands on spotted scat (*Scatophagus argus*) and grass carp (*Ctenopharyngodon idella*) MC4Rs. *Mol Cell Endocrinol.* 2020;515, 110929.
326. Cerdá-Reverter JM, Agulleiro MJ, R RG, Sanchez E, Ceinos R, Rotllant J. Fish melanocortin system. *Eur J Pharmacol.* 2011;660(1):53–60.
327. Zhang Y, Wen HS, Li Y, et al. Melanocortin-4 receptor regulation of reproductive function in black rockfish (*Sebastes schlegelii*). *Gene.* 2020;741, 144541.
328. Lampert KP, Schmidt C, Fischer P, et al. Determination of onset of sexual maturation and mating behavior by melanocortin receptor 4 polymorphisms. *Curr Biol.* 2010;20 (19):1729–1734.
329. Volff JN, Selz Y, Hoffmann C, et al. Gene amplification and functional diversification of melanocortin 4 receptor at an extremely polymorphic locus controlling sexual maturation in the platyfish. *Genetics.* 2013;195(4):1337–1352.
330. Smith CC, Harris RM, Lampert KP, Schartl M, Hofmann HA, Ryan MJ. Copy number variation in the melanocortin 4 receptor gene and alternative reproductive tactics the swordtail *Xiphophorus multilineatus*. *Environ Biol Fish.* 2015;98(1):23–33.
331. Jiang DN, Li JT, Tao YX, et al. Effects of melanocortin-4 receptor agonists and antagonists on expression of genes related to reproduction in spotted scat, *Scatophagus argus*. *J Comp Physiol B.* 2017;187:603–612.
332. Liu RQ, Du K, Ormanns J, Adolfi MC, Schartl M. Melanocortin 4 receptor signaling and puberty onset regulation in *Xiphophorus* swordtails. *Gen Comp Endocrinol.* 2020;295, 113521.
333. Irani BG, Xiang Z, Moore MC, Mandel RJ, Haskell-Luevano C. Voluntary exercise delays monogenetic obesity and overcomes reproductive dysfunction of the melanocortin-4 receptor knockout mouse. *Biochem Biophys Res Commun.* 2005;326 (3):638–644.
334. Sandrock M, Schulz A, Merkwitz C, Schoneberg T, Spanel-Borowski K, Ricken A. Reduction in corpora lutea number in obese melanocortin-4-receptor-deficient mice. *Reprod Biol Endocrinol.* 2009;7:24.
335. Zhang C, Forlano PM, Cone RD. AgRP and POMC neurons are hypophysiotropic and coordinately regulate multiple endocrine axes in a larval teleost. *Cell Metab.* 2012;15(2):256–264.
336. Aspiras AC, Rohner N, Martineau B, Borowsky RL, Tabin CJ. Melanocortin 4 receptor mutations contribute to the adaptation of cavefish to nutrient-poor conditions. *Proc Natl Acad Sci U S A.* 2015;112(31):9668–9673.
337. Cummings DE, Clement K, Purnell JQ, et al. Elevated plasma ghrelin levels in Prader Willi syndrome. *Nat Med.* 2002;8(7):643–644.
338. Patton EE, Mitchell DL, Nairn RS. Genetic and environmental melanoma models in fish. *Pigment Cell Melanoma Res.* 2010;23(3):314–337.
339. Kallman KD, Schreibman MP, Borkoski V. Genetic control of gonadotrop differentiation in the platyfish, *Xiphophorus maculatus* (Poeciliidae). *Science.* 1973;181(4100):678–680.
340. Kallman KD, Schreibman MP. A sex-linked gene controlling gonadotrop differentiation and its significance in determining the age of sexual maturation and size of the platyfish, *Xiphophorus maculatus*. *Gen Comp Endocrinol.* 1973;21(2):287–304.
341. Schreibman MP, Kallman KD. The genetic control of the pituitary-gonadal axis in the platyfish, *Xiphophorus maculatus*. *J Exp Zool.* 1977;200(2):277–293.
342. Siegrist W, Drozdz R, Cotti R, Willard DH, Wilkison WO, Eberle AN. Interactions of α-melanotropin and agouti on B16 melanoma cells: evidence for inverse agonism of agouti. *J Recept Signal Transduct Res.* 1997;17(1-3):75–98.
343. Chai BX, Neubig RR, Millhauser GL, et al. Inverse agonist activity of agouti and agouti-related protein. *Peptides.* 2003;24(4):603–609.

344. Sanchez-Mas J, Hahmann C, Gerritsen I, Garcia-Borron JC, Jimenez-Cervantes C. Agonist-independent, high constitutive activity of the human melanocortin 1 receptor. *Pigment Cell Res.* 2004;17(4):386–395.
345. Navarro S, Guillot R, Crespo D, et al. Enhanced growth without accelerated puberty in fish: a role for the melanocortin system. *Aquaculture.* 2021;540, 736721.
346. Liu R, Kinoshita M, Adolfi MC, Schartl M. Analysis of the role of the Mc4r system in development, growth, and puberty of medaka. *Front Endocrinol (Lausanne).* 2019; 10:213.
347. Sanchez E, Rubio VC, Thompson D, et al. Phosphodiesterase inhibitor-dependent inverse agonism of agouti-related protein on melanocortin 4 receptor in sea bass (*Dicentrarchus labrax*). *Am J Phys Regul Integr Comp Phys.* 2009;296(5):R1293–R1306.

CHAPTER EIGHT

Patient-derived organoids as a model for tumor research

Jia Wang[a], Xiaoying Feng[a], Zhichao Li[b,c,d], Yongsong Chen[a], and Weiren Huang[a,b,c,d,e,*]

[a]The First Affiliated Hospital of Shantou University, Shantou University Medical College, Shantou, China
[b]Department of Urology, Shenzhen Institute of Translational Medicine, Shenzhen Second People's Hospital, The First Affiliated Hospital of Shenzhen University, Shenzhen, China
[c]Guangdong Key Laboratory of Systems Biology and Synthetic Biology for Urogenital Tumors, Shenzhen, China
[d]International Cancer Center of Shenzhen University, Shenzhen, China
[e]Shenzhen Institute of Synthetic Biology, Shenzhen Institutes of Advanced Technology, Chinese Academy of Sciences, Shenzhen, China
*Corresponding author: e-mail address: pony8980@163.com

Contents

Progress in Molecular Biology and Translational Science, Volume 189
ISSN 1877-1173
https://doi.org/10.1016/bs.pmbts.2022.03.004

Abstract

Cancer represents a leading cause of death, despite the rapid progress of cancer research, leading to urgent need for accurate preclinical model to further study of tumor mechanism and accelerate translational applications. Cancer cell lines cannot fully recapitulate tumors of different patients due to the lack of tumor complexity and specification, while the high technical difficulty, long time, and substantial cost of patient-derived xenograft model makes it unable to be used extensively for all types of tumors and large-scale drug screening. Patient-derived organoids can be established rapidly with a high success rate from many tumors, and precisely replicate the key histopathological, genetic, and phenotypic features, as well as therapeutic response of patient tumor. Therefore, they are extensively used in cancer basic research, biobanking, disease modeling and precision medicine. The combinations of cancer organoids with other advanced technologies, such as 3D bio-printing, organ-on-a-chip, and CRISPR-Cas9, contributes to the more complete replication of complex tumor microenvironment and tumorigenesis. In this review, we discuss the various methods of the establishment and the application of patient-derived organoids in diverse tumors as well as the limitations and future prospects of these models. Further advances of tumor organoids are expected to bridge the huge gap between bench and bedside and provide the unprecedented opportunities to advance cancer research.

Abbreviations

2D	two-dimensional
3D	three-dimensional
5-FU	5-fluorouracil
AdSC	adult stem cell
ALI	air–liquid interface
AR	androgen receptor
BC	breast cancer
CAF	cancer-associated fibroblast
CAR-T	chimeric antigen receptor T
CI	confidence interval
COVID-19	coronavirus disease 2019
CRC	colorectal cancer
CRISPR-Cas9	clustered regularly interspaced short palindromic repeats-Cas9
CRPC-NE	castration-resistant neuroendocrine prostate cancer
CTC	circulating tumor cell
ECM	extracellular matrix
EGF	epidermal growth factor
EGFR	epidermal growth factor receptor
ER	estrogen receptor
FGF	fibroblast growth factor
FGFR	fibroblast growth factor receptor
GBM	glioblastoma
GBO	glioblastoma organoid
HCC	hepatocellular carcinoma
HER	human epidermal growth factor receptor
IC_{50}	half-maximal inhibitory concentrations
ICC	intrahepatic cholangiocarcinoma
IFN	interferon

IL	interleukin
InSMAR-chip	integrated superhydrophobic microwell array chip
IPMN	intraductal papillary mucinous neoplasm
iPSCs	induced pluripotent stem cells
LADC	lung adenocarcinoma
LCOs	lung cancer organoids
MDSC	myeloid-derived suppressor cells apoptosis
MIBC	muscle-invasive bladder cancer
MSI	microsatellite Instability
MSS	microsatellite Stable
NMBIC	non-muscle invasive bladder cancer
NSCLC	non-small-cell lung cancer
opT	organoid-primed cytotoxic T
PanINs	pancreatic intraepithelial neoplasms
PC	pancreatic cancer
PD-1	programmed cell death protein 1
PDAC	pancreatic ductal adenocarcinoma
PD-L1	programmed death-ligand 1
PDO	patient derived organoid
PDOX	patient-derived orthotopic xenograft
PDX	patient-derived xenograft
PFS	progression-free survival
PLC	primary liver cancer
PSC	pluripotent stem cell
PTC	papillary thyroid cancer
SCLC	small-cell lung cancer
SRCC	signet-ring cell carcinoma
TCGA	The Cancer Genome Atlas
TGF	transforming growth factor
TKI	tyrosine kinase inhibitor
TME	tumor microenvironment
TNF	tumor necrosis factor

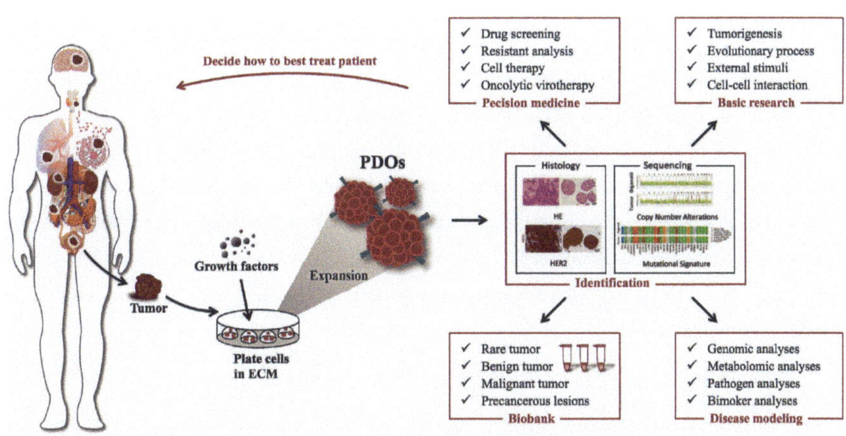

1. Introduction

Cancer represents a huge global burden, ranking as the first or second leading cause of death before the age of 70 in 112 of 183 countries and third or fourth in a further 23 countries.[1] Despite the rapid advancement of cancer research and economy, there is an urgent need for accurate preclinical model to further study tumor mechanism and accelerate their translational applications. The well-known eight hallmarks of cancer acquired during the multistep development of human tumors were proposed 10 years ago, including the acquired capabilities for sustaining proliferative signaling, evading growth suppressors, resisting cell death, enabling replicative immortality, inducing/accessing vasculature, activating invasion and metastasis, reprogramming cellular metabolism, and avoiding immune destruction.[2] As knowledge of cancer mechanisms has progressed, four additional characteristics, including unlocking phenotypic plasticity, non-mutational epigenetic reprogramming, polymorphic microbiomes, and senescent cells, were added recently as new hallmarks of cancer.[3] Cancer is daunting in the breadth and scope of its diversity, spanning genetics, cell and tissue biology, pathology, and response to therapy.[3]

Tumor heterogeneity is highly associated with resistance to cancer therapies, and is often considered as the main hinderance to develop effective treatment, causing the dire need of patient-specific treatment.[4] Tumor heterogeneity can be subdivided into two types: inter-patient heterogeneity and intra-tumor heterogeneity.[5] Specifically, the various tumor genotypes between patients are called interpatient heterogeneity, which can predict the response of patients to targeted treatments. The multiple unique subpopulations of cancer cells across different geographical regions in the same patient or evolving over time are called intra-tumor heterogeneity, which poses great challenges for effective treatment and the prevention of recurrence.[5] Pre-existing intra-tumor heterogeneity increases the probability of survival of at least some tumor cells during treatment, and the continuous diversification of tumor cell phenotypes during treatment enables tumor cells to adapt to the selective pressure imposed by treatment, resulting in *de novo* resistance and eventual recurrence.[6] Thus, personalized treatment, taking into account tumor heterogeneity, may improve the long-term treatment response, prolong the remission period, and increase the probability of tumor elimination.

Tumor heterogeneity is a complex phenomenon, which integrates genetic, epigenetic, and microenvironmental factors.[6] The tumor microenvironment (TME), primarily comprising immune cells, fibroblasts, pericytes, endothelial cells, mesenchymal stroma/stem cells, and extracellular matrix (ECM) components, has been recognized as an important component of malignancy, leading to not only disease initiation but also tumor progression, metastasis, therapy resistance, and disease recurrence through complex bi-directional signaling mechanism.[7] The complexity of TME and its evolution throughout the course of cancer progression has brought significant challenge to the study of tumor mechanisms and the determination of the most effective strategy for curing patients. One reason for the rather slow progress in understanding the link between genetic heterogeneity, TME, and tumor evolution, is the lack of suitable experimental models to replicate the heterogeneity of human tumors.

Historically, the exploration of disease mechanisms in animal models has progressed along a common discovery pipeline, whereby biological processes were initially investigated by genetic screens in invertebrates, such as yeast, fruit flies and *Caenorhabditis elegans*, followed by an analysis of evolutionary conservation in vertebrate systems, such as *Danio rerio* and mice, eventually leading to clinical translation to humans.[8] For many decades, classical cell lines and mouse models have made great achievements as the most frequently used preclinical models in tumor research, such as improving the understanding of molecular signaling pathways, identifying critical drug targets, and guiding the design and screening of antitumor drugs.

However, extrapolating the results to tumor patients has been the biggest bottleneck for these traditional models, demonstrated by the surprisingly low success rate of expensive and time-consuming human clinical trials (3.4%),[9] implying that results from cancer cell lines and mouse models cannot be used to accurately predict the response of cancer patients to anti-tumor drugs. Therefore, the continued efforts and attempts of many researchers recently have focused on the patient-derived organoids (PDOs), which are expected to bridge the huge bench-to-bedside gap and provide the unprecedented opportunities for further exploration of the mechanism of tumorigenesis and development as well as precision treatment. In this review, we will discuss the various methods of PDO establishment, the application of PDOs in diverse tumors, and the limitations and future prospects of PDOs.

2. The comparison between the organoid model and previous models

When establishing an *in vitro* cancer model, cancer cell lines are the first choice for many researchers because they are easy to obtain and maintain in culture medium and can reproduce rapidly and almost indefinitely. Since the first stable strain of HeLa cells, human cervical carcinoma cells derived from the cancer patient Henrietta Lacks, was established in 1953,[10] thousands of cancer cell lines derived from a variety of cancers have been produced and characterized.[11] Although they are useful model systems for cancer biology and translational medicine,[12] it has been demonstrated that established cancer cell lines are in fact highly genetically heterogeneous across strains, which results from clone dynamics (changes in the abundance of pre-existing clones) and ongoing instability (the appearance of new genetic variants), leading to differential drug sensitivity and unrepeatable experimental results.[13] In addition, the lack of tumor complexity and specificity is still a defect that cannot be ignored to truly recapitulate tumor heterogeneity and drug response of different patients.[14]

Patient-derived xenografts (PDXs) that combine the specificity of patient's tumor with the whole organism environment of animal models have proved to be more powerful stools to recapitulate the genetics, structure, function, and drug response of human tumors.[15] However, the high technical difficulty, long time and expensive cost of PDXs makes them unable to be used extensively for all type of tumors and large-scale drug screening. The immune deficiency of host mice is the key to the successful establishment of PDXs, but it also limits its application in the evaluation of immunotherapy. Importantly, PDX models undergo mouse-specific tumor evolution, leading to genetic and phenotypic divergence from the parental tumor.[16]

PDOs are three-dimensional (3D) self-organized assemblies of neoplastic cells derived from patient tissue samples with the ability of mimicking key histopathological, genetic, and phenotypic features of patient tumor.[17] In brief, it's a "mini-tumor" generated by culturing and expanding tumor tissue *in vitro*. PDOs have incomparable specificity, because they are derived directly from the patient's tumor, with the additional advantage of rapid expansion, long-term maintenance, and genetic operability of cell lines with the spatial structure and complexity of animal models. Therefore, they promise to provide a powerful and effective platform for the study of cancer development, specific markers, and drug screening.

3. Emergence and culture of cancer organoids

The first landmark organoid is the crypt villus organoids derived from the $Lgr5^+$ stem cells located at the bottom of small-intestinal crypts in mouse, described by Sato et al.[18] in 2009. Single isolated mouse crypts were suspended in Matrigel with the need of epidermal growth factor (EGF) and R-spondin 1 (Wnt agonist) for growth, Noggin for passaging, undergoing continuous budding events. These organoids were created within culturing for 14 days, comprising more than 40 crypt domains surrounding a central lumen lined by a villus-like epithelium, highly similar to freshly isolated small-intestinal crypts. These organoids preserved four types of mature cells of small-intestinal crypts, including Paneth cells and $Lgr5^+$ stem cells at crypt bottom, and goblet cells and enteroendocrine cells throughout the organoid structure, even after culturing for more than 8 months. Then, that team established the first human healthy organoid derived from small intestine and colon tissues in 2011 by optimizing culture conditions of mouse organoids.[19] These studies have the epic-making significance and laid the foundation for the subsequent establishment of PDOs from various tumors.

Since these studies, PDOs have been widely established from various tissues, including primary tumors,[20] metastatic lesions,[21] circulating tumor cells[22] and liquid exudates.[23] The process usually includes the following steps. First, the patient tissue is shredded or tumor cells from blood and liquid exudates isolated. Second, the tumor cells are dripped into a drop of solid colloid materials, such as Matrigel or Basement Membrane Extract, the most important component for forming 3D spatial structure. Third, nutrient medium with different growth factors, depending on the type of tumor, is added around the drops to allow PDO to grow continuously. Successful PDO, a sphere with multilayer cell structure, is often defined by the ability to expand, passage and cryopreserve cells that sustainably retain the histological, genetic, proteomic and pharmaco-typic characteristics of parental tumor, although it has not been standardized.

The success rate of PDO establishment is different depend on types of parental tumor tissue and cancer as well as the culture condition. Overall, the majority of PDOs have been established at a significantly higher success rate (>70%) than classical cancer cell lines (20–30%)[24] in many tumors, including glioblastoma,[25] lung,[26] breast,[27] pancreatic,[28] colorectal,[29] liver,[20] bladder[30] cancer. PDOs have been extensively applied in basic research, biobanking, disease model, and precision medicine fields of cancer. In this review, we discuss the various methods of the establishment and the application of PDOs in diverse tumor types as well as the limitations and future prospects of this model.

4. The applications of PDOs in various cancers

4.1 Gliomas

Glioma is the most common primary brain tumor with resistance to chemotherapy and poor prognosis, especially glioblastoma (GBM), the most aggressive type of glioma, representing approximately 57% of all gliomas and 48% of all primary malignant central nervous system tumors. GBM is a devastating disease that has seen scant progress in treatment options over the last several decades, is prone to recurrence because of the rapid growth in the brain despite surgical removal.[31] Further elucidation of GBM biology and hierarchy will be pivotal for developing more effective treatment, but the genetic/epigenetic heterogeneity and plasticity across cellular states of the disease[32] makes it exceedingly difficult to model. PDO culture systems permit a complex structure to develop, mimicking tumor development, providing an unprecedented platform for studying the mechanism of glioma.

4.1.1 Glioma organoids derived by genetically manipulating and coculturing

The establishment of glioma organoids include three methods: the first one is obtained from normal cerebral organoids by genetically manipulating to develop oncogenic properties[33,34]; the second one is coculturing with tumor spheres; the last one is derived from patients with glioma.[34,35] The introduction of oncogenic mutations in cerebral organoids *via* transposon- and CRISPR-Cas9-mediated mutagenesis caused the establishment of a neoplastic cerebral organoid, which recapitulates brain tumorigenesis and defines mutation combinations that result in glioblastoma-like and central nervous system primitive neuroectodermal tumor-like neoplasms.[36]

The use of human cerebral organoids as a scaffold for patient-derived GBM cell invasion and combination with single-cell transcriptomics successfully uncovers transcriptional changes implicated in the invasion process as well as potential ligand–receptor interactions between GBM and normal cerebral organoid cells, providing novel therapeutic targets for GBM.[37] However, these mutant organoids cannot truly reproduce glioblastoma because there are only normal nerve cells and glioma cells, lacking key elements of glioma cell components. Therefore, great efforts have been made to establish organoids by the third method, directly derived from tumor materials, to solve these problems.

4.1.2 The establishment and characterization of glioma PDOs

Glioma PDOs from biopsies and fresh resected tumor tissue in different tumor regions and tumor grades have been well established and characterized. In 2016, Hubert et al.[38] established glioblastoma organoids (GBOs) derived from multiple patient biopsies and tumor regions and showed that PDOs recapitulate hypoxic gradients and stem cell heterogeneity of parental tumors *in vivo*. Specifically, cancer stem cells near the oxygen rich surface of organoids often split and die, while cells in the hypoxic core of organoids are static. Notably, non-stem cells within organoids are sensitive to radiotherapy, whereas adjacent cancer stem cells are radio-resistant. The heterogeneity of cells and microenvironment in tumor organoids makes it possible to study the heterogeneous cell–cell relationships including hypoxic and non-hypoxic cancer cells as well as stem and non-stem cancer cells *in vivo*, which is impossible to achieve using other culture methods.

Furthermore, Jacob et al.[39] developed an improved protocol to generate GBOs without mechanical or enzymatic dissociation of the resected tumor tissue into single cells to preserve the local cytoarchitecture and native cell-cell interactions of the original tumors, and to avoid clonal selection of specific cell populations in culture. The resected glioblastoma tumor tissue obtained from patients was cut into 1 mm diameter pieces with debris and red blood cell removed, and then further cultured in a defined culture medium on an orbital shaker to facilitate organoid formation and increase nutrient and oxygen diffusion. They established GBOs within 1–2 weeks, radically reducing the time needed to establish PDOs. Further comprehensive histological, molecular, and genomic analyses showed that GBOs recapitulate inter- and intra-tumoral heterogeneity and retain many key features of parental tumors, including the histological features, cellular diversity, and genomic profiles. These GBOs could be efficiently xenografted into the adult mouse brain, displaying rapid and aggressive infiltration and maintaining expression of key driver mutations.[40] Interestingly, GBOs have been shown to reproduce a large number of microRNAs of parental tumors, contributing to construct the diagnostic model based on the expression of the selected microRNAs.[41]

Recently, the modeling approaches of PDOs from high grade glioma have been successfully applied to lower grade glioma with a success rate of 87%, while the stemness, proliferative, and tumor-stromal composition profiles, cytoarchitectural, mutational, and metabolic traits of parental tumors are also conserved in PDOs.[25] Golebiewska et al.[42] reported the establishment of a large cohort of unique PDOs including 107 GBMs and 25 grades II–III gliomas with a success rate of 79% and 68%, respectively. They further generated patient-derived orthotopic xenografts (PDOX) of PDOs, including seven paired longitudinal PDOXs from primary and recurrent tumors of the same patient, and showed that glioma PDOXs enable long-term propagation of patient tumors and represent clinically relevant patient avatars that retain histopathological, genetic, epigenetic, and transcriptomic features of parental tumors. Matched longitudinal PDOXs recapitulate the limited genetic evolution of gliomas observed in patients following treatment.

4.1.3 The comparison of four GBM models

Although expanding GBM models to PDOs has enabled improved modeling of tumor heterogeneity, how well different patient-derived GBM models are able to recreate the full heterogeneity of primary tumors is unclear. To address this issue, Pine et al.[43] compared the transcriptomes of more than 60,000 single tumor cells from five primary GBM patients across four model types by single-cell RNA sequencing, including 2D glioma sphere cultures, PDX, and 3D tumor organoids derived from PDX, PDO. The combined principal component analysis and unsupervised hierarchical clustering of the models' average transcriptomes showed that PDO and PDX tended to cluster closer together, more similar to primary samples, indicating PDO may be able to model the same transcriptional programs as PDX, such as interactions with nonmalignant brain cells, whereas the other models are unable to do so.

Furthermore, they recapitulated all four glioblastoma stem cell cellular states of the original tumor by combining cells across all models and found that PDOs have a higher percentage of neural progenitor cell-like cells, and along with the PDX, have a higher proportion of oligodendrocyte progenitor cell-like cells, and are the most similar to primary GBM samples with respect to cellular states composition. Then, they identified relevant GBM and neural developmental/stem cell genes that are differentially expressed in PDOs as compared with 2D glioma sphere and 3D tumor organoids derived from PDX, including SOX4, BCAN, and genes encoding members of the NOTCH pathway. Finally, they replated PDO cells in 2D culture conditions and showed that the subgroups of neural progenitor cell-like and astrocyte-like are present only in the PDO model and lost without the organoid microenvironment. Overall, their work reveals important differences between patient-derived models and shows that the microenvironment is crucial for recapitulating stem cell cellular states, highlighting the incomparable advantages of PDOs for GBM stem cell research.

4.1.4 The multicellular bio-printed models of organoids

Organoids cultured in submerged Matrigel usually lack immune microenvironment components, such as endothelial cells, stromal cells, and immune cells. Bioprinting technology can assemble these cells with organoids to simulate TME. The bio-printed reconstituted human glioblastoma-on-a-chip, consisting of patient-derived tumor cells, vascular endothelial cells, and decellularized extracellular matrix from brain tissue in a compartmentalized cancer-stroma concentric-ring structure that sustains a radial oxygen gradient, recapitulate the structural, biochemical and biophysical properties of the native tumors.[43]

Tang et al. further integrated macrophage into bio-printed glioblastoma constructs, which recapitulate patient-derived transcriptional profiles predictive of patient survival, maintenance of stemness, invasion, and drug resistance.[44] Maloney et al. reported a novel immersion bioprinting method to block the bioink from interacting with the well walls and provide support to maintain a spherical form by bioprinting the hydrogel bioink comprised of hyaluronic acid and collagen into a viscous gelatin bath.[45] Multicellular bio-printed models can serve as a scalable and physiologic platform to interrogate drug sensitivity, cellular crosstalk, invasion, context-specific functional dependencies, as well as immunologic interactions in a species-matched neural environment.

4.1.5 The application of PDOs in treatment

PDOs has been widely used for drug screening. Jacob et al.[40] established an organoid biological bank to test personalized therapies by correlating mutational profiles with response to specific drugs and by modeling chimeric antigen receptor T (CAR-T) cell immunotherapy. They showed that PDOs recapitulate patient-specific responses to treatment by exposing PDOs to similar post-operative treatments as patients (radiation and temozolomide (TMZ)). They assessed tumor progression in PDOs by calculating the percentage of $Ki67^+$ cells after treatment and observed that those PDOs most resistant to treatment correspond to patients with below median survival. Likewise, PDOs that respond best to treatment correspond to patients with extended survival. They also showed that CAR-T cell immunotherapy targeting epidermal growth factor receptor (EGFRvIII) is more effective for PDOs with positive EFGRvIII.

In addition, they further described procedures for investigating patient-specific responses to immunotherapies by coculturing PDOs with CAR-T cells.[39] PDOX-derived standardized glioma organoids that were amenable to high-throughput drug screens can be validated in mice, with highly clinically relevant responses to TMZ and to targeted treatments such as EGFR and CDK4/6 inhibitors. Dianhydrogalactitol (VAL-083), a promising bifunctional alkylating agent currently in clinical trial, display high therapeutic efficacy, and is able to overcome TMZ resistance in glioblastoma.[42] Patient-specific glioblastoma-on-a-chip reproduces clinically observed patient-specific resistances to treatment with concurrent chemoradiation and temozolomide, and can be used to determine drug combinations associated with superior tumor killing.[46] The immersion printing technique that bio-printed tissue organoids in 96-well plates highly increased the throughput of 3D drug screening.[45] GBM organoids have been used for testing the oncolytic activity of Zika virus, an oncolytic virus that can preferentially target glioblastoma stem cells,[47] identifying integrin αβ as a functional cancer stem cell marker essential for Zika virus infection.[48] Together, these results highlight the potential utility of patient-derived glioma organoid models and biobank for testing personalized therapies.

4.2 Lung cancer

Lung cancers are broadly classified into two subtypes with distinct histopathological and molecular characteristics, including small-cell lung cancer (SCLC), which is the cause of 15% of cases, and non-small-cell lung cancer (NSCLC), which accounts for 85% of cases, including adenocarcinoma, squamous cell carcinoma, and large cell carcinoma.[49] Despite the increasing availability of therapeutic drugs, lung cancer remains the leading cause of cancer death, with an estimated 1.8 million deaths (18%) and accounts for the second cancer morbidity (11.4%) worldwide,[1] urgently requiring preclinical models that accurately recapitulate the biology of lung cancer to further investigation of biomarkers and drug screening to allow more individualized therapy. Lung cancer organoids (LCOs) have been successfully established in NSCLC, including lung adenocarcinoma (LADC),[26] squamous cell carcinoma, adeno-squamous carcinoma,[50] large cell carcinoma,[51] and SCLC.[52]

4.2.1 The establishment and characterization of PDOs from NSCLC

Following experience with generating organoids from other adult human tissues, Sachs et al.[53] first established 18 long-term-expanding human airway organoids derived from lung tissue in patients with NSCLC with a success rate of 94% under optimized conditions. The organoids, termed airway organoids, were composed of a polarized, pseudostratified airway epithelium containing basal, secretory, and multiciliated cells and retain major characteristics of the *in vivo* epithelium. Then, they successfully generated 16 LCO lines with a success rate of 88% and 18 pure LCO lines from needle biopsies of metastatic NSCLCs with a success rate of 28%, which both recapitulate tumor histopathology and mutation status. In addition, they established eight airway organoid lines from fresh or cryopreserved broncho-alveolar lavage fluids, five of which from cystic fibrosis patients, with a 62% success rate, that present a much thicker layer of apical mucus compared to control organoids and recapitulate aspects of the *in vivo* cystic fibrosis phenotype.

Kim et al.[50] established organoids derived from primary lung cancer tissue and five paired non-neoplastic airway tissues and created a biobank of 80 LCOs from five subtypes of lung cancer (adenocarcinoma, squamous cell carcinoma, adeno-squamous carcinoma, large cell carcinoma, and SCLC) with a success rate of 87%. LCOs sufficiently represented the diverse histological features of different lung cancer subtypes, including PD-L1, the lung adenocarcinoma markers (napsin A, thyroid transcription factor 1, and cytokeratin 7), squamous cell carcinoma markers (p63 and CK5/6), large cell carcinoma markers (cytokeratin 7 and a stem cell marker, CD133), SCLC markers (CD56, synaptophysin 37, and thyroid transcription factor 1), and maintained the genomic heterogeneity and up to 77% concordance of the mutations that were present in the original cancer tissues, while normal bronchial organoids maintained bronchial epithelial organization. Interestingly, not only the distribution of cell types, but also the type of cilia in normal bronchial organoids and LCOs, was different.

Our team further established 12 LADC organoid lines from tumor resections of 15 untreated patients with a success rate of 80% and provided a biobank of LADC organoids that cover a large spectrum of different LADC subtypes, including acinar, solid, papillary, and enteric adenocarcinoma.[26] Although some LADC organoid lines stopped growing after passaging, we solved this problem by using the mechanical dissociation method. Extensive gene and histopathology profiling show that the tumor organoids retain the histological architectures, genomic landscapes, and gene expression profiles of their parental tumors. In addition, we also established three normal lung organoid lines from normal wild-type lung samples and demonstrated tumor heterogeneity among patients. We further identified 29 novel genes whose expression levels have never been reported to be altered in LADC samples, in which overexpression of RHOF, SLC16A3, and ANXA10 predicted poor prognosis, whereas decreased expression of CDHR1 correlated with poor overall survival.

Transcriptomic analysis of tumor tissues and organoids reveals the common and different mechanisms involved in LADC and lung squamous cell carcinoma carcinogenesis and progression, whose hub genes were further identified. Using LADC cell lines and organoid models, CDK1 and CCNB2 knockdown was found to suppress LADC proliferation, while CDC25A knockdown did not inhibit LUAD cell line proliferation but could effectively suppress LADC organoid growth,[54] indicating that an organoid could be used as an effective tool to study cancer biology in LADC.

4.2.2 The establishment, characterization and application of PDOs from SCLC

Compared with NSCLC, SCLC with high malignancy and poor prognosis is still considered one of the most lethal cancer types worldwide.[55] The development of representative preclinical models and effective anti-cancer agents for SCLC is therefore critical. Resected tumor tissue is a typical material source for PDO establishment. However, SCLC is a diffuse malignancy and patients rarely undergo surgical resection, resulting in limited utility of this technique.

Given the possibility of insufficient material from small needle aspiration samples to generate PDOs for drug testing, Gmeiner et al.[56] established PDX tumors in mice from SCLC needle aspiration to produce sufficient tumor volumes for PDO formation. Drug testing indicated that although relatively resistant to conventional 5-fluorouracil, four PDOs derived from PDX tumors were moderately sensitive to the front-line agent cisplatin and relatively more sensitive to CF10, a novel fluoropyrimidine polymer with enhanced thymidylate synthase inhibition activity.

Choi et al.[52] generated SCLC PDOs from fresh biopsied samples with a success rate of 80% (8/10) and attempted to determine the culture conditions most suitable for the long-term culture of SCLC cells. They cultured SCLC PDOs for more than 9 months using five different culture conditions. Results showed that the basal culture conditions previously determined for PDOs were not suitable for the long-term maintenance of SCLC PDOs, and the activation of the Wnt/β-catenin pathway by the addition of WNT3A or R-spondin1 is a prerequisite for the long-term expansion of SCLC PDOs, in combination with ALK receptor and BMP signaling inhibition. They confirmed that SCLC PDOs faithfully replicated genetic profiles, molecular characteristics, and morphological architectures of parent tumor during and after long-term expansion. Further comparison of the genetic profiles and morphological and molecular characteristics of organoids cultured under five different conditions with their corresponding tissues indicated that the culture conditions did not affect the molecular and morphological characteristics, genomic stability, and mutational status of tumor organoids.

The heterogeneous SCLC organoids with different response to cisplatin or etoposide would be useful for overcoming the current problem of SCLC recurrence in patients after the current standard etoposide–platinum doublet. The further drug testing in two SCLC PDOs with resistant to cisplatin and etoposide indicated that YPN-005, a potent inhibitor of cyclin-dependent kinase 7, has the potential to be used to treat patients with SCLC who are non-responsive to etoposide–platinum chemotherapy.[57] As ideal avatar models of SCLC patient with high-risk, PDOs can help for investigating the mechanisms underlying tumor recurrence after chemotherapy and the development of personalized medicine.

4.2.3 The establishment and characterization of PDOs from pleural effusion

Interestingly, most PDOs are established from tumor tissue of patients. Mazzocchi et al. fabricated lung cancer organoids using a rare cell source, pleural effusion aspirate.[23] They found that isolated patient cells embedded directly into 3D lung cancer organoids created anatomically relevant structures and displayed lung cancer specific behaviors compared to cultures in 2D conditions by tacking the phenotypic change that developed in short-term culturing and characterizing the cell population within the organoids. Although the 2D cultures were more chemosensitive than the 3D organoids, the organoids that preserve the phenotype of the cancer cells as well as maintain cancer and stromal cell interaction for an extended period were more accurate for drug screening.

Recently, 83 PDOs without changes in morphology for 2–3 months were established from patients with advanced lung adenocarcinoma, including 77 from malignant effusions, 3 from brain metastasis, 1 from bone metastasis, and 2 from lung primary tumor, with a success rate of 83.0% (83/100). But 17 samples failed to expand due to lack of tumor cells in effusions, fungi contaminations, and excessive numbers of immune cells. In addition to recapitulating morphologic, histologic, and genetic characteristics of original tumor tissues, PDOs could predict a progression-free survival (PFS) in patients receiving clinically approved tyrosine kinase inhibitors (TKIs) at an accuracy of 75.0%, in which four of five matching PDOs with patients achieved a PFS of >9 months were sensitive to TKI treatment ($IC_{50} < 100$ nM), two of three matching PDOs with patients had a PFS of <9 months were resistant to TKI treatment ($IC_{50} > 100$ nM). Comparing with 2D cultures generated from nine PDOs with known clinical responses to TKIs, 3D PDOs could capture unique information such as Src activation, which is not represented by genetic test. Especially, some 3D PDOs (4/9, 44.4%) even cannot be cultured in monolayer condition, suggesting that some LADC may grow only as suspension cells or require extracellular matrix for optimal growth.[58]

This study indicates that pleural effusion aspirates removed by a routine, required, and relatively non-invasive procedure can be used to successfully establish PDOs, providing a unique opportunity to researchers and clinicians to monitor disease progression and response to treatment and study the evolution of tumor and drug resistance *in vitro*.

4.2.4 Tumor microenvironment simulation and immune test of LCOs

Although typically containing exclusively neoplastic epithelium, LCOs can be used to test immunotherapy and enable immune-oncology investigations within the TME by coculture with immunocyte[59] and propagation of primary tumor epithelium *en bloc* with endogenous immune stroma.[60]

Dijkstra et al.[59] established and validated a platform to induce and analyze tumor-specific T cell response to NSCL in a personalized manner. They exposed peripheral blood lymphocytes to NSCLC organoids generated from six patients, resulting in the expansion of tumor-reactive CD8 populations in two of these patients after 2 weeks of coculture. The tumor organoid-reactive CD8 T cells, whose reactivity had been induced in the 2-week coculture with peripheral blood lymphocytes and was restricted to the tumor organoids, were specific for tumor antigens, without response to normal organoids. Furthermore, they described the detailed protocol for the generation and functional assessment of tumor-reactive cells based on cocultures of NSCLC organoids and autologous peripheral blood lymphocytes, allowing the establishment of *ex vivo* test systems for T cell-based immunotherapy at the level of the individual patient.[61]

Compared with adding T cells to organoid, an air–liquid interface organoid model containing tightly integrated epithelial and stromal compartments notably preserves diverse endogenous immune cell types in addition to T cells, including macrophages, B, and natural killer cells. Neal et al.[60] successfully established air–liquid interface PDO cultures from 100 individual patient tumors representing 14 distinct tissue sites and 28 unique disease subtypes, including common tumor sites, such as 20 lung, 20 colorectum, 20 kidney, 11 pancreas, 9 thyroid, and rarer histology, including bile duct ampullary adenocarcinoma, brain schwannoma and salivary gland pleomorphic adenoma for which cell lines are scarce. Although organoid could grow with fetal calf serum alone, they utilized optimized culture protocol with supplementation of WNT3A, EGF, NOGGIN, and RSPO1 to improve tumor epithelia-organoid growth in submerged Matrigel.

Air–liquid interface PDOs typically recapitulate histology and genetic characterization of the parental tumor, with an overall 73% success rate at 1-month culture across tumor histology. Not only fibroblast stroma, but also macrophages, CD8 and CD4 T cells, B cells, natural killer and natural killer T cells as well as infiltrating CD3 T cells expressing the immune checkpoint surface receptor programmed cell death protein-1 (PD-1), were preserved in air-liquid interface PDOs. They demonstrated accurate preservation of PDO tumor-infiltrating lymphocytes to the original tumor T cell receptor spectrum by using robust droplet-based, single cell simultaneous determination of gene expression and immune repertoire. Importantly, human PDOs successfully model immune checkpoint blockade with anti-PD-1 and/or anti-PD-L1 expanding and activating tumor antigen-specific tumor-infiltrating lymphocytes and eliciting tumor cytotoxicity with rapid 7-day assessment.

4.2.5 The microwell array chip of PDOs

While the potential of PDOs to predict patients' responses to anti-cancer treatments has been well recognized, the relatively lengthy time in establishing enough quantities of PDOs for drug screening hampers the implementation of PDO-based drug sensitivity tests in the clinic. To overcome this technical challenge, Hu et al.[62] developed an integrated superhydrophobic microwell array chip (InSMAR-chip) for high-throughput 3D culture and analysis of PDOs to reduce the reaction volumes of PDO-based drug test. The nanoliter scale of microwells causes PDOs obtained at passage 0 enough for testing an array of clinically recommended drugs in a week without need for prolonged expansion. The results showed that the 1-week drug sensitivity test of PDOs is consistent with the clinical outcome and genetic mutations. This study demonstrates that the combination of PDOs and the InSMAR-chip may provide an efficient means for promptly predicting patient-specific drug responses, leading to significant acceleration of bench-to-bedside transformation, which is extremely important to improve prognosis, especially for patients with rapidly progressing malignant tumors.

4.2.6 The tumor purity of LCOs

For clinical implementation, an additional challenge faced when culturing PDOs from primary tumors with normal epithelial cells is efficient establishment of pure tumor organoids to avoid the overgrowth of normal epithelial cells, which may have major implications for the translational potential of PDOs. To establish pure lung cancer organoids, Kim et al.[50] developed a 3D culture protocol using minimum basal medium, which is a suboptimal media inhibiting growth of normal cells due to depletion of Wnt3a and Noggin. The MDM-2 inhibitor Nutlin-3a was also applied to selectively grow out p53 mutant cells of the sample, resulting in a pure tumor organoid population.[59] Yokota et al.[63] further verified that nutlin-3a was key to select pure lung tumoroids that harbor mutant p53. But whether this strategy results in higher tumor purity without compromising organoid establishment rate still needs to be investigated in larger cohorts.

Dijkstra et al.[64] systematically evaluated several methods to identify tumor purity of PDOs from >70 intrapulmonary NSCLC, of which genetic analysis suggested that the majority (65%) of PDOs from intrapulmonary lesions are overgrown by healthy normal airway organoids. Histomorphology alone was insufficient to determine tumor purity, but when combined with p63 immunostaining, tumor and normal airway organoids could be distinguished.

In summary, we recommend that lung cancer organoids derived from intrapulmonary lesions should be established by rigorous methods to prevent overgrowth by normal cells and routinely evaluated for tumor purity; exploring protocol of tumor organoids from extrapulmonary lesions may be useful for improvement of tumor purity.

4.2.7 The application of PDOs in treatment

LCO models preserve the sensitivity of the matched parental tumor to therapeutics,[51] and are amenable for high-throughput drug screening *in vitro*. LCOs respond to drugs based on their genomic alterations: a *BRCA2* mutant organoid to olaparib, *EGFR* mutant organoids to erlotinib, and *EGFR* mutant/*MET* amplified organoids to crizotinib,[50] *P53* mutant organoids to nutlin-3a,[26] *HER2* mutant organoids to pyrotinib,[65] *BRAF*^{G469A} mutant organoids to trametinib/erlotinib, TPM3-ROS1-fusion organoids to crizotinib/entrectinib,[63] while normal bronchial organoids can be used for estimation of drug toxicity on non-cancerous cells. Of note, some LADC organoid lines also demonstrated sensitivity to afatinib and gefitinib even though they did not have mutations in *EGFR*, emphasizing the value of whole-exome sequencing in combination with *in vitro* drug screens on LADC organoids.[26]

In addition, drugs with similar nominal targets have comparable activity across all the LADC organoid lines. For example, the tubulin inhibitors vincristine and vinorelbine display similar sensitivity patterns and all the organoid lines show a similar and concordant trend for the ALK-targeted drugs brigatinib and ceritinib.[26]

Moreover, LCOs were used to develop an evaluation system for the immune checkpoint inhibitors, nivolumab and pembrolizumab.[66] LCOs could not only capture the clinical response to the dabrafenib/trametinib combination therapy in a patient with NSCLC harboring *EGFR* plus *BRAF* mutation, but also identify effective anti-cancer therapies for molecular targets, whose results demonstrated that afatinib was more potent than first-generation and third-generation EGFR-TKIs against the *EGFR* L747P mutation; poziotinib and pralsetinib are more effective against NSCLCs harboring *ERBB2* exon 20 insertions and *RET* fusions, respectively.[58] The effective treatment of Bcl-2 inhibitor (ABT-263) and a survivin inhibitor (YM-155) to EGFR TKI-resistant LCO,[63] and the effective combination of cisplatin and Halofuginone to cisplatin-resistant LCO,[67] demonstrated that the establishment of resistant LCDs may contribute to postpone or even reverse the development of drug resistance.

4.2.8 The application of human lung organoids in coronavirus disease 2019 (COVID-19)

In the current COVID-19 pandemic, both stem cell-based organoid models[68] and primary alveolar cell-derived organoids[69] have been used to model organotropism, pathogenesis and inflammatory responses following a severe acute respiratory syndrome coronavirus-2 infection. This is a very important application of human lung organoids beyond tumor research, providing novel insights into COVID-19 pathogenesis and demonstrating the great potential of organoids as a model for understanding human respiratory diseases.

4.3 Breast cancer

Breast cancer (BC) has surpassed lung cancer as the most commonly diagnosed cancer in women, with an estimated 2.3 million new cases (11.7%) according to the Global Cancer Statistics 2020,[1] comprising more than 20 distinct subtypes that differ genetically, morphologically, and clinically.[70] It is unreasonable to rely on a few dozen cell lines as *in vitro* representations of a highly heterogeneous disease affecting millions of patients. Genome-wide analysis has already revealed the substantial genomic differences between BC cell lines and patient samples and shows that surprisingly, MDA-MB-231 cells, one of the most widely used triple-negative cell lines in metastatic BC research, bear little genomic similarities to metastatic BC patient samples. Further comparison suggests that organoids more closely resemble the transcriptome of metastatic BC samples compared to cell lines, highlighting the enormous potential of organoids in breast cancer research.[71]

4.3.1 The establishment and characterization of PDOs from BC

In order to expand the application of PDOs in BC, Sachs et al.[72] established a robust protocol for long-term culturing of human mammary epithelial organoids and generated >100 primary and metastatic BC organoid lines which broadly recapitulate the diversity of original tumor, including gene expression, histopathology, and hormone receptor (including epidermal growth receptor 2 (HER2)) status. In addition, PDOs also retained the protein expression patterns of original tissues and the ability of preserving complex stem/progenitor that can differentiate in culture to generate mature basal and luminal cells.[73]

Subsequently, Dekkers et al.[74] provided a detailed, highly versatile protocols on how organoids could be derived from both normal and tumor resections covering all major subtypes of BC (triple-negative, estrogen receptor-positive/progesterone receptor-positive and HER2-positive). They outlined the respective culture conditions and explained how to maintain and cryopreserve the cultures. Additionally, they discussed three different ways to genetically manipulate organoids, including Lipofectamine 2000, electroporation, or lentivirus, and subsequent organoid selection and clonal culture. Finally, they described xenotransplantation of BC organoids. Although their methods have been optimized for breast organoids, the section on genetic manipulation and clonal selection can be applied to a wide variety of organoid cultures with minor adjustments, thus affecting research fields beyond BC. These studies indicate that PDOs may be the most representative and robust models that can be generated efficiently.

Most breast cancers at an advanced stage with high-risk clinical features, including drug-resistant and metastatic breast cancers, exhibit an aggressive nature, and lack of effective preclinical model. Although PDO culture system as an *in vitro* cancer model has been successfully established in breast cancer, whether it can be successful with drug-treated breast cancer, especially heavily treated tumor specimens, has not been explored.

Chen et al.[27] took great effort into establishment of a large BC organoid biobank from tumor tissues with a broad spectrum of treatment statuses and disease stage. They obtained 132 BC tissues, including 77 treatment-naïve, 55 exposed to systemic anticancer therapy, and 23 resistant to treatments. They eventually established 99 organoids with an overall success rate of 75%. In addition, they obtained 17 biopsied tumors during the clinical treatments with 14 from drug-treated patients and 13 from metastatic sites, generating 11 organoids with a success rate of 65%. They found that organoid lines established from drug-treated tumors and lines from treatment-naïve tumors do not have significant morphological differences, replicating the histologic features of their parental tumor tissues, including growth patterns, cellular and nuclear atypia and biomarkers, as well as retained genomic features of parental tumors to some extent. However, PDOs had a higher number of copy number alteration and single-nucleotide variant genes compared with parental tumors, which may be due to the amplification potential of cancer initiating cells or cancer stem cells in PDOs and due to the inclusion of multiple normal cells in primary tumor tissues.

4.3.2 The application of PDOs from BC in basic research

BC organoids were used in basic research, including the regulation of extracellular matrix for collective migration and local dissemination in malignant mammary epithelium,[75] the contribution of different subpopulations of cancer cells for collective invasion and identification of the most invasive cancer cell in primary BC,[76] as well as the inhibition of apical-basal polarity on epithelial–mesenchymal transition during tumor metastasis.[77] The availability of the clustered regularly interspaced short palindromic repeats (CRISPR)/Cas9-mediated gene editing in BC organoids[72] further contributes to the application of PODs in breast oncogenesis.

Dekkers et al.[78] established mutant tumor organoids with long-term culturing capacity and response to standard therapy by targeted knockout of four breast cancer-associated tumor suppressor gene (*P53*, *PTEN*, *RB1*, *NF1*) in normal breast epithelial organoids to simulate the oncogenesis. They showed that a minimum of three tumor suppressor genes must be inactivated for oncogenesis in normal breast organoids, highlighting the potential of normal organoids combined with CRISPR/Cas9 gene editing to understand critical drivers of oncogenesis. Recently, PDOs combined with metabolomics analysis have been used to identify the potential therapeutic target for high-risk triple-negative breast cancer, showing that targeting sphingosine-1-phosphate, an intermediate of the ceramide pathway, is a promising therapy for the transcriptomic luminal androgen receptor (AR) tumors, one subtype overlapped with metabolomic C1 subtype.[79] Together, PDOs cultures provide a valuable platform for studies of differentiation, transformation, and cancer risk for BC.

4.3.3 Tumor microenvironment simulation of PDOs from BC

Given that TME plays a critical role in tumor growth and progression, 3D biochemico-mimetic and mechano-mimetic scaffold[80] and 3D bioprinting[81] have been used to culture BC organoids to more faithfully recapture ECM features and complex tumor microenvironment of BC. The hybrid mechano-mimetic scaffolds developed by decellularizing cancer-associated fibroblasts (CAFs) cultured on 3D macro-porous polymer scaffolds present a unique native extracellular environment to the primary BC cells, enhancing cellular viability, attachment and organoids formation.[80]

3D bioprinting has been used to develop a tumor tissue model similar to solid tumor architecture in which a core tumor cell bioink is surrounded on all sides by epithelial and stromal cell bioink, and shows that bio-printed cells are able to survive, self-organize, and interact with one another to form tissue-like structures. The assessments of intrinsic, extrinsic, and spatial tumorigenic phenotypes in bio-printed tissues found that cellular proliferation, extracellular matrix deposition, and cellular migration are altered in response to extrinsic signals.[81] This work demonstrates that multicell-type bio-printed tissues can recapitulate aspects of *in vivo* neoplastic tissues and provide a manipulable system for the combination of organoids and distinct tumor microenvironments.

4.3.4 The application of PDOs from BC in treatment

The high heterogeneity and effective establishment of BC organoids have greatly promoted its application for drug screening. As expected, most HER2-overexpressing organoids are sensitive to six drugs targeting the HER2 signaling pathway, while 28 HER2-negative organoids are resistant. However, a few organoid lines defied these expectations, emphasizing the value of functional *in vitro* drug tests on BC organoids. *In vitro* responses of the BC organoids to tamoxifen matched that of the respective patients, indicating the potential use of BC organoids as predictive *in vitro* surrogates for BC *in vivo*.[72] Engineered BC organoids generated from normal organoid mutations responded to tamoxifen and docetaxel, supporting the potential of this model to enhance our understanding of the molecular events that culminate in specific subtypes of BC.[78]

The BC organoids on the hybrid mechano-mimetic scaffolds display varying response to two chemotherapeutic drugs, doxorubicin and mitoxanthrone, across patient samples,[80] showing the ability of the hybrid scaffolds to efficiently serve as a culture platform to test the effectiveness of chemotherapeutic drugs in PDOs. The application of 3D bioprinting can help exploring the response and mechanism of organoids to therapies in distinct microenvironment.[78] The combination of microfluidic organ-on-chip system in PDOs for multianalyte monitoring of metabolites under dynamic conditions contributes to the utility of PDOs for drug screening experiments for personalized therapy, the standardization of *in vitro* models and fundamental research on cellular metabolism and interaction between cocultures.[82]

A high-throughput screen based on the functional interaction of BC organoids and tumor-specific cytotoxic T cells contributed to the identification of epigenetic inhibitors, which promoted antigen presentation and potentiated T cell-mediated cytotoxicity. Of these epigenetic inhibitors, BML-210 substantially sensitized breast tumors to the inhibitor of checkpoint programmed death-1.[83] The coculture system of PODs with peripheral blood mononuclear cells has been used to confirm the therapeutic potential of engineered exosomes as an *in situ* dendritic cell (DC)-primed vaccine for BC. Consistent with the observations in BC cell lines and in mouse xenograft model, the *in situ* DC vaccine formed by loading the immunogenic cell death inducers, human neutrophil elastase and Hiltonol (TLR3 agonist), into α-lactalbumin-engineered BC-derived exosomes exhibited potent antitumor activity in PDOs, because they promoted the activation of DCs *in situ* and subsequent tumor-reactive $CD8^+$ T cell priming.[84]

4.3.5 The application of PDOs in treatment of heavily treated patients with BC

Due to the urgent need for personalized treatment in heavily treated patients that show resistance to a wide range of drugs, Chen et al.[27] attempted to exploit patient-derived tumor organoids as a real-time platform to guide the clinical treatment of advanced breast cancers and achieved great success.

Results showed that cancer PDOs were commonly sensitive to treatment with mitoxantrone, epirubicin, doxorubicin and bortezomib (proteasome inhibitor), while consistently resistant to formestane and carboplatin. Compared with the primary tumors, organoids derived from metastatic tumors showed a higher frequency of resistance to many drugs, including the microtubule- and EGFR-targeting drugs, palbociclib (CDK4/6 inhibitor), and cepharanthine (inhibiting TNF-α-mediated NF-κB stimulation). However, the drug responses between the drug-treated and treatment-naïve groups showed no significant difference, while three PDOs derived from multiple drug-resistant tumors showed high-frequency resistance to a wide range of tested drugs. Notably, PDOs exhibited simultaneous sensitivity or resistance to nucleic acid synthesis inhibitors (gemcitabine, cladribine, and cytarabine) and microtubule inhibitors (docetaxel, paclitaxel, vinorelbine, ixabepilone, vincristine, and vinblastine).

Then, they performed RNA sequencing on 57 BC organoid lines that were treated with six microtubule targeting drugs: docetaxel, paclitaxel, vinorelbine, ixabepilone, vincristine, and vinblastine, and found that the responses of BC organoid lines to drugs are highly correlated with their transcriptional profile. The difference in gene expression between the microtubule-targeting drug-sensitive PDOs (sensitive to all six drugs) and -resistant PDOs (resistant to all six drugs) was defined as the sensitive response signature, which was highly correlated with survival outcomes. The clear-cut retrospective data of 35 patients showed that PDO pharmaco-phenotyping results could reflect the previous treatment responses of the corresponding patients and are highly correlated with clinical outcomes ($P=0.00009824$), with 80.10% AUC (95% CI, 69.70–90.50%), 76.67% accuracy (95% CI, 65.00–86.67%), 82.35% sensitivity (95% CI, 41.12–94.12%), and 69.23% specificity (95% CI, 53.85–100.00%).

Finally, they performed new personalized chemotherapy for 16 patients with multiple-drug resistance and metastatic BC using the effective agents identified by PDOs, and all patients exhibited efficient response as predicted by drug sensitivity test. Notably, the patient who was diagnosed with multiple-drug resistance and multiple-lung metastatic triple-negative BC, remains alive more than 2 years after personalized therapy.

This study demonstrates that PDOs not only could be successfully established from the tumor specimens of multiple drug-resistant patients but also will provide personalized therapy recommendations for patients with advanced disease.

4.4 Pancreatic cancer (PC)

PC is a highly aggressive disease and a generalized poor treatment response with an overall 5-year survival rate of 9%, the lowest among common cancers,[85] causing 530,000 incident cancer cases and 531,000 deaths across 204 countries and territories in 2019. The majority of pancreatic cancer are pancreatic ductal adenocarcinoma (PDAC); rare pancreatic neoplasms include neuroendocrine tumors which can secrete hormones such as insulin or glucagon, and acinar carcinomas which can release digestive enzymes into the circulation.[86] Effective screening is not available for PDAC, and most patients present with locally advanced (30–35%) or metastatic (50–55%) disease at diagnosis with 10–15% of patients who have the chance to undergo surgical resection. The incidence of PDAC is increasing by 0.5–1.0% per year, and it is projected to become the second leading cause of cancer-related mortality by 2030.[87] PDOs bring new hope for effective personalized treatment of the heterogeneous pancreatic cancer.

4.4.1 The establishment and characterization of PDOs from PC

PC organoids have been successfully established from PC patients and proved to accurately replicate the characteristics of original tumors. Boj et al.[28] first established normal and tumor PDOs from freshly resected specimens and biopsies of PDAC with a success rate of 75% ($n=3/4$) and 83% ($n=5/6$) in the Netherlands and United States, respectively. While the normal pancreatic organoids had a simple, cuboidal morphology, the PDAC organoids had different degrees of dysplastic tall columnar cells, resembling low-grade pancreatic intraepithelial neoplasms (PanINs), the pre-neoplastic lesions that could progress to be locally invasive and then metastatic carcinomas. Both normal and tumor PDOs effectively replicated the mutant characteristics, karyotypes, and biomarkers of normal and PDAC cells. When orthotopically transplanted in mice, PDAC organoids modeled the translation of early-grade pre-invasive PanIN to infiltrative carcinoma progression.

Huang et al.[88] induced the differentiation of human pluripotent stem cells (PSCs) into exocrine progenitor organoids that formed ductal and acinar structures. Expression of mutant KRAS or TP53 in progenitor organoids induces mutation-specific phenotypes in culture and *in vivo*. Expression of $TP53^{R175H}$ induced cytosolic SOX9 localization. In patient tumors bearing *TP53* mutations, SOX9 was cytoplasmic and negatively associated with clinical outcome. Then, they established PDAC organoids from surgical resections of primary tumor samples with a success rate of 85% (17/20); the three that failed were intraductal papillary mucinous cystic neoplasm, poorly to moderately differentiated PDACs. The PDAC organoids replicated the histological organization, differentiation status, and morphologic heterogeneity of parent tumors, as well as patient-specific traits including repressive epigenetic markers, oxygen consumption, and differential sensitivity to EZH2 inhibition.

Seino et al.[89] used surgical fine-needle aspiration and ascites specimens to establish a pancreatic tumor organoid library, including 39 PDAC organoids and 10 normal pancreatic organoids. During the culture, these PDAC organoids harbored common driver gene alterations at the expected frequencies: KRAS (36/38), CDKN2A (31/38), TP53 (29/38), and SMAD4 (14/38).

Tiriac et al.[90] generated a pancreatic cancer PDO library from 159 human samples of primary tumors and metastases with a success rate of 75% (72% for fine needle biopsies, 78% for tumor resections), covering four clinical stages of pancreatic cancer. PDOs recapitulated genetic hallmarks of pancreatic cancer and revealed new characteristics. Transcriptomic profiling showed that 70% of the PDO cultures are the Classical subtype (31/44) and 30% are Basal-like (13/44), which is a notable finding as there are very few available cell line models of the Classical PDAC subtype. They further proposed a classification schema for PDO culture maturation in three clinically relevant phases: establishment, expansion, and characterization.[91] In addition, they showed that PDOs could be generated from specimens obtained remotely with a success rate of 77%.

Romero-Calvo et al.[92] demonstrated that organoid from primary PDAC and PDX shared three morphological classifications with the primary tumor: simple, papillary, and solid/cribriform, while the architecture of the PDX models was more complex and atypical, in part due to the presence of mouse stroma within the tumor. In addition, organoids and PDX replicated protein expression and transcriptional profiles of parent tumor, as well as mutation profiles, stable even over 10 passages in PDX-derived organoids. Clustering of single-cell RNAseq identified three transcriptionally distinct groups of cells, including the largest cluster 1 comprised 97.2% of the total population of cells, the cluster 2 (0.9%) expressed markers amphiregulin and epiregulin that were associated with tumorigenesis and poor outcomes, the cluster 3 (1.9%) expressed Prominin-1 (CD133), a marker for pancreatic cancer stem cells, indicating that organoids derived from PDAC were largely clonal populations of the primary tumor.

Driehuis et al.[93] reported a pancreas tumor biobank, including 19 PDAC, 1 adeno-squamous PDAC, 3 intraductal papillary mucinous neoplasm (IPMN)-derived PDAC, and 7 other non-PDAC tumor types, such as pancreatic acinar cell carcinoma and distal cholangiocarcinoma. Tumor PDOs recapitulate tumor histology and contain genetic alterations typical of pancreatic cancer, while "healthy pancreas" organoid established from normal pancreatic tissue had the anecdotal expansion of precancerous cells with an activating G12R KRAS mutation. This precancerous lesion was detected upon pathological assessment of the biopsy specimen, showing the special application of PDOs in culture and preservation of rare precancerous cell. Furthermore, they described the detailed protocols, including the establishment, expanding, freezing, and thawing of both normal pancreatic and tumor PDOs from resections and biopsies.[94]

Recently, PDOs from adeno-squamous carcinoma of the pancreas, a rare variant of pancreatic ductal adenocarcinoma, have been successfully established and show remarkable resemblance to the parent tumor. These PDOs were a mixture of glandular and squamous differentiation and formed ducts with a cribriform growth pattern and cytonuclear atypia.[95]

Resistant PDAC organoids were also successfully generated from patients who received eight cycles of neoadjuvant FOLFIRINOX chemotherapy (a combined regimen of 5-fluorouracil (5-FU), irinotecan, and oxaliplatin) before surgery and recapitulated histologic, genetic, and transcriptional characteristics of primary tumor tissue.[96] Interestingly, extracellular vesicle miRNA profiles were shared by PDAC organoids with matched blood plasma samples and largely differed among patients.[97]

4.4.2 The improvement of culture conditions of PDOs from PC

The culture conditions of PDOs are constantly improving. Boj et al.[28] first improved the culture conditions of normal pancreatic organoids from adult murine ductal cells[98] to establish PDOs. They found that the propagation of human normal pancreatic organoids needs TGF-β pathway inhibitors (A83-01 and Noggin), R-Spondin1 and Wnt3a-conditioned media, EGF, and PGE_2 compared with murine pancreatic organoids. While the normal pancreatic organoids had limited propagation of 20 passages or approximately 6 months, the PDAC organoids had unlimited propagation. Huang et al.[88] identified the growth factors and nutrients that were critical for pancreatic organoid development, including FGF2, insulin, hydrocortisone, ascorbic acid, all-*trans* retinoic acid, and B27 serum-free supplement.

Senino et al.[89] found that serum is detrimental for the long-term maintenance of human pancreas organoids due to its triggering senescence.

The application of serum-free Afamin-stabilized Wnt3A medium instead of serum-stabilized Wnt3A-conditioned medium enabled stable culture for over 8 months. The sensitivity to EGF removal, Noggin removal/BMP4 treatment, A83-01 removal/transforming growth factor β1 (TGF-β1) treatment, and Nutlin3 treatment, were associated with *KRAS*, *SMAD4*, *TGFBR2*, and *TP53* mutations/indel alterations, respectively, indicating that driver gene alterations largely dictated the requirements for the corresponding niche factors.

However, Wnt/R-spondin dependency was mostly unrelated to Wnt signaling mutations in PDAC organoids, causing the identification of three functional PDAC organoid subtypes with distinct Wnt niche requirements. Wnt-secreting subtype autonomously created their own Wnt niche by expressing epithelial Wnts, which also predicts aggressive clinical behaviors, while Wnt-non-secreting subtype critically depended on Wnt ligands from cancer-associated fibroblasts. Wnt and R-spondin-independent subtype grew in the absence of Wnt and R-spondin. Furthermore, CRISPR-Cas9-based genome editing of PDAC driver genes (KRAS, CDKN2A, SMAD4, and TP53) demonstrated that niche independency was mainly acquired through driver gene mutations, whereas the Wnt niche independency was predominantly regulated by epigenetic mechanisms. This suggests that usage of a Wnt ligand-free culture medium can preselect more aggressive PDAC organoid subpopulations.

Driehuis et al.[93] demonstrated again that organoids with tumor-specific genetic alterations could be selected in culture by pharmacologic inhibition or removal/addition of certain growth factors. Addition of Wnt3a, R-spondin1, EGF, and FGF-10 in a conventional organoid culture system showed $CD44^+$ cancer-initiating cells were enriched >65% in PDAC organoids, while $CD44^-$ cancer cells derived from organoids were able to reform organoids and be re-programed as CD44-expressing cancer-initiating cells.[99] Inclusion of Y-27632 Rho kinase inhibitor contributed to the recovery of organoids from the freeze/thaw stress and enhanced cell survival after cryopreservation.[100]

4.4.3 The application of PDOs in cachectic patients with cancer

Cachexia, present in up to 80% of patients with PDAC, is associated with loss of skeletal muscle tissue, reduced survival, more progressive disease, and higher rates of metastasis.[101] PDOs represent a powerful novel approach to understand the inducing mechanisms and inter-individual differences of common cachexia.

Vaes et al.[95] established five PDOs from five patients diagnosed with cancer cachexia, including three PDAC, one ampullary carcinoma, and one cholangiocarcinoma, as well as three PDAC organoids without cachexia. All organoids showed typical features of malignancy corresponding to the parental tumor, including nuclear enlargement, multiple nucleoli, mitoses, and apoptosis, *KRAS* and *TP53* mutation. Comparing with non-cachectic organoids, cachectic organoids expressed lower mRNA of IL-1α and IL-1β, and higher mRNA of IL-8, GDF15, and LIF, as well as higher protein levels of IL-8 and GDF15, while IL-6 was only found in two cachectic organoids, although it is highly present in the plasma of all non-cachectic and cachectic patients.

They further used non-cachectic and cachectic tumor organoids to assess the direct effects of tumor-derived factors on C2C12 skeletal muscle cells.[102] The incubation of C2C12 myotubes with conditioned medium from four pancreatic cancer cell lines (PK-45H, PANC-1, PK-1, and KLM-1) showed that tumor-derived factors from 2D pancreatic cancer cell lines inconsistently induced NF-κB or atrophy signaling in mature C2C12 myotubes with small effects. Although tumor-derived factors from PDAC organoids did not induce skeletal muscle atrophy signaling in mature C2C12 myotubes, they significantly promoted myoblast alignment and fusion during differentiation, accompanied with significantly increased myomaker mRNA expression levels. In addition, tumor-derived factors from PDAC organoids transiently increased proliferation of C2C12 myoblast, together with significant downregulation of the satellite cell marker paired-box 7 and myogenic factor 5, and increased expression of muscle differentiation markers such as MyHC-IIB. In this study, the presence of cachectic phenotype did not impact myogenesis with tumor specificity, indicating that muscle damage may be consistent from the early stage of cancer without cachexia to the cachectic stage. Whether the cachexia phenotype is associated with more severe muscle damage needs further large study to verify due to the inclusion of only one pair of cachectic and non-cachectic organoids in their study.

4.4.4 The establishment and characterization of PDOs from intraductal papillary mucinous neoplasms (IPMN)

IPMNs that comprise nearly 50% of all resected pancreatic cysts are non-invasive and cystic pancreatic cancer precursors with the potential to progress into invasive pancreatic adenocarcinoma and the few *in vitro* models.[103] PDOs provide an ideal preclinical platform to model and study the biology of IPMNs and their progression to invasive carcinoma.

Huang et al.[104] established a living biobank of organoids derived from 7 normal pancreatic ducts and 10 IPMNs with low- and high-grade dysplasia. Whole genome sequencing and transcriptome sequencing identified prevalent somatic mutations in *KRAS*, *GNAS*, *CDKN2A*, *TP53*, and *RNF43* and 13 other genes with recurrent somatic mutations in IPMN, as well as 28 genes that were differentially expressed between normal duct organoids and IPMN organoids. The most significantly upregulated and downregulated genes were CLDN18 and FOXA1, respectively. Further immunohistochemical analysis demonstrated that the FOXA1 protein expression decreased in low-grade IPMNs, mucinous cystic neoplasms, and PDAC, indicating that the FOXA1 loss of expression occurs early in pancreatic tumorigenesis.

Subsequently, Beato et al.[100] reported another living biobank of PDOs from IPMN specimens and paired normal tissue with a success rate of 81% (13/16) and 87% (13/15), respectively, covering distinct grades of differentiation, including three low-grade IPMNs, two moderate-grade IPMNs, seven high-grade IPMNs, and three IPMNs associated with invasive carcinoma as well as distinct epithelial subtypes, including pancreatobiliary, intestinal, and mixed subtypes. Compared with frozen resected tissue, PDOs from fresh resected tissue grew at a faster rate, while PDOs from high-grade IPMNs grew at a slower rate than low and intermediate/moderate grade. IPMN organoids replicated the biomarkers and morphological characters of parental tumors, in which a low grade IPMN organoids showed a central core surrounded by neoplastic cells while a high grade IPMN organoids grew in clusters. In addition, IPMN organoids recapitulated the genomic characteristics of the parental IPMN tumor and had the somatic mutations in key genes including *KRAS*, *GNAS*, *RNF43*, and *BRAF*, showing the potential of PDOs to identify the genetic alterations commonly found in a pre-malignant disease state.

Compared with PDAC, IPMN and IPMN with an associated invasive carcinoma organoid showed different expression profiles with *GNAS*, *RNF43*, or *KLF4* mutations.[105] Transcription factor footprint analysis and functional assays demonstrated that MNX1 and HNF1B were biologically indispensable for IPMN lineages, which was specifically marked by upregulation of MNX1. The MNX1-HNF1B axis governed a set of genes, including MYC, SOX9, and OLFM4, which are known to be essential for gastrointestinal stem cells.

4.4.5 The coculture models of PDOs with various cells

Coculture models of PDOs with various cells contribute to the study of cell–cell interaction. Tsai et al.[106] developed a simple 3D organotypic microenvironment models by coculturing PDAC organoids with human cancer-associated fibroblasts and T cells, finding that T cells infiltrated the Matrigel and migrated toward the organoids in a tumor-dependent manner, while cancer-associated fibroblasts seemed to contribute to the resistance to gemcitabine. The symbiosis of fibroblasts with patient organoid cells increased collagen deposition sixfold and corresponding tissue stiffness compared with fibroblast free controls.[107]

The coculture of PDOs with peripheral blood mononuclear cells (PBMC) resulted in the generation of organoid-primed cytotoxic T (opT) cells with a memory phenotype whose expansion, cell surface markers, and cell types, had patient-specific differences.[108] Compared with naïve PBMCs, the activated features and antitumor cytotoxicity of opT cells were significantly increased. The coculture increased the expression of checkpoint inhibitors on opT and ligands for the immunomodulatory receptors on tumor cells. Compared with blocking other ligands, the IFN-γ production most potently increased with NKG2A blockade after stimulation, demonstrating the ability of the coculture platform to identify checkpoint inhibition strategies for effective immunotherapy. TCRs identified on opT cells demonstrated a dramatic clonal expansion of T and conferred tumor recognition and cytotoxic ability to engineered primary human T cells.

The coculture models of PDAC organoids with cytotoxic T lymphocyte (CTL) showed that polymorphonuclear myeloid-derived suppressor cells not only promoted tumor growth, but also suppressed CTL proliferation, resulting in diminished efficacy of checkpoint inhibition, while its depletion maximized the effect of checkpoint inhibition.[109] In the coculture model with endothelial cells, PDAC organoids maintained cancer-initiating cells by interacting with endothelial cells *via* Wnt and Notch pathways.[99] Compared with indirect coculture, the direct coculture with pancreatic stellate cells caused the loss of ductal structures in organoids, with more organoids invaded into collagen and increased invasion area of PDAC1 organoids.[110] 3-D bioprinting[81] and organ-on-a-chip system[111] can help the organoids to model the more complete tumor microenvironment and perfusable vasculature, overcoming the limitation of the lack of diverse stroma and immune cells and vascular system in PDOs.

4.4.6 The application of PDOs from PC in treatment

PDO as a model has been widely used to evaluate the therapeutic efficacy of pancreatic cancer. To assess the sensitivity of PDOs to various chemotherapeutic agents, Driehuis et al.[93] exposed 24 PDOs to a panel of 76 therapeutics, including chemotherapeutics currently used in the treatment of PDAC. They found that specific targeted therapies were effective in only a subset of patients, and highlighting the importance of personalized approaches for effective cancer treatment. In addition, they used pancreatic PDOs to confirm that protein arginine methyltransferase 5 inhibitor EZP015556 was effective in both methylthioadenosine phosphorylase $(MTAP)^{-}$ and a subset of $MTAP^{+}$ PDOs, both with high methylthioadenosine levels.

Tiriac et al.[90] demonstrated that PDO pharmaco-typing was corresponded with individual patient treatment responses and clinical outcomes. Longitudinal PDO sampling following repeat biopsies showed that PDOs replicated the spatial intra-patient heterogeneity of chemo-sensitivity for pancreatic cancer as well as temporal evolution and could be used to evaluate the acquisition of resistance mechanisms to first line chemotherapeutic regimens. They found the organoid-based gene expression signatures of chemo-sensitivity that predicted improved responses for many patients to chemotherapy in both the adjuvant and advanced disease settings. Finally, they nominated alternative treatment strategies for chemo-refractory PDOs using targeted agent therapeutic profiling. These studies indicate that combined molecular and therapeutic profiling of PDOs may predict clinical response and enable prospective therapeutic selection.

They further established a framework for rapid pharmaco-typing of PDOs across a multi-institutional consortium of academic medical centers.[91] They demonstrated that there were similar drug sensitivity between early and late passage of PDOs as well as 25 and 1000 cells per well, after exposing to gemcitabine, paclitaxel, irinotecan. Rapid pharmaco-typing with a median of 48 days can expedite the interval between tumor tissue acquisition and *ex vivo* pharmaco-typing to fit within a clinically meaningful timeframe. PDOs showed significant inter-patient heterogeneity in chemosensitivity to gemcitabine, paclitaxel, irinotecan, 5-Fluorouracil, and oxaliplatin. Due to the faster drug-induced changes in cell metabolism than tumor size or overall cell viability, non-invasive optical metabolic imaging as a fast, dynamic method has been suggested to evaluate heterogeneous drug response at the organoid to speed up the determination of optimal treatment.[112]

Although neoadjuvant FOLFIRINOX-treated organoids had a similar treatment response to 5-FU as naïve organoids, they were resistant to irinotecan and oxaliplatin.[96] Global mRNA-seq analyses showed that the main FOLFIRINOX resistance pathways involved are genomic instability, energy metabolism, and innate immune system, providing targetable pathways to overcome the chemoresistance. The establishment of triple drug resistant PDO to FOLFORINOX treatment will further help the discovery of acquired chemoresistance mechanisms.[113] Tumor-specific targeted delivery of 5-FU using EGFR aptamers has been demonstrated to overcome 5-FU resistance in PDOs, with the potential to target virtually any cancer entity sensitive to 5-FU treatment.[114]

Perfusion of gemcitabine into a stiffened matrix of 3D perfused vascular network did not have the dose-dependent effects on tumor viability as those under static conditions, highlighting the value of organ-on-a-chip vasculature in drug screening.[111] PDAC organoids have been used to identify the superior combinatorial effect of magnetic field and radiotherapy than monotherapy.[115] Recently, Raimondi et al.[116] demonstrated that PDOs of PDAC can identify therapeutic responses to oncolytic adenoviruses. They further described the detailed protocol for tumor organoid adenoviral infection and functional assessment of patient-specific responses to oncolytic adenoviruses,[117] greatly promoting the application of PDOs in preclinical testing of oncolytic adenovirus sensitivity.

4.5 Colorectal cancer

Colorectal cancer (CRC) ranks third in terms of incidence, but second in terms of mortality, estimating globally 1.9 million new cases and 935,000 deaths in 2020[1] with an increasing trend year by year, especially in patients <50 years old.[118] CRC can be divided into two subtypes: Microsatellite Stable (MSS) with chromosomal instability, accounting for 85%, Microsatellite Instability (MSI) with the defective function of mismatch repair DNA system, accounting for 15%.[119] The multiple different genetic mutations associated with CRC bring tremendous challenge for effective treatment as no single specific molecular therapy is effective for all patients.[120] Organoids may be a powerful weapon to win the fight against one of the world's deadliest cancers.

4.5.1 The establishment and characterization of PDOs from CRC

Sato et al.[19] preformed robust and long-term culture of primary human epithelial cells isolated from small intestine, colon, adenocarcinomas, and Barrett's esophagus by adding nicotinamide, A83-01 (Alk inhibitor), prostaglandin E_2, and the p38 inhibitor SB202190, after they first established crypt villus organoids from mouse small intestine.[18] Then, they established the cancer organoid biobank comprised of 22 CRC organoids and 19 normal-adjacent organoids derived from surgically resected tissues from 20 untreated CRC patients.[121] All but two organoids could be readily expanded and frozen with cell survival typically >80% upon thawing. Both normal organoids and tumor organoids replicated the histological morphology of parent tissues. Tumor organoids well represented the most commonly altered genes in CRC, such as inactivating alterations to the tumor suppressors *APC, TP53, FBXW7* and *SMAD4*, as well as activating mutations in *KRAS, PIK3CA, BRAF,* and *TGFBR1/2*. They further established a human CRC organoid expression profile by quantitative proteomics and transcriptomics, revealing strong patient-specific features in which a single patient with a hypermutated phenotype in CRC was characterized by MSI.[122]

Weeber et al.[123] demonstrated CRC organoids could be established from metastatic tumor biopsies with a success rate of 71%. PDOs replicated 90% of the somatic mutations of parental tumors and had DNA copy number profiles with a correlation of 0.89 to original tumor.

With improved culture conditions, Fujii et al.[29] were able to propagate organoids from CRC tissues with 100% efficiency and established a large colorectal tumor biobank consisting of 41 counterpart organoids from normal colorectal mucosa and 55 colorectal tumor organoid lines from a range of histological subtypes and clinical stages, including rare subtypes, such as metastatic CRC, hyperplastic polyps, sessile serrated adenoma/polyps, tubulovillous adenoma, neuroendocrine carcinoma, mucinous adenocarcinoma and poorly differentiated adenocarcinoma. *In vitro* and in xenografts, PDOs replicated the histopathological grade and differentiation capacity of parental tumors.

Distinct patterns of genetic mutations exist among CRC organoids, consistent with that of The Cancer Genome Atlas (TCGA) database. Tumor primary and metastatic CRCs organoids from different tumor regions within the same individuals shared driver gene mutations, indicating that the driver gene mutations preceded metastasis development. They evaluated the niche factor requirements for each tumor organoid and found the decreased niche factor requirements in colorectal tumors during the normal-adenoma and adenoma-carcinoma transition, which was predominantly determined by genetic mutations and contributed to local tumorigenicity. Specifically, most tumor organoids that harbored *APC, CTNNB1,* and *TCF7L2* mutations of Wnt signaling pathways and *SMAD3, SMAD4,* or *TGFBR2* mutations of TGF-β signaling pathways did not require Wnt pathway stimulation

(Wnt3A/R-spondin 1) or TGF-β pathway inhibition (A83-01). However, organoids with RAS/MAPK pathway mutations alone were insufficient for growth without EGF stimulus due to the ligand-mediated EGFR internalization, and required a p38 inhibitor as an EGF receptor-specific stabilizer for stable culture.

Ganesh et al.[124] established 65 CRC organoids from biopsies of 41 patients with an overall success rate of 77% (65/84), as well as 51 normal rectal organoids from normal adjacent tissue. They could generate PDOs from patient with or without undergoing therapeutic intervention, chemotherapy and/or radiotherapy before surgical resection, indicating that clinical therapy did not predict the success or failure of organoid establishment. CRC organoids retain molecular features of parental tumor, as well as the invasive ability of tumor formation and metastasis after transplantation into mice. In addition to surgical specimens and biopsies, PDOs can be derived from circulating,[125] ascites, and peritoneal metastases[21] CRC tumor cells. These studies show that CRC organoids can be established from a variety of tumor tissues of patients with CRC, with a high success rate, and accurately recapitulate the important characteristics of parental tumors.

4.5.2 The variety of heterogeneity in CRC recapitulated by PDOs

Although recapitulating the major histologic, genomic, and transcriptomic characteristics of parental tumor, the CRC organoids derived from multiple regions of single tumors presented genetic and transcriptional heterogeneity and strong variance in drug responses, suggesting the widespread subregional heterogeneity of CRC in patients.[126]

The metabolic analysis of PDOs showed distinct metabolic phenotypes between cancer stem cells and differentiated cancer cells in CRC. Interestingly, lactate from differentiated cancer cells promoted self-renewal of cancer stem cells and thus contributed to CRC progression.[127]

Because of the heterogeneity of CRC, PDOs cannot be established from all types of CRC. Compared with other types of CRC, tumors characterized as MSI-mutated, poorly differentiated, and/or of a mucinous type, were more difficult to establish PDOs, suggesting the biological difference in tumor growth, which may depend on factors associated with inflammation and pathogen exposure. In addition, TCGA data indicated that patients with organoid-forming tumors tend to have a worse prognosis, reflecting the clinical difference.[128]

Okamoto et al.[129] established 72 PDOs from 21 stage IV CRC patients by collecting synchronous metastatic tumors as well as recurrent lesions.

Transcriptomic analysis showed that OLFM4 was the most differentially expressed gene between primary and metastatic CRC organoids, while single cell RNA-seq analysis indicated that metastatic/recurrent PDOs had fewer stem-like cell and differentiated cell clusters and more proliferating cell clusters compared with primary organoid, showing the cellular heterogeneity of metastatic CRC. They identified OLFM4 was the gene most strongly associated with a stem-like cell cluster, and found OLFM4 cells had the capacity to initiate organoid culture growth and differentiation in primary PDOs. In addition, CRC organoids from metastatic lesions had more aggressive phenotypes, tumorigenesis, and metastatic capacity compared with those from primary lesions in the same patient, which may be associated with SOX2 gene.[130]

Although CRC organoids cannot recapitulate the tumor immune microenvironment of primary tumors, such as lymphocytes, fibroblasts, and blood vessel components, they enabled the identification of the heterogeneity of immunogenic properties.[131] HLA peptide presented great heterogeneity between clonal cell of single-cell derived tumor organoids even from the same patient with CRC, suggesting that a multipeptide vaccination approach against highly conserved tumor suppressors may only benefit patients with low tumor mutational burden.[132]

The methods to comprehensively analyze cell-type specific post-translational modification of signaling networks in organoids have also been well described.[133] The different exosomal miRNAs extracted from organoids from before and after endoscopic resections in patients with colorectal adenoma may have the potential to be a diagnostic biomarker for colorectal adenoma.[134]

4.5.3 The establishment and characterization of PDOs from rare tumor

PDOs can be used to expand, characterize and preserve the small amount of tissues that are available in rare tumors, such as malignant dedifferentiated colon signet ring cell carcinoma (SRCC) that is not noted above and difficult to directly culture as 2D cell lines. Li et al.[135] first established a human colon SRCC organoid from a surgical sample of primary tumor from a 43-years-old Chinese female patient with colon SRCC. Different with other adenocarcinoma organoids, SB202190, gastrin, PGE_2, and A83-01 did not improve the growth of SRCC organoid, while NICO obviously inhibited SRCC growth. Unexpectedly, 10% fetal bovine serum greatly promoted the growth of SRCC organoid instead of cell aging. The SRCC organoid recapitulated the histological and growth characteristics of the original tumor, with a high tumor formation rate in xenografts. Combined with DNA sequencing, the drug screening on the *in vitro* organoid and the *in vivo* xenograft identified the effective JAK2 inhibitors, AT9283 and Pacritinib. They further found that enhanced glutamine utilization mediated by SLC1A5 and GPT2 is an essential metabolic feature of colorectal signet ring cell carcinoma with therapeutic potential.[136]

4.5.4 T cells and CRC organoids

As T cells are a key player in cancer immunotherapy, strategies to expand tumor-reactive cells and explore their interaction with tumor cells at the level of an individual patient is of great significance. Dijkstra et al.[59] demonstrated that cocultures of autologous tumor organoids and peripheral blood lymphocytes can be used to enrich tumor-reactive T cells from peripheral blood of patients with mismatch repair-deficient colorectal cancer and these T cells can be used to assess the efficiency of specific killing of matched tumor organoids. They further described the detailed protocol for the generation and functional assessment of tumor-reactive cells based on cocultures of MSI CRC organoids and autologous peripheral blood lymphocytes,[61] providing a means to assess the sensitivity of tumor cells to T cell-mediated attack at the level of the individual patient. The coculture immunotherapy models of PDO and T cell showed that the heterogeneity and plasticity of carcinoembryonic antigen conferred resistance to cibisatamab, the bispecific immunotherapy antibody targeting carcinoembryonic antigen, in CRC PDOs.[137]

The ALI method enabled the CRC organoid to preserve native immune cells, such as T cells, B cells, NK cells, and macrophages.[60] By using the ALI PDO models, it was found that autocrine type I IFNs from the activated T cells inducted by anti-PD-1 act synergistically with TNF-α to induce myeloid-derived suppressor cells apoptosis (MDSC) and suppress MDSC function, while blockade of TNF-α and IFN-α/β abolished the efficacy of anti-PD-1 treatment in a CRC mouse model, indicating that reprogramming MDSC by TNF-α and IFN-α/β might be an effective therapeutic strategy to promoted the response to anti-PD-1 therapy in CRC.[138]

Combined with cell lines, syngeneic orthotopic mouse models, human CRC organoids were used to define the effects of tumor cell-secreted extracellular vesicles on antitumor immune response.[139] Data showed that these extracellular vesicles regulated T cell costimulation, leading to immune checkpoint blockade resistance, and could be manipulated to enhanced T cell-mediated antitumor immune response in CRC models to increase the immune checkpoint blockade response.

4.5.5 Fibroblast and CRC organoids

As fibroblasts contribute to tumor progression and drug resistance, exploring the exact mechanism of fibroblast in CRC is very important. The lack of fibroblasts in PDO models is a major limitation, which can be partly addressed by advanced models such as ALI mentioned above,[60] or coculture with fibroblasts.[140] The coculture of tumor organoids with CAFs demonstrated that the Wnt-induced phenotypic switch in CAFs inhibited epithelial–mesenchymal transition in CRC.[140] It was found that CAFs increased the frequency of tumor-initiating cells, and this effect was enhanced by TGF-β signaling. By using PDOs and xenografts, it was shown that blocking the cross-talk between cancer cells and the microenvironment by TGF-β signaling inhibitors halted the progression of CRC.[137]

Interestingly, fibroblast-derived extracellular vesicles could induce cell proliferation in EGF-dependent PDOs, which was mainly caused by transmitting amphiregulin.[141] In addition, fibroblast-derived extracellular vesicles induced the colony formation of CRC organoid cells, even under hypoxia, while tumor cell-derived extracellular vesicles had no significant effect on the activation of fibroblasts.[142] Compared with CAFs from mature-type desmoplastic reaction, CAFs from tumor with immature-type desmoplastic reaction significantly increased growth of CRC organoids, which may be caused by ADAM9 in CAFs, implicating deteriorating survival in CRC patients.[143]

CAFs promoted the proliferation, migration, and invasion of colon cancer cells *via* minimal glycolysis. Inhibition of glycolysis impaired the survival and growth of CRC organoids and decreased tumor growth and intraperitoneal dissemination *in vivo*, indicating that enhancing fatty acid catabolism in CAFs drove CRC metastasis.[144] By coculturing with CAFs, CRC organoids even re-expressed REG family and DUOXs genes, which are known to promote the proliferation, anti-apoptotic state, and drug resistant phenotypes of tumor cells.[145]

4.5.6 The application of PDOs from CRC in treatment

The large drug screening for 83 compound library including the anti-EGFR antibody cetuximab, used clinically for KRAS/NRAS/BRAF wild-type CRC, as well as oxaliplatin and 5-FU, first line chemotherapeutics for CRC treatment, showed huge differences in the response of heterogeneous

PDOs to treatment, highlighting the value of PDOs in personalized treatment.[121] None of the mutations were found exclusively in either the tumor or organoids were in genes amenable for drug targeting, further supported to use PDOs as an *ex vivo* platform in personalized treatment.[123] The *ex vivo* responses of CRC organoids to clinically relevant chemotherapy and radiation treatment were correlated with the clinical responses noted in individual patients' tumors.[124]

Optical metabolic imaging of these organoids, which contributed to resolve the therapeutic differences between various patient samples by providing cell-level quantification of treatment response and tumor heterogeneity, allowed prospectively predicting treatment response for patient with CRC.[146] The combination of PDO tests and clinical trial data confirmed that the chemo-radiation responses in CRC organoid were highly matched to that in patients with 84.43% accuracy, 78.01% sensitivity, and 91.97% specificity.[147]

However, in a prospective clinical study, PDO tests predicted response of the biopsied lesion in more than 80% of patients treated with irinotecan-based therapies, while they failed to predict outcome for treatment with 5-FU plus oxaliplatin.[148] The drug resistance of 5-FU may be associated with the upregulation of revival stem cell marker genes of CRC organoids following chemotherapeutic treatment,[149] which may be involved in the p53-mediated Wnt/β-catenin pathway activation,[150] demonstrating that the combination therapy of Wnt inhibitor and 5-FU may be an effective strategy to improve the poor outcomes of current 5-FU-based therapies for CRC patients.

Application of oxaliplatin-resistant CRC organoids found that Lnc-RP11-536K7.3/SOX2/HIF-1α signaling axis regulated oxaliplatin resistance by promoting proliferation, glycolysis, and angiogenesis.[151] Integrated chromatin and transcriptomic profiling of CRC PDOs contributed to identify personalized drug targets to overcome oxaliplatin resistance.[152] Unexpectedly, DNA mismatch repair deficiency status seemed to be not associated with the response of PDOs to any of the drugs tested, including 5-FU, oxaliplatin, SN-38, binimetinib, encorafenib, and cetuximab, although mutant KRAS and BRAF impacted the intrinsic sensitivity of tumor cells to chemotherapy and targeted therapy.[153]

Quantifying single cell extracellular signal-regulated kinase dynamics in CRC organoids by biosensor demonstrated that EGFR was an amplifier of oncogenic MAPK pathway signaling, providing a mechanistic understanding of the effectiveness of EGFR inhibitors within combination therapies against KRAS and BRAF mutant CRC.[154] As EGFR blockade reverted resistance to KRAS inhibition in colorectal cancer, the combinatorial targeting of EGFR and KRAS was highly effective in CRC organoids.[155] Moreover, RAC1B inhibition sensitized cetuximab resistant human CRC organoids to the effects of EGFR inhibition, providing a potential therapeutic target for improving the clinical efficacy of EGFR inhibitors in CRC.[156] PDOs can provide a sensitive platform to evaluate CAR efficacy and tumor specificity in a personalized manner for patients with CRC.[157] Combined with murine intestinal tumorigenesis models, patient-derived CRC organoids were used to demonstrate that *Lactobacillus gallinarum* modulated gut microbiota and produced anti-cancer metabolite, indole-3-lactic acid, to protect against colorectal tumorigenesis, providing a new line of thinking for the treatment of CRC.[158]

4.6 Liver cancer

Primary liver cancer (PLC) is the sixth most commonly diagnosed cancer and the third leading cause of cancer death worldwide in 2020, with approximately 906,000 new cases and 830,000 deaths.[1] Approximately 72% of all liver cancer arises in Asia, with China accounting for 47% of the world's burden.[159] Primary liver cancer includes hepatocellular carcinoma (HCC) (comprising 75–85% of cases) and intrahepatic cholangiocarcinoma (ICC) (10–15%), as well as other rare types.[1] Currently available treatment strategies for PLC are unsatisfactory. Given the limited efficacy of current PLC treatment strategies, there is an urgent need for new therapies for PLC. Indeed, the generation of PDOs from patients with PLC has been a major breakthrough in cancer biology. The application of PDOs can help to gain a better understanding and more personalized treatment of PLC.

4.6.1 The establishment and characterization of PDOs from adult PLC

After the realization of long-term expansion of epithelial organoids derived from single mouse $Lgr5^{+}$ liver stem cells[160] and healthy human liver organoid,[161] Broutier et al.[20] first established PLC organoids derived from surgically resected liver tumor tissue by optimizing their established human liver expansion protocol. The PLC organoids, including three HCC, three ICC, and two hepatocellular cholangiocarcinoma, with a success rate of 100%, recapitulated the histological characteristics, expression profile, genomic profile, and *in vivo* tumorigenesis of the parental tumor, even after long-term expansion in culture. They reported *C19ORF48*, *UBE2S*, *DTYMK* (for HCC) and *C1QBP* and *STMN1* (for ICC) as novel genes associated with poor prognosis for primary liver cancer. By using human ICC organoid models, it was found that inhibition of Wnt signaling could induce ICC cells to differentiate into functional hepatocytes, which reduced the malignant potential *in vitro* and *in vivo*, while Wnt3a derived from macrophages recruited to the liver by inflammation could promote the malignant transformation of mature hepatocytes to ICC cells.[162]

Nuciforo et al.[163] generated PDOs from tumor needle biopsies of HCC patients with various etiologies, such as viral hepatitis, nonalcoholic fatty liver disease, and alcoholic liver disease, and different tumor stages, with a success rate of 26% (10/38) and 33% (8/24) based on the number of biopsies or patient, respectively, as well as generating three PDOs from 5 patients with ICC. The success rate of HCC organoid generation was not associated with clinically relevant patient characteristics, while it was strongly correlated with the histopathological grades of the HCCs as all PDOs were derived from poorly differentiated tumors (Edmondson grades III and IV).

In addition, the KI-67 staining of tumor biopsies predicted a higher HCC cell proliferation. HCC organoids without immune cell infiltrates and tumor stromal cells recapitulated the growth pattern, differentiation grade and biomarkers of the parental tumors, as well as the genetic alterations, mutational signatures, intratumoral mutational heterogeneity and the ability to form *bona fide* tumors in mice. The establishment of PDOs from ICC needle biopsies overcomes a major limitation of that from surgically resected specimens as biopsies can be obtained not only from patients without treatment in early stages, but also from patients with frequent systemic therapies in intermediate and advanced tumor stages. The lower success rate here than that in other cancers above might be because the origin hepatocytes of organoids lack the features of epithelial stem cells that favor their propagation in the organoid culture system.

Zhao et al.[164] established seven organoids from different primary hepatobiliary tumor patients, including four HCC organoids, two ICC organoids, and one gallbladder cancer organoid, by the addition of EGF, FGF2, FGF10, hepatocyte growth factor, Wnt agonist R-spondin 1, TGF-β inhibitor Noggin, and ROCK inhibitor Y-27632. As expected, cancer stem cell-like markers, such as CD133, CD44, EPCAM, CD13, SOX9, OCT4, varied greatly among different organoids, of which HCC272 with high status of epithelia–mesenchymal transition had broad spectrum drug resistance due to the abnormal enrichment of CD44 cells. By exploring coexpressed genes, trajectory, and pseudo-time analysis, they defined the CTNNB1-enriched subpopulation as the proliferation advantage cluster and the GAPDH-enriched as the metabolism advantage cluster, which contributed to HCC272 drug resistance by accelerating the usage of glucose, enhancing hypoxia-induced HIF-1 signaling, and causing the upregulation of NEAT1 in CD44 high cells. This study indicates that PDO can help us to characterize the cancer stem cell heterogeneity and understand the distinctive metabolic circuitry in resistant subpopulations, which may contribute to predict the therapeutic response.

4.6.2 The establishment and characterization of PDOs from childhood hepatoblastoma

Hepatoblastoma is the most common childhood liver cancer, accounting for 67% of malignant liver tumors in children, with more than 90% occurring in children under 5 years of age.[165] Saltsman et al.[166] successfully established six tumor organoid lines from three patients with aggressive hepatoblastoma. Hepatoblastoma tumor organoids recapitulated the key features of patient tumors, including tumor architecture, mutational profile, gene expression patterns, and features of Wnt/β-catenin signaling, features central to the pathogenesis of hepatoblastoma. In addition, they developed two murine patient-derived xenografts from patient samples and found that hepatoblastoma organoids shared similar biology with PDX models of hepatoblastoma. Drug screening of PDOs showed that cisplatin was not effective for hepatoblastoma organoids, but one drug, JQ1, had therapeutic potential for hepatoblastoma, while some drugs (a674563 and AZD3463) resulted in greater reduction in tumor organoid as well as significant death in normal organoids.

4.6.3 The coculture models of ICC organoids with CAFs

Liu et al.[167] successfully established the coculture models of ICC organoids with CAFs after the establishment of four human ICC organoids from resected tumors and the isolation of CAFs from tumors of HCC and ICC patients. After coculturing for 14 days, CAFs enlarged the size of ICC organoids instead of the number, at least partially through paracrine signaling, while ICC organoids led to the increased expression of CAF markers in CAFs, such as CAV1, Alpha-SMA, CD29, FAP, and COL1A1. In addition, CAFs promoted the growth of ICC organoid-formed tumors *in vivo*, as well as the resistance to sorafenib, regorafenib and 5-FU *in vitro*. The coculture model partially improves the limitation of PLC organoids lack of stroma components, and lays a foundation for the coculture of other stromal cells, immune cells, and PLC organoids.

4.6.4 The application of PDOs from PLC in treatment

Sorafenib reduced HCC organoid growth *in vitro* in a dose-dependent manner as well as rare lympho-epithelioma-like ICC organoids, indicating PDOs from PLC can be used to test tumor-specific sensitivities of drug.[163] The sensitivity test of PLC organoids for 29 anti-cancer compounds showed that they were resistant to the majority of the compounds, while some drug sensitivities were associated with mutational profiles in PLC organoids, such as *CTNNB1* gene mutations caused the porcupine inhibitor LGK974 resistant, *KRAS* mutations caused AZD8931 resistant. PLC-derived organoids not only were amenable for drug screening testing, but also for new targets identification, leading to the identification of the ERK inhibitor SCH772984 as a potential therapeutic agent for PLC.[20]

The screening results of 129 cancer drugs in 27 PCL organoid lines revealed great intratumor and interpatient drug response heterogeneity.[168] Although the vast majority of drugs were ineffective, 13 drugs showed more than 90% killing across all 27 liver cancer PDO lines, of which four drugs, including daunorubicin, ceritinib, doxorubicin, and mitomycin, had at least one PDO with suboptimal response, while nine drugs (histone deacetylase inhibitors, proteasome inhibitors, DNA topoisomerase II inhibitors, protein translation inhibitors, and RNA synthesis inhibitors) were effective for all PDO lines. They further expanded and established 40 HCC organoids, which were all sensitive to omacetaxine (an FDA-approved drug for chronic myelogenous leukemia) at nanomolar concentrations, providing a proof-of-principle basis for further clinical trials to evaluate omacetaxine in patients with HCC.[169]

CD44-positive organoid was not only resistance to trametinib,[164] but also to sorafenib due to hedgehog signaling.[170] GANT61 reversed the upregulation of CD44 and Hedgehog signaling on HCC organoids, indicating that the combination of sorafenib and Hedgehog signaling inhibitors might be effective in HCC patients with high CD44 levels.

4.7 Gastric cancer

Gastric cancer is a common health problem, ranking fourth deadliest disease worldwide.[1] It is anatomically divided into non-cardia gastric cancer and cardia gastric cancer (gastro-esophageal-junction adenocarcinoma),[171] mainly caused by *H. pylori* infection, Epstein-Barr virus infection, heritable pathogenic mutations, and environmental factors.[172] According to the analysis of TCGA research network, gastric cancer is divided into four tumor subgroups by deep sequencing analyses: chromosomal unstable tumors, microsatellite unstable tumors, genome stable tumors, and Epstein-Barr virus tumors.[173] Therefore, classification of gastric cancer based on molecular subtypes may be helpful in determining prognosis and personalized therapy.

4.7.1 The establishment of gastric organoids

Gastric organoids can be derived from adult stem cells of surgical specimens as well as induced pluripotent stem cells (iPSC). iPSC-derived organoids are induced by a step-by-step differentiation program from iPSC, while gastric organoids derived from adult stem cells always require the same identical medium. Therefore, iPSC-derived organoid establishment needs about 1–2 months, while the differentiation of adult stem cells into organoids takes only 7–14 days.[174]

The first human stomach organoid model was reported by McCracken et al.,[175] who induced human pluripotent stem cells to differentiate into gland-type and pit-type gastric organoids, which contained LGR5-expressing stem cells, surface cells, antral mucous cells, and a diversity of gastric endocrine cells. Based on the analysis of temporal and spatial expression of molecular markers, the organoids and human fetal stomach tissue shared several critical stages, such as posterior foregut patterning, specification of the fundic and antral domains of the stomach, and the growth of a pseudo-stratified epithelium into an elaborate glandular epithelium.

Furthermore, Bartfeld et al.[176] proposed a modified organoid medium for gastric cancer organoids that could maintain stable passage for more than 1 year with staying cystic and without morphological abnormality, chromosomal aberration or marker loss. They demonstrated that EGF, Noggin, R-spondin 1, and WNT were essential components in traditional organoid media, and the inhibition of FGF10, gastrin, and TGFβ could greatly extend the lifespan of organoids to 10–20 weeks, while nicotinamide, which contributed to organoid establishment but later limited the lifespan of the cultures, was recommended to be removed after successful establishment of organoids.

Meanwhile, addition of nicotinamide and withdrawal of WNT could guide gastric stem cells to differentiate into gland or pit lineages. The four complete types of gastric lineages ($MUC5AC^{+}$, $MUC6^{+}$, PGC^{+}, SST^{+}) organized into gland and pit domains after adding all the above modified medium, while only gland type ($MUC5AC^{-}$) were generated by adding only nicotinamide into the above modified medium, or a large number of pit cells (high expression of MUC5AC and low expression of MUC6 and PGC) were generated after removing only WNT from the above modified medium.

Overall, the establishment and improvement of gastric normal and cancer organoid systems is conducive to understand the mechanism of gastric development and the occurrence and development of gastric cancer.

4.7.2 The biobank of PDOs from gastric cancer

In recent years, biobanks of patient-derived gastric cancer organoids have been well reported. Seidlitz et al.[177] successfully established a biobank containing 20 human gastric cancer organoid cultures and conducted the deep genomic characterization of four organoid lines. Different molecular genotype organoids had different morphological phenotypes. For example, corpus-like organoids had a cystic structure with a multilayered epithelium while antrum-like organoids had a compact morphology without lumen.

In addition, Nanki et al.[178] generated a biobank consisting of 37 patient-derived organoid lines, including rare genome-stable tumors. Furthermore, Yan et al.[179] established a large and unique biobank, covering nearly all molecular subtypes and subtype-specific mutational profiles, including organoids carrying EBV, RHOA mutations, CLDN18-ARHGAP6 fusion and CTNND1-ARHGAP26 fusion, all of which had been rarely described in cell lines or organoids so far. Importantly, although most newly diagnosed gastric cancer patients have advanced or metastatic disease, this biobank collected organoids from four stage I and one pre-invasive lesion, as well as from different regions of the tumor.[173] Parallel exome sequencing analysis confirmed that these organoids reproduced most of the key driver markers of the original tumor. These studies suggest that organoid biobank can serve as valuable tools for providing a human cancer resource to inquire about the functional roles of key driver genes and their consequences.

4.7.3 The infection model of PDOs

It has been found that there are 228 items of human diseases (particularly gastric cancer) related to *H. pylori* infection, which not only directly damages host cells, but also disorganizes the gastric environment, triggers an inflammatory cascade and alters the microbiome.[175] These findings highlight the importance of studying the mechanism between *H. pylori* infection and gastric cancer.

Gastric organoids can used to study the relationship between *H. pylori* infection and different lineages of gastric cells. It was found that after microinjection, *H. pylori* was more inclined to locate and colonize in the gastric gland domains rather than the pit domains as well as provoking more strong immune response with high expression of IL8.[176] iPSC-derived gastric organoids were used to simulate the acute response of the gastric epithelium to *H. pylori* infections in 24 h, which caused a robust epithelial responses including c-Met phosphorylation and epithelial cell proliferation increasing through virulence factor CagA of *H. pylori*.[175]

Moreover, gastric organoids were used to elucidate how *H. pylori* is located in the gastric epithelium. Huang et al.[176] demonstrated that urea, decomposition product of gastric epithelial metabolites, activated the chemical receptor TlpB of *H. pylori*, which caused *H. pylori* to move to a favorable ecological niche and rapidly destroy the environment of the colonization area because of the ability of TlpB to detect the concentrations of urea to as low as 50 nM and the effective degradation of urease secreted by *H. pylori*.

All of these advantageous characteristics make organoids a useful technology. Combined with genome sequencing, transcriptome sequencing, and proteome analysis, the molecular characteristics of organoids can be clearly understood in temporal and spatial expression. Organoids can also be genetically manipulated by lentivirus, bacterial artificial chromosome,[180] or CRISPR/Cas9 systems.[181] The application of organoids in bacterial or viral infection plays an important role to study the association between infectious diseases and cancer.

4.7.4 The application of PDOs in drug response testing

Large-scale drug screening revealed the heterogeneity of drug response in different regions from the same tumor. PDOs from gastric cancer were sensitive to unexpected drugs that were recently approved or in clinical trials, including Napabucasin, Abemaciclib, and the ataxia telangiectasia and Rad3-related protein (ATR) inhibitor VE-822, indicating that gastric cancer organoids have the ability to recapitulate the similar response from clinical trials and can be used for *in vitro* drug screening.[179] Trastuzumab was identified to effectively treat gastric cancer organoids with abnormally amplified and mutated *ERBB2*, producing an accumulative therapeutic effect by combining with 5-FU treatment.[177] Lapatinib inhibited MAPK and PI3K/AKT signaling downstream of EGFR/ERBB2 in *ERBB2*-amplified PDOs and induced apoptosis of organoids.[182] Two AKT inhibitors, MK-2206 and GSK690693, produced a strong response in *AKT1*-mutated gastric organoids.[182] Importantly, about 50% of cancer biopsies can be developed in organoids within 20 days,[174] which allows to test response to standard chemotherapy drugs in a short time, and quickly adopt further personalized treatment.

4.7.5 Organoids-on-a-chip

Various types of mechanical forces (e.g., traction forces between cells, fluid shear stress, and solid mechanical forces generated from cell masses[183]) act synergistically with signaling molecules and extracellular matrix to regulate organ development and maturation *in vivo*.[184] The development and application of organoid models are limited by the lack of biomechanical research.

To overcome this difficulty, Lee et al.[185] established a microfluidic imaging platform which embedded single human iPSCs in the central chamber. The micropipettes were inserted into both sides of organoid separately and peristaltic pump was used to deliver the media containing visual FITC-dextran, nutrients or drugs for a long time *via* flexible tubing for simulating stomach contraction and luminal flow *in vivo*.

Jin et al.[186] developed a organoids-on-a-chip to coculture gastric, liver, and intestinal organoids in different compartment to evaluate interactions between organs in drug metabolism. By adding the primary bile acid chenodeoxycholic acid, it was found that intestinal organoids secreted paracrine factors, which resulted in low expression of bile acid synthase CYP7A1 in liver organoids, suggesting that there was effective cross-talk between these organoids. This platform relies on highly simple setups using rockers-induced media flow without pumps or tubes compared to other microfluidic chips which involve complex fluidic connections for creating dynamic flow,[187] so this microfluidic system can be developed flexibly into a testing platform enabling high-throughput culture and analysis. In addition, this chip simulated ECM environments by applying liver extracellular matrix hydrogel, facilitating organoid maturation and improving their functions. These multiorganoids-on-a-chips will play an important role in disease research and drug development.

4.8 Prostate cancer

Prostate cancer (PC) is the second most frequent cancer and the fifth leading cause of cancer death among men in 2020, with an estimated about 1.4 million new cases and 375,000 deaths worldwide.[1] For a disease as common as PC, relatively little is known about its etiology. PDO can help us more accurately model PC and better understand the disease pathogenesis and therapy response of PC.

4.8.1 The establishment and characterization of PDOs from PC

Gao et al.[188] first established seven PDOs from distinct tumor samples (L2 vertebral body, acetabulum, two retroperitoneal lymph node, pleural effusion, circulating tumor cells, salvage prostatectomy) of seven patients with advanced PC, with a low success rate of 15–20% (6/32 cancer bearing samples, except for circulating tumor cells). These PDOs were reproducibly maintained for 1–2 months for 70% of soft tissue tumor biopsies and 30% of bone biopsy biopsies, but many of these tumor cultures were overtaken by tumor-associated spindle cells or normal epithelial cells present in the biopsy material. These PDOs recapitulated the molecular diversity of PC subtypes, including *TMPRSS2-ERG* fusion, *SPOP* mutation, *FOXA1* and *PIK3R1* mutations, *SPINK1* overexpression and *CHD1* loss, as well as *in situ* histology of PC *in vitro* and *in vivo*. Loss of PTEN, TP53, and RB tumor suppressor pathways are the most common characteristic across these PDOs.

The low success rate of organoids establishment in their study may be mainly due to the small amount of input material and the presence of excess normal epithelial and stromal cells. Strategies to selectively isolate tumor from biopsies might partly overcome this problem. They further reported a detailed protocol to generate 3D prostate organoid cultures from healthy mouse and human prostates, metastatic prostate cancer lesions and circulating tumor cells, including the stepwise establishment and the fully defined serum-free conditioned medium, providing a reference for further exploring the establishment of PC organoids.[189] Whether this protocol can be used to effectively establish PC organoids remains to be further tested.

Beshiri et al.[190] established 20 PC organoid models derived from the LuCaP mCRPC PDX cohort. The PC organoids recapitulated the genotypic and phenotypic characteristics of PDXs, as well as the defining feature of AR-dependent gene expression and growth dependence. By using RNA and whole-exome sequencing, Karkampouna et al.[191] further demonstrated that both PDX tissues and organoids derived from PDX recapitulated the transcriptomic and genomic of parental tumors, including mutations in *BRCA2*, *CHD1*, *ACVR2A*, *RNF43*, *APC*, and *KMT2D*.

4.8.2 The improvement of culture conditions of PC organoids

Karthaus et al.[192] described an R-spondin 1-based culture method for long-term propagation of murine and human prostate epithelium and identified that EGF, FGF2, nicotinamide, PGE_2, and the TGF-β pathway inhibitor A83-01, were all essential to sustain prolonged growth of human prostate organoids. The culture conditions are another possible reason that cannot be ignored for their low success rate of PC organoid establishment.[188] Given this possibility, Beshiri et al.[190] systematically tested more than 20 modifications to prostate organoid conditions mentioned above using PC organoid models derived from PDXs. They found that none of the PC organoids could be maintained long-term under previously reported conditions, but 11 PC organoids were grown for >10 generations (~6 months) by removing SB202190 (a p38 MAPK inhibitor) and *N*-acetylcysteine.

Other modifications specific to individual LC organoid might support continuous growth, such as the addition of a GSK3β inhibitor and an LSD1/MAO inhibitor to organoids that expressed high levels of the Wnt pathway antagonist GSK3β and weakly expressed neuroendocrine-associated transcripts. Altogether, these studies highlight the crucial need to optimize culture conditions for the successful establishment of PDOs.

4.8.3 The establishment and characterization of PDOs from rare prostate cancer

Given the lack of existing preclinical models for rare prostate cancer phenotypes, Puca et al.[193] established 4 organoids from fresh tumor tissues from 25 patients with castration-resistant neuroendocrine prostate cancer (CRPC-NE), an uncommon and aggressive histologic variant of prostate cancer, with a success rate of 16%. CRPC-NE organoids had the typical pathology of neuroendocrine prostate cancer, characterized by the presence of small- to medium-sized round cells with fine chromatin pattern and nuclear molding, with the expression of classical neuroendocrine markers and lack of AR protein expression. In addition, PDOs recapitulated the genomic, transcriptomic, and epigenomic concordance of corresponding patient tumors, including overexpression of MYCN, PEG10, SRRM4, EZH2, SOX2, BRN2, and FOXA2, and low expression of AR signaling genes. By knockdown of EZH2 on organoids with short hairpin RNAs targeting EZH2, it was found that EZH2 was associated with CRPC-NE program dysregulation, but the suppression of EZH2 alone was insufficient to re-express AR in this late-stage AR-negative CRPC-NE state. This study indicated that PDOs can be used not only to model and preserve the rare cancer, but also to better understand the pathogenesis of rare cancer.

4.8.4 The establishment and characterization of PDOs from circulating tumor cells (CTCs)

As shown above, established PDO models of PC have been notoriously difficult, presumably because of the low proliferative capacity of PC cells as well as overgrowth of benign epithelial and stromal cells. Given the advantages of tumor cells obtained from liquid biopsy lacking normal epithelial cells, Mout et al.[22] were committed to the establishment of PDOs from CTCs in patient with PC. They prospectively screened 102 patients with metastatic prostate cancer (mPCa) and recovered 46% of CTCs from 40 patients with ≥5 CTCs/7.5 mL blood using optimized methods. Then they established 14 PDOs from the isolated and enriched CTC samples with a success rate of 35%.

The majority of these PDOs were of short-term, maintained for 6–8 weeks. One of the two organoids that could be maintained for over 6 months successfully yielded a stable organoids line. The established organoid line harbored genomic features that were frequently identified in mCRPC tumors, such as alterations of *AR*, *PTEN* and *ERG* genes. However, single cell sequencing suggested that the established organoid line probably did not recapitulate the genetic heterogeneity of parental tumor, which might be due to a selective pressure that resulted in *ex vivo* expansion of one clone. Their work suggested that blood samples provide a rich source of viable tumor cells, providing an important step forward in the application of CTCs to establish PDOs models. Nevertheless, it also identifies several challenges that require further optimization.

4.8.5 The determinants of PDO establishment outcome

Given that the factors that caused the low success rate of PC organoid establishment were unclear, which greatly impeded their use for modeling, characterization, and personalized medicine of PC, Servant et al.[194] reported the detailed outcome of the PC PDO establishment from a prospective cohort of 81 diverse PC samples, including radical prostatectomy samples (RP, $n=56$), transurethral resection samples of the prostate (TUR-P, $n=14$), and metastasis samples (MET, $n=9$). Compared with the combination of mechanical and enzymatic digestion, mechanical digestion alone led to more accurate phenotypic recapitulation of parental tumors and better maintenance for PDOs. They found that organoids derived from PCa patients initially grew with variable efficiency (55/56 PR *versus* 4/14 TUR-P *versus* 4/9 MET), which may depend on the source and the intrinsic characteristics of the samples. RP organoids initially grew with high efficiency and 63% of them were maintained long-term, while TUR-P organoids initially grew with moderate success and could not be maintained over long periods of time, MET were established with poor efficiency and low stability.

The histological parameters such as Gleason score, percentage of tumor cells, and proliferative index, were not associated with sustained organoid growth in the RP and MET cohort, while high percentage of tumor cells seemed to predict the better growth of TUR-P organoid. Organoids with clear PCa hallmarks were initially present but were progressively lost and overgrown by organoids with a benign-like phenotype in culture. Finally, they generated a long-term maintenance organoid model derived from a hormone-naïve lung metastasis, which displayed alterations in the PI3K/Akt and Wnt/β-catenin pathways and responded to androgen deprivation. This comprehensive study explored determinants of PDO establishment outcome, which may contribute to improvement of the low success rate of establishing PDO from patients with PC.

4.8.6 The application of PDOs in treatment

Although the low success rate of organoid establishment hinders the development of PDOs for personalized medicine, some achievements have been made. Gao et al.[188] confirmed that drug sensitivity was associated with the mutation profiles: AR amplified organoid was exquisitely sensitive to enzalutamide, and organoids with both *PTEN* loss and *PIK3R1* mutation were sensitive to both everolimus and BKM-120. *CHD1* (a homozygously deleted putative tumor suppressor gene) status might be a predictive biomarker for DNA damaging agents such as carboplatin or PARPi for mCRPC (metastatic castration resistant prostate cancers).[195] In addition, *BRCA2* loss contributed to the sensitivity of PC organoids to olaparib and carboplatin.[190] As shown that a modest number of drugs such as pozotinib and vandetanib were effective in killing CRPC-NE organoids, PDOs were also used for high-throughput drug screening for patients with CRPC-NE by a manner of patient-specific sensitivities.[193] The medium-throughput automated drug screening of 74 compounds, including routinely used PC standard-of-care drugs and different FDA-approved drugs with indications for other cancer types, showed that PC organoids were resistant to the majority of the tested drugs, while 14 drug compounds were significantly effective for them, in which multikinase inhibitors (ponatinib, sunitinib, sorafenib) were broadly effective for both PDX and PC organoids.[191]

4.9 Bladder cancer

Bladder cancer is the tenth most common malignancy worldwide, with approximately 600,000 new cases and 200,000 deaths in 2020.[1] Urothelial carcinoma accounts for 90% of bladder cancers and can be categorized two subtypes: non-muscle invasive bladder cancer (NMBIC) and muscle-invasive bladder cancer (MIBC), with unique pathological features and different molecular characteristics.[196] Bladder cancer is a very heterogeneous disease with a large number of mutations, in which NMIBC occurs mainly as papillary disease with frequent *FGFR3* mutations, whereas MIBC has a more diverse mutation spectrum as well as copy-number instability.[197]

Overall, 75% of patients with high-risk bladder cancer will experience the recurrence and advancement of cancer, or decease within 10 years of their diagnosis.[198] The common tumor models of bladder cancer are cell lines generated from MIBC, while NMBIC, the most common type of bladder cancer, has not been derived as cell lines successfully.[198] PDXs cannot really recapitulate the host tumor microenvironment, leading to the limitation of their application to simulate the immune response and test immunotherapy. Therefore, PDOs are the perfect models for both NMIBC and NIBC to study their disease progression and drug preclinical studies, with the urgent need to their establishment.

4.9.1 The establishment of PDOs from bladder cancer

Pauli et al.[199] established 56 PDO lines from a wide range of tumor types, including eight bladder tumor organoid lines that were not extensively characterized. Lee et al.[30] established 12 PDOs from different bladder cancer with a success rate of70% (12/17). These PDOs recapitulated the histopathological and molecular diversity of human bladder cancer, as well as tumor heterogeneity of parental tumors, and showed a spectrum of genomic changes that were consistent with tumor evolution in culture. Mullenders et al.[200] created a living biobank consisting of 77 bladder cancer organoid lines derived from 53 patients. They identified that both FGF7 and FGF10 were sufficient for the growth of human bladder cancer organoids. These bladder cancer organoids had vast histological and functional heterogeneity, with both basal and luminal bladder cancer subtypes as well as common bladder cancer mutations like *TP53* and *FGFR3*.

Our team established three bladder cancer organoid lines of basal and luminal subtypes and confirmed that these organoids retained the key characteristics of the original tumors, including histopathological morphology, genomic alteration, and specific marker expression profiles of parental tumors.[201] Based on 3D printing, Kim et al.[202] reconstructed bladder assembloids-organoids containing stromal fibroblasts, endothelial cells, immune cells, and muscle layers, exhibiting the characteristics of mature bladder in terms of cell composition and gene expression at the level of single-cell transcriptome as well as reproducing the dynamics of regenerative response of tissues *in vivo* to injury. These experiments show that PDOs from bladder cancer provide an effective tool to model and study oncogenic mutations.

4.9.2 The application of PDOs in drug test

Although bladder cancer has high mortality and carries a huge social and medical pressure, systematic treatment of bladder cancer has not changed essentially since 1990.[203] Cisplatin is the common chemotherapy for bladder cancer patients, but only 30–40% of them exhibit a response with this therapy.[204] Analyses of drug response using PDOs from bladder tumor showed partial correlations with mutational profiles, as well as changes associated with treatment resistance, and specific responses could be validated using xenografts *in vivo*.[30]

Tan et al.[205] reported SRT1720, an activator of Sirtuin 1, is the first identified new drug based on small molecule screening of bladder cancer organoids. It had dramatic anti-tumor effect by inhibiting the hypoxia pathway through deacetylating HIF1α by activating Sirtuin 1. Given the importance of the predictive biomarkers for drug sensitivity, Kong et al.[206] developed the first machine-learning framework to predict patient-drug responses by using pharmacogenomic data derived from bladder and colorectal cancer organoids with combination of gene modules and network-based approaches. The biomarkers identified by this method accurately predicted the drug responses of 77 bladder cancer patients treated with cisplatin. This system pointed out that tolerance to cisplatin is associated with low expression of genes in "amino acid synthesis and interconversion" pathway. These studies indicate that PDOs from bladder tumor represent a faithful model system for treatment response in the context of precision cancer medicine.

4.9.3 The application of PDOs in immunotherapy

The coculture system containing PDO and immune cells can be used to evaluate the efficacy of cancer immunotherapy and precision therapy, such as chimeric antigen receptor (CAR-T). Our team demonstrated that patient-derived bladder cancer organoids could be used to test the *in vitro* immune response of CAR-T cells targeting MUC1 by coculturing CAR-T cells with organoids, providing new insights for improving personalized immunotherapy for bladder cancer and other solid tumors.[201] Gong et al.[207] developed an acoustic droplet printer used in combination with a hydrophobic matrix to generate a large number of homogeneous and highly viable PDO, including various immune elements and tumor cells. Then tumor-responsive T cells were induced, followed by enhancing the killing efficiency of derived organoids. In summary, 3D bladder cancer organoid models provide a platform for testing and optimizing immunotherapies for solid tumors *in vitro*. Importantly, because bladder cancer organoids recapitulate the complexity and variety of tumor-associated antigens, the accurate assessment would be assessed for target reactivity, response threshold, and specificity of immune killer cells.

4.10 Other types of cancer

Ovarian cancer organoids have been successfully established from diverse samples including cisplatin-resistant tissues,[208] covering all main subtypes of ovarian cancer and recapitulating histological and genomic features of parental tumors.[209] In addition to assess DNA damage repair defects and their influence on immune-oncologic agents,[210] these organoids are used for drug screening and capture different tumor subtype responses to the gold standard platinum-based chemotherapy, including acquisition of chemoresistance in recurrent disease.[209] These PDO drug responses had a significant correlation with the clinical response[211] as well as the clinical outcome.[212]

Recently, our team has described a 3D culture system for the long-term expansion of patient-derived papillary thyroid cancer (PTC) organoid lines.[213] These PTC organoids preserve the histopathological profiles and genomic heterogeneity of the originating tumors. Drug sensitivity assays of PTC organoids demonstrate patient-specific drug responses, and large correlations with the respective mutational profiles. Estradiol was shown to promote cell proliferation of PTC organoids in the presence of estrogen receptor α (ERα), regardless of the expression of ERβ and G protein-coupled estrogen receptor, GPR30. In addition, it has been found that the iodine uptake pathway is functionally conserved in organoids derived from thyroid carcinomas,[214] which can be used to identify early the iodine-resistant patients to reduce radioactive iodine overtreatment and its many side effects for thyroid cancer patients.[215]

Head and neck cancer organoids have been generated from CTCs[216] and resected tumor specimens,[217] showing similar histological features and stem cell, epithelial, and mesenchymal marker expression of the original tumors.

Organoids from oropharyngeal and esophageal squamous cell cancer,[218] gallbladder cancer,[219] biliary tract cancer,[220] malignant rhabdoid tumor,[221] clear cell renal cell carcinoma[222] and childhood kidney cancers[223] have been established, awaiting further exploration.

5. Limitations and prospects

Although research during the past decade has made explosive progress and obtained many achievements, PDOs still have several limitations that hinder their (pre-)clinical implementation.

5.1 Standardization

There is a lack of standardized and defined culture protocols across cancer organoid studies, which significantly affected the recapitulation of patient tissues and the repeatability of experimental results. These non-standardizations include cancer tissue sources, subsequent processing, medium formulations, matrices sources, and identification of successful establishment. Standardization of PDOs is a problem that requires interdisciplinary efforts from clinicians, biologists, and engineers all over the world. The transparency of culture protocols and strengthening exchange and cooperation may help to accelerate the solution of this problem.

5.2 Cost

As a new drug screening model, PDOs cost less than PDX, but it is still much higher than cell lines. The expensive growth factors and conditioned medium that must be used to promote cancer cell growth as well as matrix gels that must be used to support the 3D structure of PDOs are a considerable burden for long-term maintenance and high-throughput applications. Exploring the growth factors necessary for the growth of different PDOs and reducing redundant growth factors will help to reduce costs. The development of non-animal source Matrigel is key problem that need to be solved in the standardization and later implication of PDOs.

5.3 Efficiency

At present, the studies of PDOs are mainly focused on common and highly malignant tumors, whose tissues are easier to obtain and the success rate of PDO establishment is quite high. However, the efficiency of PDO establishment and expansion is still very low in some other tumors, such as prostate cancer. Rare subtypes of cancer and non-epithelial cancer such as rhabdomyosarcoma are much less studied, and the efficiency of organoid derivation is unpredictable. Exploring the determinants of PDO establishment outcome is necessary and important for improving the efficiency of organoid derivation and expansion, especially in prostate cancer and other cancers with low efficiency.

5.4 Immunization

Submerge cultured PDOs usually contain only tumor cells and lack stromal cells and immune cells, which limit the exploration of the mechanism of TME in driving cancer evolution and drug sensitivity. The ALL culture, 3D bioprinting technology, organoids-on-a-chip and the coculture with stromal and immune cells can help us partly remodel TME and test immunotherapy.

5.5 Vascularization

Most PDOs do not have vascularized structure, leading to the limitation for the growth of large organoids and the evaluation of the impact of blood vessels and hemodynamics on tumors and drug response. Organ-on-a-chip technology can help to realize the vascularization of organoids and evaluate drug response.[107]

5.6 Systematization

Compared with a single organoid, the construction of multiorganoid system can make a more complete and comprehensive evaluation of drug efficacy and potential toxicity. A organoids-on-a-chip to coculture gastric, liver, and intestinal organoids in different compartment has been successfully developed to evaluate interactions between organs in drug metabolism.[186]

5.7 Automatization

The development of automatization in organoids can decrease the labor-intensive and time-consuming tasks of 3D cell culture, human error, and minimize the consumption of expensive reagents, greatly improve the speed, cost, and reproducibility of culture, and accelerate the realization of standardization. Automated microfluidic platform for dynamic and combinatorial drug screening of tumor organoids have been established and well described.[224] The further enhanced automatization of organoids will become an essential tool for both basic and translational research.

6. Conclusion

In the past decade, the development of organoids derived from patients with different cancers has been a great breakthrough in cancer medicine. Tumor organoids precisely recapitulate the key histopathological, genetic, and phenotypic features, therapeutic response of parental tumor, constantly advancing our understanding of tumor spatial and temporal heterogeneity. In addition, the implication of organoid culture techniques in precancerous lesions and rare tumor types have provided a powerful support to explore previous understudied diseases. Although there are several limitations of organoids, the combination with other advance technologies, for example, 3D bio-printing, organ-on-a-chip, CRISPR-Cas9, and other future improvements will accelerate their (pre-)clinical implementation. Further advancement of patient-derived organoids has great potential to revolutionize tumor research in a profound way.

Acknowledgments

This work was supported by the National Key R&D Program of China (2019YFA0906000), the Shenzhen Municipal Government of China (JCYJ20200109120016553 , GJHZ 20180926165202081), the Sanming Project of Shenzhen Health and Family Planning Commission (SZSM202011017), the Shenzhen High-level Hospital Construction Fund.

References

1. Sung H, Ferlay J, Siegel R-L, et al. Global cancer statistics 2020: GLOBOCAN estimates of incidence and mortality worldwide for 36 cancers in 185 countries. *CA Cancer J Clin*. 2021;71(3):209–249.
2. Hanahan D, Weinberg R-A. Hallmarks of cancer: The next generation. *Cell*. 2011;144 (5):646–674.
3. Hanahan D. Hallmarks of cancer: New dimensions. *Cancer Discov*. 2022;12(1):31–46.
4. Dagogo-Jack I, Shaw A-T. Tumor heterogeneity and resistance to cancer therapies. *Nat Rev Clin Oncol*. 2018;15(2):81–94.
5. Bedard P-L, Hansen A-R, Ratain M-J, et al. Tumor heterogeneity in the clinic. *Nature*. 2013;501(7467):355–364.
6. Marusyk A, Janiszewska M, Polyak K. Intratumor heterogeneity: The Rosetta stone of therapy resistance. *Cancer Cell*. 2020;37(4):471–484.
7. Runa F, Hamalian S, Meade K, et al. Tumor microenvironment heterogeneity: Challenges and opportunities. Curr. *Mol Biol Rep*. 2017;3(4):218–229.
8. Kim J, Koo B-K, Knoblich J-A. Human organoids: Model systems for human biology and medicine. *Nat Rev Mol Cell Biol*. 2020;21(10):571–584.
9. Wong C-H, Siah K-W, Lo A-W. Estimation of clinical trial success rates and related parameters. *Biostatistics*. 2019;20(2):273–286.
10. Scherer W-F, Syverton J-T, Gey G-O. Studies on the propagation in vitro of poliomyelitis viruses. IV. Viral multiplication in a stable strain of human malignant

epithelial cells (strain HeLa) derived from an epidermoid carcinoma of the cervix. *J Exp Med.* 1953;97(5):695–710.
11. Barretina J, Caponigro G, Stransky N, et al. The cancer cell line encyclopedia enables predictive modelling of anticancer drug sensitivity. *Nature.* 2012;483(7391):603–607.
12. Mirabelli P, Coppola L, Salvatore M. Cancer cell lines are useful model systems for medical research. *Cancers (Basel).* 2019;11(8):1098.
13. Ben-David U, Siranosian B, Ha G, et al. Genetic and transcriptional evolution alters cancer cell line drug response. *Nature.* 2018;560(7718):325–330.
14. Gillet J-P, Varma S, Gottesman M-M. The clinical relevance of cancer cell lines. *J Natl Cancer Inst.* 2013;105(7):452–458.
15. Lai Y, Wei X, Lin S, et al. Current status and perspectives of patient-derived xenograft models in cancer research. *J Hematol Oncol.* 2017;10(1):106.
16. Ben-David U, Ha G, Tseng Y-Y, et al. Patient-derived xenografts undergo mouse-specific tumor evolution. *Nat Genet.* 2017;49(11):1567–1575.
17. LeSavage B-L, Suhar R-A, Broguiere N, et al. Next-generation cancer organoids. *Nat Mater.* 2022;21(2):143–159.
18. Sato T, Vries R-G, Snippert H-J, et al. Single Lgr5 stem cells build crypt-villus structures in vitro without a mesenchymal niche. *Nature.* 2009;459(7244):262–265.
19. Sato T, Stange D-E, Ferrante M, et al. Long-term expansion of epithelial organoids from human colon, adenoma, adenocarcinoma, and Barrett's epithelium. *Gastroenterology.* 2011;141(5):1762–1772.
20. Broutier L, Mastrogiovanni G, Verstegen M-M, et al. Human primary liver cancer-derived organoid cultures for disease modeling and drug screening. *Nat Med.* 2017;23(12):1424–1435.
21. Ubink I, Bolhaqueiro ACF, Elias S-G, et al. Organoids from colorectal peritoneal metastases as a platform for improving hyperthermic intraperitoneal chemotherapy. *Br J Surg.* 2019;106(10):1404–1414.
22. Mout L, van Dessel L-F, Kraan J, et al. Generating human prostate cancer organoids from leukapheresis enriched circulating tumor cells. *Eur J Cancer.* 2021;150:179–189.
23. Mazzocchi A, Dominijanni A, Soker S. Pleural effusion aspirate for use in 3D lung cancer modeling and chemotherapy screening. *ACS Biomater Sci Eng.* 2019;5(4):1937–1943.
24. Kodack D-P, Farago A-F, Dastur A, et al. Primary patient-derived cancer cells and their potential for personalized cancer patient care. *Cell Rep.* 2017;21(11):3298–3309.
25. Abdullah K-G, Bird C-E, Buehler J-D, et al. Establishment of patient-derived organoid models of lower grade glioma. *Neuro Oncol.* 2021;noab273.
26. Li Z, Qian Y, Li W, et al. Human lung adenocarcinoma-derived organoid models for drug screening. *iScience.* 2020;23(8), 101411.
27. Chen P, Zhang X, Ding R, et al. Patient-derived organoids can guide personalized-therapies for patients with advanced breast cancer. *Adv Sci (Weinh).* 2021;8(22), e2101176.
28. Boj S-F, Hwang C-I, Baker L-A, et al. Organoid models of human and mouse ductal pancreatic cancer. *Cell.* 2015;160(1–2):324–338.
29. Fujii M, Shimokawa M, Date S, et al. A colorectal tumor organoid library demonstrates progressive loss of niche factor requirements during tumorigenesis. *Cell Stem Cell.* 2016;18(6):827–838.
30. Lee S-H, Hu W, Matulay J-T, et al. Tumor evolution and drug response in patient-derived organoid models of bladder cancer. *Cell.* 2018;173(2):515–528.
31. Tan A-C, Ashley D-M, Lopez G-Y, et al. Management of glioblastoma: State of the art and future directions. *CA Cancer J Clin.* 2020;70(4):299–312.
32. Neftel C, Laffy J, Filbin M-G, et al. An integrative model of cellular states, plasticity, and genetics for glioblastoma. *Cell.* 2019;178(4):835–849.

33. Ogawa J, Pao G-M, Shokhirev M-N, et al. Glioblastoma model using human cerebral organoids. *Cell Rep.* 2018;23(4):1220–1229.
34. Da S-B, Mathew R-K, Polson E-S, et al. Spontaneous glioblastoma spheroid infiltration of early-stage cerebral organoids models brain tumor invasion. *SLAS Discov.* 2018;23 (8):862–868.
35. Linkous A, Balamatsias D, Snuderl M, et al. Modeling patient-derived glioblastoma with cerebral organoids. *Cell Rep.* 2019;26(12):3203–3211.
36. Bian S, Repic M, Guo Z, et al. Genetically engineered cerebral organoids model brain tumor formation. *Nat Methods.* 2018;15(8):631–639.
37. Krieger T-G, Tirier S-M, Park J, et al. Modeling glioblastoma invasion using human brain organoids and single-cell transcriptomics. *Neuro Oncol.* 2020;22(8):1138–1149.
38. Hubert C-G, Rivera M, Spangler L-C, et al. A three-dimensional organoid culture system derived from human glioblastomas recapitulates the hypoxic gradients and cancer stem cell heterogeneity of tumors found in vivo. *Cancer Res.* 2016;76 (8):2465–2477.
39. Jacob F, Ming G-L, Song H. Generation and biobanking of patient-derived glioblastoma organoids and their application in CAR T cell testing. *Nat Protoc.* 2020;15 (12):4000–4033.
40. Jacob F, Salinas R-D, Zhang D-Y, et al. A patient-derived glioblastoma organoid model and biobank recapitulates inter- and intra-tumoral heterogeneity. *Cell.* 2020;180(1):188–204.
41. Kitano Y, Aoki K, Ohka F, et al. Urinary microRNA-based diagnostic model for central nervous system tumors using nanowire scaffolds. *ACS Appl Mater Interfaces.* 2021;13(15):17316–17329.
42. Golebiewska A, Hau A-C, Oudin A, et al. Patient-derived organoids and orthotopic xenografts of primary and recurrent gliomas represent relevant patient avatars for precision oncology. *Acta Neuropathol.* 2020;140(6):919–949.
43. Pine A-R, Cirigliano S-M, Nicholson J-G, et al. Tumor microenvironment is critical for the maintenance of cellular states found in primary glioblastomas. *Cancer Discov.* 2020;10(7):964–979.
44. Tang M, Xie Q, Gimple R-C, et al. Three-dimensional bio-printed glioblastoma microenvironments model cellular dependencies and immune interactions. *Cell Res.* 2020;30(10):833–853.
45. Maloney E, Clark C, Sivakumar H, et al. Immersion bioprinting of tumor organoids in multi-well plates for increasing chemotherapy screening throughput. *Micromachines (Basel).* 2020;11(2):208.
46. Yi H-G, Jeong Y-H, Kim Y, et al. A bio-printed human-glioblastoma-on-a-chip for the identification of patient-specific responses to chemoradiotherapy. Nat. *Biomed Eng.* 2019;3(7):509–519.
47. Zhu Z, Gorman M-J, McKenzie L-D, et al. Zika virus has oncolytic activity against glioblastoma stem cells. *J Exp Med.* 2017;214(10):2843–2857.
48. Zhu Z, Mesci P, Bernatchez J-A, et al. Zika virus targets glioblastoma stem cells through a SOX2-integrin $\alpha_v\beta_5$ axis. *Cell Stem Cell.* 2020;26(2):187–204.
49. Oser M-G, Niederst M-J, Sequist L-V, et al. Transformation from non-small-cell lung cancer to small-cell lung cancer: Molecular drivers and cells of origin. *Lancet Oncol.* 2015;16(4):e165–e172.
50. Kim M, Mun H, Sung C-O, et al. Patient-derived lung cancer organoids as in vitro cancer models for therapeutic screening. *Nat Commun.* 2019;10(1):3991.
51. Shi R, Radulovich N, Ng C, et al. Organoid cultures as preclinical models of non-small cell lung cancer. *Clin Cancer Res.* 2020;26(5):1162–1174.
52. Choi S-Y, Cho Y-H, Kim D-S, et al. Establishment and long-term expansion of small cell lung cancer patient-derived tumor organoids. *Int J Mol Sci.* 2021;22(3):1349.

53. Sachs N, Papaspyropoulos A, Ommen-DD Z-v, et al. Long-term expanding human airway organoids for disease modeling. *EMBO J*. 2019;38(4), e100300.
54. Ma X, Yang S, Jiang H, et al. Transcriptomic analysis of tumor tissues and organoids reveals the crucial genes regulating the proliferation of lung adenocarcinoma. *J Transl Med*. 2021;19(1):368.
55. Yang S, Zhang Z, Wang Q. Emerging therapies for small cell lung cancer. *J Hematol Oncol*. 2019;12(1):47.
56. Gmeiner W-H, Miller L-D, Chou J-W, et al. Dysregulated pyrimidine biosynthesis contributes to 5-FU resistance in SCLC patient-derived organoids but response to a novel polymeric fluoropyrimidine, CF10. *Cancers (Basel)*. 2020;12(4):788.
57. Choi Y-J, Lee H, Kim D-S, et al. Discovery of a novel CDK7 inhibitor YPN-005 in small cell lung cancer. *Eur J Pharmacol*. 2021;907, 174298.
58. Kim S-Y, Kim S-M, Lim S, et al. Modeling clinical responses to targeted therapies by patient-derived organoids of advanced lung adenocarcinoma. *Clin Cancer Res*. 2021;27 (15):4397–4409.
59. Dijkstra K-K, Cattaneo C-M, Weeber F, et al. Generation of tumor-reactive T cells by co-culture of peripheral blood lymphocytes and tumor organoids. *Cell*. 2018;174 (6):1586–1598.
60. Neal J-T, Li X, Zhu J, et al. Organoid modeling of the tumor immune microenvironment. *Cell*. 2018;175(7):1972–1988.
61. Cattaneo C-M, Dijkstra K-K, Fanchi L-F, et al. Tumor organoid-T-cell coculture systems. *Nat Protoc*. 2020;15(1):15–39.
62. Hu Y, Sui X, Song F, et al. Lung cancer organoids analyzed on microwell arrays predict drug responses of patients within a week. *Nat Commun*. 2021;12(1):2581.
63. Yokota E, Iwai M, Yukawa T, et al. Clinical application of a lung cancer organoid (tumoroid) culture system. *NPJ Precis Oncol*. 2021;5(1):29.
64. Dijkstra K-K, Monkhorst K, Schipper L-J, et al. Challenges in establishing pure lung cancer organoids limit their utility for personalized medicine. *Cell Rep*. 2020;31(5), 107588.
65. Wang Y, Jiang T, Qin Z, et al. HER2 exon 20 insertions in non-small-cell lung cancer are sensitive to the irreversible pan-HER receptor tyrosine kinase inhibitor pyrotinib. *Ann Oncol*. 2019;30(3):447–455.
66. Takahashi N, Hoshi H, Higa A, et al. An in vitro system for evaluating molecular targeted drugs using lung patient-derived tumor organoids. *Cell*. 2019;8(5):481.
67. Li H, Zhang Y, Lan X, et al. Halofuginone sensitizes lung cancer organoids to cisplatin via suppressing PI3K/AKT and MAPK signaling pathways. *Front Cell Dev Biol*. 2021;9, 773048.
68. Katsura H, Sontake V, Tata A, et al. Human lung stem cell-based alveolospheres provide insights into SARS-CoV-2-mediated interferon responses and pneumocyte dysfunction. *Cell Stem Cell*. 2020;27(6):890–904.
69. Huang J, Hume A-J, Abo K-M, et al. SARS-CoV-2 infection of pluripotent stem cell-derived human lung alveolar type 2 cells elicits a rapid epithelial-intrinsic inflammatory response. *Cell Stem Cell*. 2020;27(6):962–973.
70. Frank G-A, Danilova N-V, Andreeva II, et al. WHO classification of tumors of the breast, 2012. *Arkh Patol*. 2013;75(2):53–63.
71. Liu K, Newbury P-A, Glicksberg B-S, et al. Evaluating cell lines as models for metastatic breast cancer through integrative analysis of genomic data. *Nat Commun*. 2019;10(1):2138.
72. Sachs N, de Ligt J, Kopper O, et al. A living biobank of breast cancer organoids captures disease heterogeneity. *Cell*. 2018;172(1–2):373–386.
73. Rosenbluth J-M, Schackmann RCJ, Gray G-K, et al. Organoid cultures from normal and cancer-prone human breast tissues preserve complex epithelial lineages. *Nat Commun*. 2020;11(1):1711.

74. Dekkers J-F, van Vliet E-J, Sachs N, et al. Long-term culture, genetic manipulation and xenotransplantation of human normal and breast cancer organoids. *Nat Protoc.* 2021;16 (4):1936–1965.
75. Nguyen-Ngoc K-V, Cheung K-J, Brenot A, et al. ECM microenvironment regulates collective migration and local dissemination in normal and malignant mammary epithelium. *Proc Natl Acad Sci U S A.* 2012;109(39):E2595–E2604.
76. Cheung K-J, Gabrielson E, Werb Z, et al. Collective invasion in breast cancer requires a conserved basal epithelial program. *Cell.* 2013;155(7):1639–1651.
77. Jung H-Y, Fattet L, Tsai J-H, et al. Apical-basal polarity inhibits epithelial-mesenchymal transition and tumor metastasis by PAR-complex-mediated SNAI1 degradation. *Nat Cell Biol.* 2019;21(3):359–371.
78. Dekkers J-F, Whittle J-R, Vaillant F, et al. Modeling breast cancer using CRISPR-Cas9-mediated engineering of human breast organoids. *J Natl Cancer Inst.* 2020;112(5):540–544.
79. Xiao Y, Ma D, Yang Y-S, et al. Comprehensive metabolomics expands precision medicine for triple-negative breast cancer. *Cell Res.* 2022. https://doi.org/10.1038/s41422-022-00614-0.
80. Nayak B, Balachander G-M, Manjunath S, et al. Tissue mimetic 3D scaffold for breast tumor-derived organoid culture toward personalized chemotherapy. *Colloids Surf B Biointerfaces.* 2019;180334–180343.
81. Langer E-M, Allen-Petersen B-L, King S-M, et al. Modeling tumor phenotypes in vitro with three-dimensional bioprinting. *Cell Rep.* 2019;26(3):608–623.
82. Dornhof J, Kieninger J, Muralidharan H, et al. Microfluidic organ-on-chip system for multi-analyte monitoring of metabolites in 3D cell cultures. *Lab Chip.* 2022;22 (2):225–239.
83. Zhou Z, Van der Jeught K, Fang Y, et al. An organoid-based screen for epigenetic inhibitors that stimulate antigen presentation and potentiate T-cell-mediated cytotoxicity. Nat. *Biomed Eng.* 2021;5(11):1320–1335.
84. Huang L, Rong Y, Tang X, et al. Engineered exosomes as an in situ DC-primed vaccine to boost antitumor immunity in breast cancer. *Mol Cancer.* 2022;21(1):45.
85. Siegel R-L, Miller K-D, Jemal A. Cancer statistics, 2020. *Cancer J Clin.* 2020;70 (1):7–30.
86. Kleeff J, Korc M, Apte M, et al. Pancreatic cancer. *Nat Rev Dis Primers.* 2016;2:16022.
87. Park W, Chawla A, O'Reilly E-M. Pancreatic cancer: A review. *JAMA.* 2021;326 (9):851–862.
88. Huang L, Holtzinger A, Jagan I, et al. Ductal pancreatic cancer modeling and drug screening using human pluripotent stem cell- and patient-derived tumor organoids. *Nat Med.* 2015;21(11):1364–1371.
89. Seino T, Kawasaki S, Shimokawa M, et al. Human pancreatic tumor organoids reveal loss of stem cell niche factor dependence during disease progression. *Cell Stem Cell.* 2018;22(3):454–467.
90. Tiriac H, Belleau P, Engle D-D, et al. Organoid profiling identifies common responders to chemotherapy in pancreatic cancer. *Cancer Discov.* 2018;8(9):1112–1129.
91. Seppala T-T, Zimmerman J-W, Sereni E, et al. Patient-derived organoid pharmaco-typing is a clinically tractable strategy for precision medicine in pancreatic cancer. *Ann Surg.* 2020;272(3):427–435.
92. Romero-Calvo I, Weber C-R, Ray M, et al. Human organoids share structural and genetic features with primary pancreatic adenocarcinoma tumors. *Mol Cancer Res.* 2019;17(1):70–83.
93. Driehuis E, van Hoeck A, Moore K, et al. Pancreatic cancer organoids recapitulate disease and allow personalized drug screening. *Proc Natl Acad Sci U S A.* 2019;116 (52):26580–26590.

94. Driehuis E, Gracanin A, Vries RGJ, et al. Establishment of pancreatic organoids from normal tissue and tumors. *STAR Protoc*. 2020;1(3), 100192.
95. Vaes RDW, van Dijk DPJ, Welbers TTJ, et al. Generation and initial characterization of novel tumor organoid models to study human pancreatic cancer-induced cachexia. *J Cachexia Sarcopenia Muscle*. 2020;11(6):1509–1524.
96. Farshadi E-A, Chang J, Sampadi B, et al. Organoids derived from neoadjuvant FOLFIRINOX patients recapitulate therapy resistance in pancreatic ductal adenocarcinoma. *Clin Cancer Res*. 2021;27(23):6602–6612.
97. Zeold A, Sandor G-O, Kiss A, et al. Shared extracellular vesicle miRNA profiles of matched ductal pancreatic adenocarcinoma organoids and blood plasma samples show the power of organoid technology. *Cell Mol Life Sci*. 2021;78(6):3005–3020.
98. Huch M, Bonfanti P, Boj S-F, et al. Unlimited in vitro expansion of adult bi-potent pancreas progenitors through the Lgr5/R-spondin axis. *EMBO J*. 2013;32(20): 2708–2721.
99. Choi J-I, Jang S-I, Hong J, et al. Cancer-initiating cells in human pancreatic cancer organoids are maintained by interactions with endothelial cells. *Cancer Lett*. 2021; 49842–49853.
100. Beato F, Reveron D, Dezsi K-B, et al. Establishing a living biobank of patient-derived organoids of intraductal papillary mucinous neoplasms of the pancreas. *Lab Invest*. 2021;101(2):204–217.
101. Baracos V-E, Martin L, Korc M, et al. Cancer-associated cachexia. *Nat Rev Dis Primers*. 2018;4:17105.
102. Vaes RDW, van Dijk DPJ, Farshadi E-A, et al. Human pancreatic tumor organoid-derived factors enhance myogenic differentiation. *J Cachexia Sarcopenia Muscle*. 2022. https://doi.org/10.1002/jcsm.12917.
103. Farrell J-J. Prevalence, diagnosis and management of pancreatic cystic neoplasms: Current status and future directions. *Gut Liver*. 2015;9(5):571–589.
104. Huang B, Trujillo M-A, Fujikura K, et al. Molecular characterization of organoids derived from pancreatic intraductal papillary mucinous neoplasms. *J Pathol*. 2020;252(3):252–262.
105. Kato H, Tateishi K, Fujiwara H, et al. MNX1-HNF1B axis is indispensable for intraductal papillary mucinous neoplasm lineages. *Gastroenterology*. 2021; [S0016-5085(21) 04090-7].
106. Tsai S, McOlash L, Palen K, et al. Development of primary human pancreatic cancer organoids, matched stromal and immune cells and 3D tumor microenvironment models. *BMC Cancer*. 2018;18(1):335.
107. Benjamin-FL Lai L, Rick-X HY, et al. Recapitulating pancreatic tumor microenvironment through synergistic use of patient organoids and organ-on-a-chip vasculature. *Adv Funct Mater*. 2020;30(48):2000545.
108. Meng Q, Xie S, Gray G-K, et al. Empirical identification and validation of tumor-targeting T cell receptors from circulation using autologous pancreatic tumor organoids. *J Immunother Cancer*. 2021;9(11), e003213.
109. Holokai L, Chakrabarti J, Lundy J, et al. Murine- and human-derived autologous organoid/immune cell co-cultures as pre-clinical models of pancreatic ductal adenocarcinoma. *Cancers (Basel)*. 2020;12(12):3816.
110. Koikawa K, Ohuchida K, Ando Y, et al. Basement membrane destruction by pancreatic stellate cells leads to local invasion in pancreatic ductal adenocarcinoma. *Cancer Lett*. 2018;42565–42577.
111. Lai BFL, Lu RXZ, Huyer-L D, et al. A well plate-based multiplexed platform for incorporation of organoids into an organ-on-a-chip system with a perfusable vasculature. *Nat Protoc*. 2021;16(4):2158–2189.

112. Sharick J-T, Walsh C-M, Sprackling C-M, et al. Metabolic heterogeneity in patient tumor-derived organoids by primary site and drug treatment. Front. *Oncologia.* 2020; 10:553.
113. Hadj B-E, Poiraud C, Paget S, et al. A new pancreatic adenocarcinoma-derived organoid model of acquired chemoresistance to FOLFIRINOX: First insight of the underlying mechanisms. *Biol Cell.* 2022;114(1):32–55.
114. Mahajan U-M, Li Q, Alnatsha A, et al. Tumor-specific delivery of 5-fluorouracil-incorporated epidermal growth factor receptor-targeted aptamers as an efficient treatment in pancreatic ductal adenocarcinoma models. *Gastroenterology.* 2021;161(3):996–1010.
115. Nicosia L, Alongi F, Andreani S, et al. Combinatorial effect of magnetic field and radiotherapy in PDAC organoids: A pilot study. *Biomedicine.* 2020;8(12):609.
116. Raimondi G, Mato-Berciano A, Pascual-Sabater S, et al. Patient-derived pancreatic tumor organoids identify therapeutic responses to oncolytic adenoviruses. *EBioMedicine.* 2020;56, 102786.
117. Pascual-Sabater S, Raimondi G, Mato-Berciano A, et al. Preclinical testing of oncolytic adenovirus sensitivity in patient-derived tumor organoids. *STAR Protoc.* 2021;2(4), 101017.
118. Hofseth L-J, Hebert J-R, Chanda A, et al. Early-onset colorectal cancer: Initial clues and current views. *Nat Rev Gastroenterol Hepatol.* 2020;17(6):352–364.
119. Hong S-N. Genetic and epigenetic alterations of colorectal cancer. *Intest Res.* 2018;16 (3):327–337.
120. Ciardiello D, Vitiello P-P, Cardone C, et al. Immunotherapy of colorectal cancer: Challenges for therapeutic efficacy. *Cancer Treat Rev.* 2019;7622–7632.
121. van de Wetering M, Francies H-E, Francis J-M, et al. Prospective derivation of a living organoid biobank of colorectal cancer patients. *Cell.* 2015;161(4):933–945.
122. Cristobal A, van den Toorn HWP, van de Wetering M, et al. Personalized proteome profiles of healthy and tumor human colon organoids reveal both individual diversity and basic features of colorectal cancer. *Cell Rep.* 2017;18(1):263–274.
123. Weeber F, van de Wetering M, Hoogstraat M, et al. Preserved genetic diversity in organoids cultured from biopsies of human colorectal cancer metastases. *Proc Natl Acad Sci U S A.* 2015;112(43):13308–13311.
124. Ganesh K, Wu C, O'Rourke K-P, et al. A rectal cancer organoid platform to study individual responses to chemoradiation. *Nat Med.* 2019;25(10):1607–1614.
125. Zhang L, Mo S, Hu X, et al. Establishment and identification of organoids from human circulating colorectal cancer cells. *Clin Transl Med.* 2020;10(8), e247.
126. Kim S-C, Park J-W, Seo H-Y, et al. Multifocal organoid capturing of colon cancer reveals pervasive intratumoral heterogenous drug responses. *Adv Sci (Weinh).* 2022;9 (5), e2103360.
127. Zhao H, Yan C, Hu Y, et al. Differentiated cancer cell-originated lactate promotes the self-renewal of cancer stem cells in patient-derived colorectal cancer organoids. *Cancer Lett.* 2020;493:236–244.
128. Li X, Larsson P, Ljuslinder I, et al. Ex vivo organoid cultures reveal the importance of the tumor microenvironment for maintenance of colorectal cancer stem cells. *Cancers (Basel).* 2020;12(4):923.
129. Okamoto T, DuVerle D, Yaginuma K, et al. Comparative analysis of patient-matched PDOs revealed a reduction in OLFM4-associated clusters in metastatic lesions in colorectal cancer. *Stem Cell Rep.* 2021;16(4):954–967.
130. Li H, Dai W, Xia X, et al. Modeling tumor development and metastasis using paired organoids derived from patients with colorectal cancer liver metastases. *J Hematol Oncol.* 2020;13(1):119.

131. Cho E-J, Kim M, Jo D, et al. Immuno-genomic classification of colorectal cancer organoids reveals cancer cells with intrinsic immunogenic properties associated with patient survival. *J Exp Clin Cancer Res*. 2021;40(1):230.
132. Demmers L-C, Kretzschmar K, Van Hoeck A, et al. Single-cell derived tumor organoids display diversity in HLA class I peptide presentation. *Nat Commun*. 2020;11(1):5338.
133. Qin X, Sufi J, Vlckova P, et al. Cell-type-specific signaling networks in heterocellular organoids. *Nat Methods*. 2020;17(3):335–342.
134. Handa T, Kuroha M, Nagai H, et al. Liquid biopsy for colorectal adenoma: Is the exosomal miRNA derived from organoid a potential diagnostic biomarker? *Clin Transl Gastroenterol*. 2021;12(5), e356.
135. Li Y, Wang R, Huang D, et al. A novel human colon signet-ring cell carcinoma organoid line: Establishment, characterization and application. *Carcinogenesis*. 2020; 41(7):993–1004.
136. Wang R, Xiang W, Xu Y, et al. Enhanced glutamine utilization mediated by SLC1A5 and GPT2 is an essential metabolic feature of colorectal signet ring cell carcinoma with therapeutic potential. *Ann Transl Med*. 2020;8(6):302.
137. Calon A, Lonardo E, Berenguer-Llergo A, et al. Stromal gene expression defines poor-prognosis subtypes in colorectal cancer. *Nat Genet*. 2015;47(4):320–329.
138. Chen J, Sun H-W, Yang Y-Y, et al. Reprogramming immunosuppressive myeloid cells by activated T cells promotes the response to anti-PD-1 therapy in colorectal cancer. *Signal Transduct Target Ther*. 2021;6(1):4.
139. Zhao X, Yuan C, Wangmo D, et al. Tumor-secreted extracellular vesicles regulate T-cell costimulation and can be manipulated to induce tumor-specific T-cell responses. *Gastroenterology*. 2021;161(2):560–574.
140. Mosa M-H, Michels B-E, Menche C, et al. A Wnt-induced phenotypic switch in cancer-associated fibroblasts inhibits EMT in colorectal cancer. *Cancer Res*. 2020;80 (24):5569–5582.
141. Oszvald A, Szvicsek Z, Papai M, et al. Fibroblast-derived extracellular vesicles induce colorectal cancer progression by transmitting amphiregulin. Front cell. *Dev Biol*. 2020;8558.
142. Szvicsek Z, Oszvald A, Szabo L, et al. Extracellular vesicle release from intestinal organoids is modulated by Apc mutation and other colorectal cancer progression factors. *Cell Mol Life Sci*. 2019;76(12):2463–2476.
143. Ao T, Mochizuki S, Kajiwara Y, et al. Cancer-associated fibroblasts at the unfavorable desmoplastic stroma promote colorectal cancer aggressiveness: Potential role of ADAM9. *Int J Cancer*. 2022. https://doi.org/10.1002/ijc.33947.
144. Peng S, Chen D, Cai J, et al. Enhancing cancer-associated fibroblast fatty acid catabolism within a metabolically challenging tumor microenvironment drives colon cancer peritoneal metastasis. *Mol Oncol*. 2021;15(5):1391–1411.
145. Naruse M, Ochiai M, Sekine S, et al. Re-expression of REG family and DUOXs genes in CRC organoids by co-culturing with CAFs. *Sci Rep*. 2021;11(1):2077.
146. Pasch C-A, Favreau P-F, Yueh A-E, et al. Patient-derived cancer organoid cultures to predict sensitivity to chemotherapy and radiation. *Clin Cancer Res*. 2019;25 (17):5376–5387.
147. Yao Y, Xu X, Yang L, et al. Patient-derived organoids predict chemoradiation responses of locally advanced rectal cancer. *Cell Stem Cell*. 2020;26(1):17–26.
148. Ooft S-N, Weeber F, Dijkstra K-K, et al. Patient-derived organoids can predict response to chemotherapy in metastatic colorectal cancer patients. *Sci Transl Med*. 2019;11(513):eaay2574.

149. Engel R-M, Chan W-H, Nickless D, et al. Patient-derived colorectal cancer organoids upregulate revival stem cell marker genes following chemotherapeutic treatment. *J Clin Med*. 2020;9(1):128.
150. Cho Y-H, Ro E-J, Yoon J-S, et al. 5-FU promotes stemness of colorectal cancer via p53-mediated WNT/beta-catenin pathway activation. *Nat Commun*. 2020;11(1):5321.
151. Li Q, Sun H, Luo D, et al. Lnc-RP11-536 K7.3/SOX2/HIF-1alpha signaling axis regulates oxaliplatin resistance in patient-derived colorectal cancer organoids. *J Exp Clin Cancer Res*. 2021;40(1):348.
152. Tung K-L, Chen K-Y, Negrete M, et al. Integrated chromatin and transcriptomic profiling of patient-derived colon cancer organoids identifies personalized drug targets to overcome oxaliplatin resistance. *Genes Dis*. 2021;8(2):203–214.
153. Kucukkose E, Wensink G-E, Roelse C-M, et al. Mismatch repair status in patient-derived colorectal cancer organoids does not affect intrinsic tumor cell sensitivity to systemic therapy. *Cancers (Basel)*. 2021;13(21):5434.
154. Ponsioen B, Post J-B, Des-Amorie-JR B, et al. Quantifying single-cell ERK dynamics in colorectal cancer organoids reveals EGFR as an amplifier of oncogenic MAPK pathway signaling. *Nat Cell Biol*. 2021;23(4):377–390.
155. Amodio V, Yaeger R, Arcella P, et al. EGFR blockade reverts resistance to $KRAS^{G12C}$ inhibition in colorectal cancer. *Cancer Discov*. 2020;10(8):1129–1139.
156. Gudino V, Pohl S-O, Billard C-V, et al. RAC1B modulates intestinal tumorigenesis via modulation of WNT and EGFR signaling pathways. *Nat Commun*. 2021;12(1):2335.
157. Schnalzger T-E, de Groot M-H, Zhang C, et al. 3D model for CAR-mediated cytotoxicity using patient-derived colorectal cancer organoids. *EMBO J*. 2019;38(12), e100928.
158. Sugimura N, Li Q, Chu ESH, et al. Lactobacillus gallinarum modulates the gut microbiota and produces anti-cancer metabolites to protect against colorectal tumorigenesis. *Gut*. 2021. gutjnl-2020-323951.
159. Petrick J-L, Florio A-A, Znaor A, et al. International trends in hepatocellular carcinoma incidence, 1978-2012. *Int J Cancer*. 2020;147(2):317–330.
160. Huch M, Dorrell C, Boj S-F, et al. In vitro expansion of single Lgr5 + liver stem cells induced by Wnt-driven regeneration. *Nature*. 2013;494(7436):247–250.
161. Huch M, Gehart H, van Boxtel R, et al. Long-term culture of genome-stable bipotent stem cells from adult human liver. *Cell*. 2015;160(1–2):299–312.
162. Saito Y, Nakaoka T, Muramatsu T, et al. Induction of differentiation of intrahepatic cholangiocarcinoma cells to functional hepatocytes using an organoid culture system. *Sci Rep*. 2018;8(1):2821.
163. Nuciforo S, Fofana I, Matter M-S, et al. Organoid models of human liver cancers derived from tumor needle biopsies. *Cell Rep*. 2018;24(5):1363–1376.
164. Zhao Y, Li Z-X, Zhu Y-J, et al. Single-cell transcriptome analysis uncovers intratumoral heterogeneity and underlying mechanisms for drug resistance in hepatobiliary tumor organoids. *Adv Sci (Weinh)*. 2021;8(11), e2003897.
165. Darbari A, Sabin K-M, Shapiro C-N, et al. Epidemiology of primary hepatic malignancies in U.S. children. *Hepatology*. 2003;38(3):560–566.
166. Saltsman J-A, Hammond W-J, Narayan NJC, et al. A human organoid model of aggressive hepatoblastoma for disease modeling and drug testing. *Cancers (Basel)*. 2020;12(9):2668.
167. Liu J, Li P, Wang L, et al. Cancer-associated fibroblasts provide a stromal niche for liver cancer organoids that confers trophic effects and therapy resistance. *Cell Mol Gastroenterol Hepatol*. 2021;11(2):407–431.
168. Li L, Knutsdottir H, Hui K, et al. Human primary liver cancer organoids reveal intratumor and interpatient drug response heterogeneity. JCI. *Insight*. 2019;4(2), e121490.

169. Li L, Halpert G, Lerner M-G, et al. Protein synthesis inhibitor omacetaxine is effective against hepatocellular carcinoma. JCI. *Insight*. 2021;6(12), e138197.
170. Wang S, Wang Y, Xun X, et al. Hedgehog signaling promotes sorafenib resistance in hepatocellular carcinoma patient-derived organoids. *J Exp Clin Cancer Res*. 2020;39 (1):22.
171. Colquhoun A, Arnold M, Ferlay J, et al. Global patterns of cardia and non-cardia gastric cancer incidence in 2012. *Gut*. 2015;64(12):1881–1888.
172. Smyth E-C, Nilsson M, Grabsch H-I, et al. Gastric cancer. *Lancet*. 2020;396 (10251):635–648.
173. The Cancer Genome Atlas Research Network. Comprehensive molecular characterization of gastric adenocarcinoma. *Nature*. 2014;513(7517):202–209.
174. Seidlitz T, Koo B-K, Stange D-E. Gastric organoids-an in vitro model system for the study of gastric development and road to personalized medicine. *Cell Death Differ*. 2021;28(1):68–83.
175. McCracken K-W, Cata E-M, Crawford C-M, et al. Modelling human development and disease in pluripotent stem-cell-derived gastric organoids. *Nature*. 2014;516 (7531):400–404.
176. Bartfeld S, Bayram T, van de Wetering M, et al. In vitro expansion of human gastric epithelial stem cells and their responses to bacterial infection. *Gastroenterology*. 2015;148 (1):126–136.
177. Seidlitz T, Merker S-R, Rothe A, et al. Human gastric cancer modelling using organoids. *Gut*. 2019;68(2):207–217.
178. Nanki K, Toshimitsu K, Takano A, et al. Divergent routes toward Wnt and R-spondin niche independency during human gastric carcinogenesis. *Cell*. 2018;174(4):856–869.
179. Yan HHN, Siu H-C, Law S, et al. A comprehensive human gastric cancer organoid biobank captures tumor subtype heterogeneity and enables therapeutic screening. *Cell Stem Cell*. 2018;23(6):882–897.
180. Schwank G, Andersson-Rolf A, Koo B-K, et al. Generation of BAC transgenic epithelial organoids. *PLoS One*. 2013;8(10), e76871.
181. Lo Y-H, Kolahi K-S, Du Y, et al. A CRISPR/Cas9-engineered ARID1A-deficient human gastric cancer organoid model reveals essential and nonessential modes of oncogenic transformation. *Cancer Discov*. 2021;11(6):1562–1581.
182. Vlachogiannis G, Hedayat S, Vatsiou A, et al. Patient-derived organoids model treatment response of metastatic gastrointestinal cancers. *Science*. 2018;359(6378):920–926.
183. Vining K-H, Mooney D-J. Mechanical forces direct stem cell behaviour in development and regeneration. *Nat Rev Mol Cell Biol*. 2017;18(12):728–742.
184. Mammoto T, Ingber D-E. Mechanical control of tissue and organ development. *Development*. 2010;137(9):1407–1420.
185. Lee K-K, McCauley H-A, Broda T-R, et al. Human stomach-on-a-chip with luminal flow and peristaltic-like motility. *Lab Chip*. 2018;18(20):3079–3085.
186. Jin Y, Jin K, Seung L-J, et al. Drug screening: Vascularized liver organoids generated using induced hepatic tissue and dynamic liver-specific microenvironment as a drug testing platform. *Adv Funct Mater*. 2018;28(37) [1870266].
187. Park S-E, Georgescu A, Huh D. Organoids-on-a-chip. *Science*. 2019;364(6444): 960–965.
188. Gao D, Vela I, Sboner A, et al. Organoid cultures derived from patients with advanced prostate cancer. *Cell*. 2014;159(1):176–187.
189. Drost J, Karthaus W-R, Gao D, et al. Organoid culture systems for prostate epithelial and cancer tissue. *Nat Protoc*. 2016;11(2):347–358.
190. Beshiri M-L, Tice C-M, Tran C, et al. A PDX/organoid biobank of advanced prostate cancers captures genomic and phenotypic heterogeneity for disease modeling and therapeutic screening. *Clin Cancer Res*. 2018;24(17):4332–4345.

191. Karkampouna S, La Manna F, Benjak A, et al. Patient-derived xenografts and organoids model therapy response in prostate cancer. *Nat Commun.* 2021;12(1):1117.
192. Karthaus W-R, Iaquinta P-J, Drost J, et al. Identification of multipotent luminal progenitor cells in human prostate organoid cultures. *Cell.* 2014;159(1):163–175.
193. Puca L, Bareja R, Prandi D, et al. Patient derived organoids to model rare prostate cancer phenotypes. *Nat Commun.* 2018;9(1):2404.
194. Servant R, Garioni M, Vlajnic T, et al. Prostate cancer patient-derived organoids: Detailed outcome from a prospective cohort of 81 clinical specimens. *J Pathol.* 2021;254(5):543–555.
195. Shenoy T-R, Boysen G, Wang M-Y, et al. CHD1 loss sensitizes prostate cancer to DNA damaging therapy by promoting error-prone double-strand break repair. *Ann Oncol.* 2017;28(7):1495–1507.
196. Sanli O, Dobruch J, Knowles M-A, et al. Bladder cancer. *Nat Rev Dis Primers.* 2017;3:17022.
197. Robertson A-G, Kim J, Al-Ahmadie H, et al. Comprehensive molecular characterization of muscle-invasive bladder cancer. *Cell.* 2017;171(3):540–556.
198. Minkler S, Lucien F, Kimber M-J, et al. Emerging roles of urine-derived components for the management of bladder cancer: One man's trash is another man's treasure. *Cancers (Basel).* 2021;13(3):422.
199. Pauli C, Hopkins B-D, Prandi D, et al. Personalized in vitro and in vivo cancer models to guide precision medicine. *Cancer Discov.* 2017;7(5):462–477.
200. Mullenders J, de Jongh E, Brousali A, et al. Mouse and human urothelial cancer organoids: A tool for bladder cancer research. *Proc Natl Acad Sci U S A.* 2019;116 (10):4567–4574.
201. Yu L, Li Z, Mei H, et al. Patient-derived organoids of bladder cancer recapitulate antigen expression profiles and serve as a personal evaluation model for CAR-T cells in vitro. *Clin Transl Immunology.* 2021;10(2), e1248.
202. Kim E, Choi S, Kang B, et al. Creation of bladder assembloids mimicking tissue regeneration and cancer. *Nature.* 2020;588(7839):664–669.
203. Patel V-G, Oh W-K, Galsky M-D. Treatment of muscle-invasive and advanced bladder cancer in 2020. *CA Cancer J Clin.* 2020;70(5):404–423.
204. Cancer Genome Atlas Research Network. Comprehensive molecular characterization of urothelial bladder carcinoma. *Nature.* 2014;507(7492):315–322.
205. Tan P, Wang M, Zhong A, et al. SRT1720 inhibits the growth of bladder cancer in organoids and murine models through the SIRT1-HIF axis. *Oncogene.* 2021;40 (42):6081–6092.
206. Kong J, Lee H, Kim D, et al. Network-based machine learning in colorectal and bladder organoid models predicts anti-cancer drug efficacy in patients. *Nat Commun.* 2020;11(1):5485.
207. Gong Z, Huang L, Tang X, et al. Acoustic droplet printing tumor organoids for modeling bladder tumor immune microenvironment within a week. *Adv Healthc Mater.* 2021;10(22), e2101312.
208. Sun H, Wang H, Wang X, et al. Aurora-a/SOX8/FOXK1 signaling axis promotes chemoresistance via suppression of cell senescence and induction of glucose metabolism in ovarian cancer organoids and cells. *Theranostics.* 2020;10(15):6928–6945.
209. Kopper O, de Witte C-J, Lohmussaar K, et al. An organoid platform for ovarian cancer captures intra- and interpatient heterogeneity. *Nat Med.* 2019;25(5):838–849.
210. Hill S-J, Decker B, Roberts E-A, et al. Prediction of DNA repair inhibitor response in short-term patient-derived ovarian cancer organoids. *Cancer Discov.* 2018;8(11):1404–1421.
211. de Witte C-J, Valle-Inclan-J E, Hami N, et al. Patient-derived ovarian cancer organoids mimic clinical response and exhibit heterogeneous inter- and intrapatient drug responses. *Cell Rep.* 2020;31(11), 107762.

212. Nanki Y, Chiyoda T, Hirasawa A, et al. Patient-derived ovarian cancer organoids capture the genomic profiles of primary tumors applicable for drug sensitivity and resistance testing. *Sci Rep*. 2020;10(1):12581.
213. Chen D, Tan Y, Li Z, et al. Organoid cultures derived from patients with papillary thyroid cancer. *J Clin Endocrinol Metab*. 2021;106(5):1410–1426.
214. Khan S, Shin J-H, Ferri V, et al. High-resolution positron emission microscopy of patient-derived tumor organoids. *Nat Commun*. 2021;12(1):5883.
215. Sondorp LHJ, Ogundipe VML, Groen A-H, et al. Patient-derived papillary thyroid cancer organoids for radioactive iodine refractory screening. *Cancers (Basel)*. 2020;12 (11):3212.
216. Lin K-C, Ting L-L, Chang C-L, et al. Ex vivo expanded circulating tumor cells for clinical anti-cancer drug prediction in patients with head and neck cancer. *Cancers (Basel)*. 2021;13(23):6076.
217. Tanaka N, Osman A-A, Takahashi Y, et al. Head and neck cancer organoids established by modification of the CTOS method can be used to predict in vivo drug sensitivity. *Oral Oncol*. 2018;8749–8757.
218. Kijima T, Nakagawa H, Shimonosono M, et al. Three-dimensional organoids reveal therapy resistance of esophageal and oropharyngeal squamous cell carcinoma cells. Cell Mol. *Gastroenterol Hepatol*. 2019;7(1):73–91.
219. Wang Z, Guo Y, Jin Y, et al. Establishment and drug screening of patient-derived extrahepatic biliary tract carcinoma organoids. *Cancer Cell Int*. 2021;21(1):519.
220. Saito Y, Muramatsu T, Saito H. Establishment and long-term culture of organoids derived from human biliary tract carcinoma. *STAR Protoc*. 2020;1(1), 100009.
221. Calandrini C, Drost J. Normal and tumor-derived organoids as a drug screening platform for tumor-specific drug vulnerabilities. *STAR Protoc*. 2022;3(1), 101079.
222. Esser L-K, Branchi V, Leonardelli S, et al. Cultivation of clear cell renal cell carcinoma patient-derived organoids in an air-liquid interface system as a tool for studying individualized therapy. Front. *Oncologia*. 2020;10:1775.
223. Calandrini C, Schutgens F, Oka R, et al. An organoid biobank for childhood kidney cancers that captures disease and tissue heterogeneity. *Nat Commun*. 2020;11(1):1310.
224. Schuster B, Junkin M, Kashaf S-S, et al. Automated microfluidic platform for dynamic and combinatorial drug screening of tumor organoids. *Nat Commun*. 2020;11(1):5271.

Index

Note: Page numbers followed by "*f*" indicate figures and "*t*" indicate tables.

D

P

FSC
www.fsc.org
MIX
Papier | Fördert gute Waldnutzung
FSC® C083411

Kolibri 360 GmbH
Ferdinand-Jühlke-Straße 7
99095 Erfurt, Deutschland
produktsicherheit@kolibri360.de